International Conference on Recent Trends in Physics

(ICRTP 2014)

AA002691

Journal of Physics: Conference Series Volume 534

Indore, India
22 – 23 February 2014

Editors:

S. N. Kane
A. Mishra
A. Gaur

ISBN: 978-1-63439-462-8
ISSN: 1742-6588

Printed from e-media with permission by:

Curran Associates, Inc.
57 Morehouse Lane
Red Hook, NY 12571

Printed by Curran Associates, Inc. (2014)

Institute of Physics
Dirac House, Temple Back
Bristol BS1 6BE UK

Phone: 44 1 17 929 7481
Fax: 44 1 17 920 0979

techtracking@iop.org

Additional copies of this publication are available from:

Curran Associates, Inc.
57 Morehouse Lane
Red Hook, NY 12571 USA
Phone: 845-758-0400
Fax: 845-758-2634
Email: curran@proceedings.com
Web: www.proceedings.com

International Conference on Recent Trends in Physics

(ICRTP 2014)

Journal of Physics: Conference Series Volume 534

Indore, India
22 – 23 February 2014

ISBN: 978-1-63439-462-8
ISSN: 1742-6588

IOP Institute of Physics

TABLE OF CONTENTS

INVITED TALKS

CONDENSED MATTER AND MATERIAL PHYSICS

LASER AND PLASMA PHYSICS

OTHERS

Preface

The International Conference on Recent Trends in Physics (ICRTP 2014) was held at Indore, India, during 22–23 February 2014. The conference was hosted by the School of Physics, Devi Ahilya University, Indore – 452001.

The objective of the conference was to provide a platform for interaction among scientists, teachers, researchers and students, and to share their ideas, thoughts and scientific findings in various areas of physics, including condensed matter and materials physics, laser and plasma physics.

ICRTP 2014 attracted a total of 103 abstracts submitted by scientists from France, Germany, Hungary, India, Italy and Portugal. The conference included an inaugural talk and 17 invited talks. PhD students presented their work in the form of posters. Presented posters were judged by a panel of five experts. Two best posters were awarded prizes.

It is our pleasure to thank the members of the Advisory Committee and Local Organizing Committee for their invaluable help, especially for their proposals for invited talks. A total of 80 papers were submitted to be considered for publication and 68 papers have been accepted for inclusion in the proceedings. All the papers were reviewed, and we wish to thank all the referees for their support and prompt reviewing of the papers.

We are grateful to the University Grants Commission (UGC), New Delhi, India for their financial support.

Personally, we would like to thank all the volunteers without whom the conference would not have been functional.

We express our sincere thanks to our university administration for their continuous support. Special thanks go to all the faculty members, administrative staff and students of the School of Physics for their tireless efforts in organizing ICRTP 2014.

Finally, we deeply appreciate the assistance of Ms Sarah Toms, Conference Publishing Coordinator, Journal of Physics Conference Series (JPCS), IOP Publishing Ltd, UK, for all help in getting ICRTP 2014 published in JPCS.

Indore, 28 July 2014

Shashank N Kane*, Ashutosh Mishra
School of Physics, Devi Ahilya University, Indore – 452001, India
Abhijeet Gaur
School of Studies in Physics, Vikram University, Ujjain – 456010, India

Guest Editors

* E-mail address: kane_sn@yahoo.com

Organized by

School of Physics, Devi Ahilya University, Indore 452001, India
Tel: +91 – 731- 2467028

Web site: www.icrtp2014.com
www.physics.dauniv.ac.in

Patron
Professor (Dr) D P Singh
Honourable *Vice Chancellor*

Conference Chairs
Professor (Dr) A Mishra
Professor (Dr) Shashank N Kane

Advisory committe
F Alves, LGEP, Orsay, France
J P Araujo, UP, Porto, Portugal
V Basso, INRIM, Italy
K S Bhatt, DRDO, Bangalore, India
M Coisson, INRIM, Torino, Italy
A K Dutta, DAVV, Indore, India
E Fleury, UM, Metz, France
V Ganesan, UGC-DAE CSR, Indore, India
A Gupta, UGC-DAE CSR, Indore, India
S N Joshi, CEERI, Pilani, India
S N Kane, DAVV, Indore, India
S N Kaul,UOH, Hyderabad, India
T Kulik, WUT, Warsaw, Poland
M Lo-Bue, ENSC, Cachan, France
F Mazaleyrat, ENSC, Cachan, France
A Mishra, DAVV, Indore, India
A K Singa, RRCAT, Indore, India
B D Shrivastava, VU, Ujjain, India
P Tiberto, INRIM, Torino, Italy
L K Varga, RISSPO, Budapest, Hungary
M Vazquez, CENIM-CSIC, Cantoblanco, Madrid, Spain
F Vinai, INRIM, Torino, Italy

Program Committee
A Mishra, DAVV, Indore (*Conference Chair)*
A K Dutta, DAVV, Indore
P Sen, DAVV, Indore
D Varshney, DAVV, Indore
G Mishra, DAVV, Indore
M Banerjee, DAVV, Indore
S N Kane, DAVV, Indore (*Conference Chair)*
Y Choyal, DAVV, Indore
A Sheikh, DAVV, Indore
R P Sharma, DAVV, Indore

Local Organizing Committee
A Sheikh, R P Sharma, S Ninama, M Satalkar,
P Tapkir, S Patel, S Bisen and K Mahalaxmi

Peer review statement

All papers published in this volume of *Journal of Physics: Conference Series* have been peer reviewed through processes administered by the proceedings Editors. Reviews were conducted by expert referees to the professional and scientific standards expected of a proceedings journal published by IOP Publishing.

Bottom-up nanofabrication using self-organized porous templates

A. Apolinário*, P. Quitério*, C.T. Sousa, M. P. Proença, J. Azevedo, M. Susano, S. Moraes, P. Lopes, J. Ventura and J.P. Araújo
IFIMUP-IN, Departamento de Física e Astronomia, Faculdade de Ciências, Universidade do Porto, 4169-007 Porto, Portugal
* These authors have contributed equally.

E-mail: jearaujo@fc.up.pt

Abstract. The great advances in nanoscience and nanotechnology in the last decade have led to the development of new platforms where all physical properties such as size, porosity, geometry and surface can be controlled at the nanoscale. Self-organized nanostructuring using template synthesis is a very promising and rapidly expanding field for the preparation of many different ordered structures ranging from the micrometer to the nanometer size. Porousaluminium, titanium, iron or hafnium oxide templates demonstrated real potential as well-controlled self-organized nanostructured materials for various applications. They are also an elegant tool to fabricate two- or three-dimensional arrays of periodic nanostructures. This work presents detailed methods to implement and fabricate nanoporous templates by anodic oxidation of Al and Ti metal substrates (that can be further used for Fe or Hf anodization). The characterization of nanowires grown in the pores of the synthesized Al_2O_3 and TiO_2 templates is also discussed.

1. Introduction

Precise control of the physical and chemical properties of materials at the nanometer scale is a central issue of current nanotechnology research. For practical nanoscience and nanotechology applications, nanostructures with one, two, or three-dimensional architectures must be synthesized with controlled shape and size. In fact, since a broad range of material's properties depend critically on the chemical composition, size, shape and regularity at the sub-nanometer scale, it is of increasing importance to develop structurally and dimensionally well-defined nanomaterials [1-4].

Basically, there are two nanofabrication approaches to create materials with the desired structural, mechanical, optical, magnetic and/or electronic properties: the top-down and the bottom-up. In the former one, large structures are machined in an attempt to build practical devices at the nanoscale [1]. Such nanostructures are produced from a bulk material or a continuous sheet-film with subsequent milling or lithography of the desired pattern [5-7]. Recent advances in various lithographic techniques, such as X-ray, ion-beam or electron-beam [1, 8,9], have provided access to the sub-100 nm size scale. However, these techniques have a crucial disadvantage: they are expensive, requiring state-of-the-art facilities and time consuming processes for the production of nanostructures on large scales.

On the other hand, the bottom-up approach starts with individual sub-nm or nm-sized building units such as atoms, molecules, polymers and colloids and assembles them, in a controlled way, creating the desired nanostructured materials. In the last two decades, a wide variety of techniques have been developed to control the size, shape, and crystalline structure of nanoparticles, including co-precipitation in solution [10], chemical vapour deposition [11], sol-gel process [12] and electrochemical deposition [13]. The chemical reactions involved in these growth techniques can be performed in confined nanoreactors, where the nucleation and growth processes occur inside small spaces, such as micelles [14] or porous templates [15], increasing the nanoparticles homogeneity and organization. The bottom-up growth process from small molecules to big particles involves several steps. All processes start with the nucleation stage, which takes place in a very short period of time. Nuclei activation can be achieved after supersaturation of the reagents or with temperature or external electric field application. Subsequent growth needs to be diffusion-controlled, which is achieved with the efficient domain of all experimental parameters. These techniques often result in nanomaterials with a broad size distribution and limited control of the nanoparticles shape. Nevertheless, the exhaustive research in this field showed that nanomaterials obtained using the bottom-up approach have promising applications in a diversified set of fields as a result of their ability to form organized one-, two- and three-dimensional periodic nanostructures.

Among the various methods used for the fabrication of nanostructures using a bottom-up approach, template assisted fabrication is a practical physical-chemical method and a reliable alternative to top-down techniques [16-18]. In this methodology, a membrane with micro or nanometer pores acts as template for subsequent material deposition, adopting its shape. Several natural and synthetic materials with pores or voids in the micro- or nanoscale have been used as templates, from which Al_2O_3 and TiO_2 templates are excellent examples [19, 20].

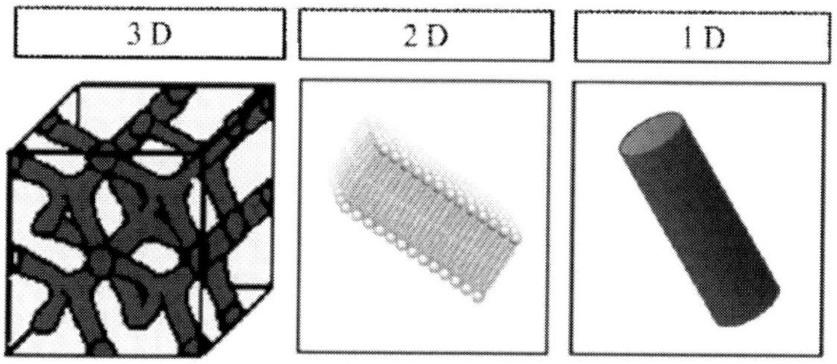

Figure 1. Three dimensional (3D), two-dimensional (2D) and one-dimensional (1D) structures.

Self-assembled templates, depending on their architecture, are typically classified taking into account the dimensionality of the resulting template structure: three-, two- and one-dimensional, as illustrated in Fig. 1. Three-dimensional templates include micelles, zeolites, glasses and block copolymers [17]. Two-dimensional templates are layered materials [21] being some minerals as the Langmuir-Blodgett films and ionic polymer multilayer, the most representative examples. One-dimensional templates are typically membranes with one-dimensional pores running through the sizeable thickness of the membrane. One of the most used one-dimensional template is the porous anodic oxide of metals like Al, Ti, Fe and Hf [15, 22-24]. When grown under appropriated conditions, cylindrical pores are formed perpendicularly to the surface of the membrane. The pores resulting from the electrochemical anodization of these metals present highly monodispersed pore diameters varying from 10 to 350 nm depending on the applied voltage. Obtainable pore density ranges from 10^9 to 10^{12} pores/cm^{2}whereas porosity varies in the range of 3 - 70%. By carefully adjusting the anodization conditions, these templates can also present a highly ordered hexagonal array of pores.

Furthermore, it is possible to manipulate the templates as to grow different nanostructures inside their pores. Figure 2 illustrates some of the possible fabrication steps proposed for porous anodic alumina (PAA) templates, which can also be used for other metal oxides. Several post-treatments can be employed after PAA fabrication in order to change the pore dimensions and architecture, such as pore widening (with chemical etching), barrier layer removal, branched pores or dendrites formation (Fig. 2). Afterwards, several types of nanostructures, such as nanotubes, nanowires, dots and anti-dots, can be grown inside (or at the surface) of PAA templates using for example electrodeposition, sol-gel or sputtering techniques, as also illustrated in Fig. 2.Such wide diversity of applications of the PAA membranes, as templates or replication masters, provides flexible and realistic opportunities to create well-defined nanostructured materials.

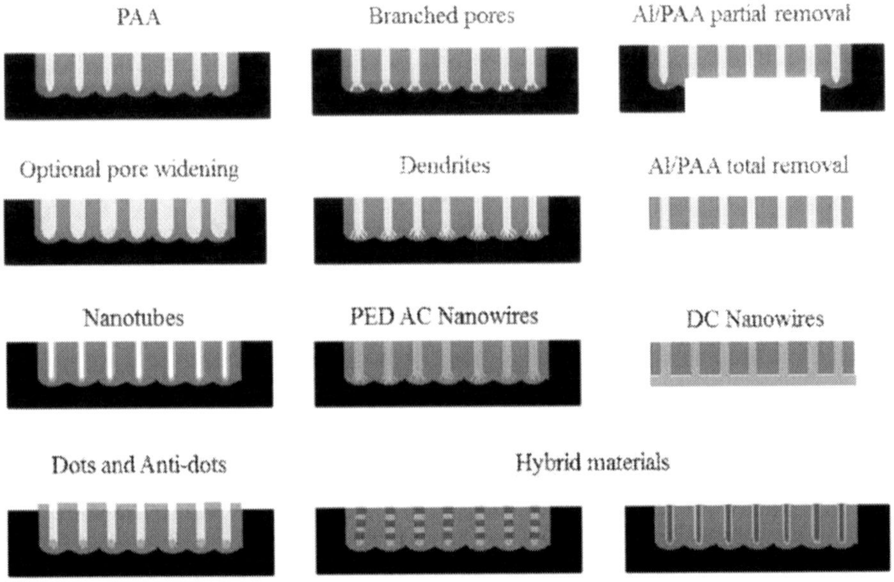

Figure 2. Processing steps in the PAA manipulation and fabrication of arrays composed of high aspect ratio nanoparticles.

2. Experimental

2.1 Anodization of Al and Ti

Both porous anodic alumina (PAA) and porous anodic titania (PAT) films were grown under regulated anodic bias using a Pt mesh as cathode. The anodization potentials (V), electrolyte concentration (C), anodization temperature (T), pore diameter (D_p), interpore distance (D_{int}), porosity (P) and anodization rates (AR) are displayed in Table 1.

Table 1. Anodization conditions and pore characteristics.

	Electrolyte	C(M)	V (V)	T (°C)	D_p (nm)	D_{int}(nm)	P (%)	AR (μm/h)
PAA	H_2SO_4 (aq)	0.3	25	2	20	65	10	2
	$C_2O_4H_2$ (aq)	0.3	40	4	35	105	10	2.5
	H_3PO_4(aq)	0.1	195	1	200	500	9	2
PAT	Ethyleneglycol +NH_4F (0.3 wt%) + H_2O (2 wt%)		60	21	37	160	5	7.9 (3 hours)
	Ethyleneglycol +NH_4F (0.3 wt%) + H_2O (2 wt%)		60	21	38	160	5	4.1 (17 hours)

A typical electrochemical cell used in the anodization process is depicted in Fig. 3. The anodization cell consists of a teflon container with a hole at the bottom, where the metal substrate is placed (working electrode), and a Pt mesh inserted on top of the teflon cell (counter electrode). The anodization was conducted under constant potential and the current monitored as a function of time using a digital source meter (Keithley 2400C) controlled by a home-developed LabVIEW program.

During the anodization, the temperature of the electrolyte solution was maintained constant using a copper cooling plate. The electrolyte was gently stirred at 200 rpm in the case of the PAA and 75 rpm for PAT.

In order to fabricate PAA templates with ordered nanopore arrays, in which the holes are straight and regularly arranged throughout the oxide film, a two-step anodization process, similar to that reported by Masuda *et al.*, was employed [15]. During the first anodization process, the hexagonally ordered pores were formed only at the bottom layer by a self-organization process. Subsequently, a chemical dissolution of the resulting PAA was achieved by submerging the sample in an aqueous solution of H_3PO_4 (0.4 M) and $H_2Cr_2O_7$(0.2 M) at 70 °C for about 12 hours. The resulting periodic concave patterns on the Al substrate act as pore nucleation centers for the second anodization process. An ordered nanopore array isthen obtained after anodizing for the second time under the same parameters as in the first step. In the case of the PAT, a two-step anodization process can also be employed to improve the structural order of the pore arrays [25, 26].However, we developed a simple and efficient method to induce a high degree of organization in PATs using an electropolishing pretreatment of the Ti foil [26]. Before starting the anodization, the Ti foils were electropolished in a H_2SO_4/HF (8:3) solution, with an applied potential of 10 V between the Ti foil and an inert Pt mesh (distance 2.5 cm) for 4 min; the solution was mechanically stirred (300 rpm) and kept at room temperature.

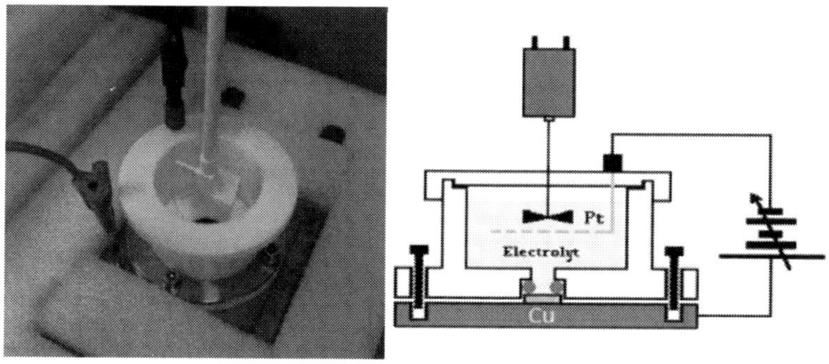

Figure 3. Home-made anodization cell.

2.2 *Electrodeposition in PAA and PAT pores*

To electrodeposit inside PAA templates, the alumina barrier-layer present at the bottom of each nanopore (intrinsic to the anodization process; Fig. 2) needs to be removed or reduced [27]. To perform DC potentiostatic electrodeposition, the barrier-layer is totally removed in a chemical bath of 0.5 M H_3PO_4 at 21 °C for 2 h (after chemically etching the Al substrate in an aqueous solution of 0.2 M $CuCl_2$ and 4.1 M HCl at room temperature). Finally, a layer of 100 nm of Au was sputtered on the bottom side of the open through porous membrane to serve as the working electrode (WE).

In order to use a pulsed electrodeposition method (PED) in PAA, the alumina barrier layer was thinned down to enable electrodeposition using the Al substrate as the cathode. To this end, the voltage was exponentially decreased from the second anodization potential until a final anodization potential corresponding to the desired thickness of the barrier layer is reached. In this method the applied potential was stepwise reduced to avoid discrete constant voltage drops, according to:

$$V_f = V_{anod} \cdot e^{-t/\eta}$$

where η was phenomenological determined and depends on the acid solutions used[28]. The pulses of V_{anod} last only for a pre-defined amount of time, also depending on the used acid. This restrain prevents the anodization current density [$j(t)$] to reach 90% of the corresponding stable value, where V_{anod} is again reduced and a new non-steady-state anodization step occurs.

In this work, Ni nanowires (NWs) were deposited in PAA and PAT pores using two different electrodeposition techniques: potentiostatic DC and pulsed electrodeposition. In both cases the depositions were performed in the so-called Watts bath [29], consisting of a mixture of 1.14 M $NiSO_4 \cdot 6H_2O$, 0.19 M $NiCl_2 \cdot 6H_2O$ and 0.73 M H_3BO_3. During the electrodeposition, the solution was magnetically stirred at 250 rpm and kept at a constant temperature of 35 °C with a pH ranging between 4 and 5.

Potentiostatic DC electrodeposition: A three-electrode cell was used, where the porous membranes serve as the WE, a Pt mesh was used as counter electrode (CE) and Ag/AgCl (in 4 M KCl) as reference electrode (0.197 V vs. standard hydrogen electrode – SHE). To deposit Ni nanowires, the electrodeposition potential of -1.5 V vs Ag/AgCl was optimized in a previous work [30].

Pulsed electrodeposition: To electrodeposit, a two-electrode cell was used. In this case the Al substrate acts as cathode and a Pt mesh as counter electrode. The PED was carried out by applying a modulated signal constituted by three consecutive pulses in the millisecond range supplied by a Source Metter Keithley 2400[27]. First, a pulse of 70 $mAcm^{-2}$is applied during 8 ms to deposit the material, followed by a second one with opposite polarity to discharge the insulator barrier layer capacitance in 1 ms at 8 V (corresponding to the final applied potential in the dendrite formation process[31]), and finally a delay time of 0.7 s.

3. Results and discussion

3.1 Influence of the electrochemical conditions on PAA and PAT formation

The most important parameters that affect the anodic oxidation of Al and the oxide film properties are the applied voltage or current, pH, type/concentration of the electrolyte and temperature. Both types of Al oxide, barrier-type and porous-type, can be obtained at constant potential or current[32]. The shapes of the current and potential-time transient curves are well established and can give insightsof the growth process and its kinetics [32]. The thickness of the barrier-type alumina is mainly determined by the applied voltage, although there is a small dependence on the electrolyte and temperature [33]. The anodizing ratio, which is defined as oxide thickness formed per volt, demonstrates that the barrier-type films are also strongly influenced by the type of metal which is anodized. The maximum attainable thickness of the barrier-type alumina was reported to be less than 1μm, corresponding to voltages lower than 500 V, since above this value the breakdown of the barrier was observed [32].

For the PAT formation, also several anodization parameters, such as electrolyte type and concentration, pH, temperature and applied potential, influence the template structure and geometry: interpore distance, diameter, wall thickness, length, quality and organization [34-36]. In the case of PAT templates, the thickness of the oxide barrier layer, diameter and interpore distance depends linearly on the applied voltage [25, 26].But differently from the Al case, the anodizing ratio is not constant, being strongly influenced by additional chemical dissolution processes during the anodization, resulting in a limitedPAT thickness[35, 37].Thus, the maximum attainable PAT thickness was reported to be hundreds of microns(220 μm in 17h),obtained in fluoride and ethylene glycol-based electrolyte that has, for that reason, become the main choice for the growth of long and ordered arrays of PAT with TiO_2nanotubes (NTs) [34, 35].

The porous-type oxide growth can be divided in several stages (Fig. 4) that are easily identified by monitoring the current transients during the anodization process (for anodizations at constant potential) [32] or that can be directly observed by high-resolution electron microscopy [38].

For both PAA and PAT cases, immediately after switching-on the anodic potential a barrier oxide layer starts to grow (stage 1). Relatively fine-featured pathways are then revealed in the outer regions of the barrier oxide prior to any true pore formation (stage 2). Further anodizing results in the propagation of individual paths through the barrier oxide with their heads becoming enlarged (stage 3). Finally, a steady-state pore structure is formed by cylindrical cells, each containing a pore at the center and separated from the valve metal by a layer of scalloped hemispherical barrier oxide (stage 4).

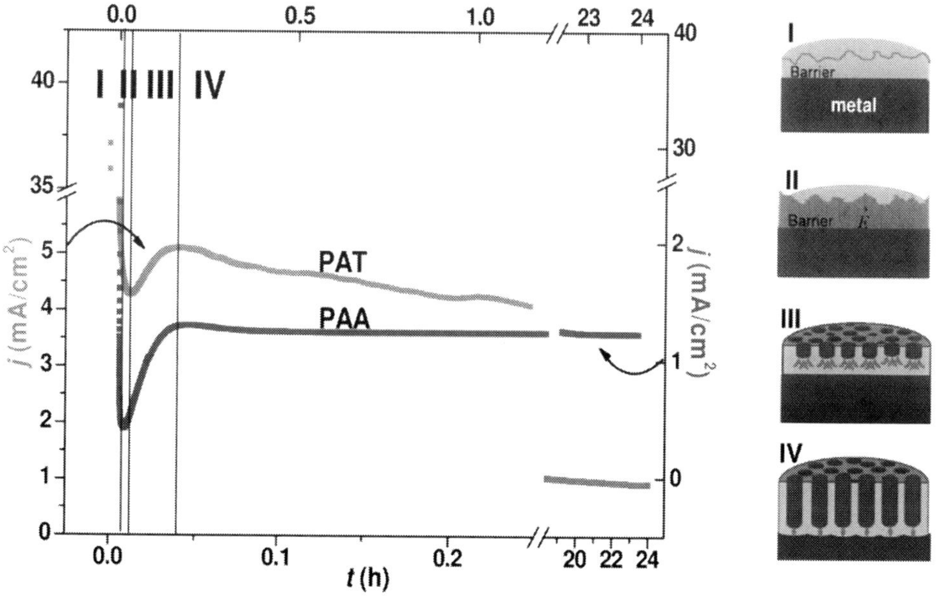

Figure 4. Current density vs time behavior during PAA and PAT formation and schematic diagram of porous structure growth.

The PAT growth mechanisms, evaluated from the current density transient curves, present the typical behaviour when compared to the PAA formation. However, at stage 4, the oxidation/dissolution processes are not quite in equilibrium, where the oxidation is seen to be stronger than the dissolution, leading to a slow decay of $j(t)$ during the Ti anodization (differently from the constant $j(t)$for the PAA case). This is due to the progressively failure of F^- ions from the electrolyte, at the NT bottoms, which are difficult to replace by new ones giving the high viscosity of the electrolyte and increasing length of the pores (diffusion limited process). Consequently, the oxidation process is faster than the dissolution (j slowly decrease with time), leading to an increase of the oxide barrier layer thicknessδ_b at the NTs' bottom. This significantly extends theionic diffusion paths along the oxide barrier[39], thus inhibiting the transport of F^-, Ti^{4+} and O^{2-} ions across δ_b which subsequently limits the PAT growth. The result is a typical PAT non-steady-state anodization equilibrium between oxide formation and dissolution rate and a deceleration of the template growth [40, 41].

3.2 Nanoporous structures of PAA and PAT

Regarding the PAA structure, based on the SEM analysis, a direct dependence is observed between the applied voltage and the pore diameter (D_p), interpore distance (D_{int}) and barrier layer thickness (δ_b), with a relationship of 0.85 nm/V, 1.29 nm/V and 2.77 nm/V, respectively[42], although these values depend on the pH, temperature and electrolyte used. Recently it was showed that the porosity also depends on the applied potential. Careful measurements and simulations showed a decrease in porosity with voltage, which implies a weaker dependence of the pore diameter on the applied voltage[25, 43]. Only at high voltages and in a self-ordered regime, was the porosity found independent of the applied voltage[25, 44]. The content of incorporated anionic impurities also increases with increasing voltage, as the driving force for anion adsorption and transport in the oxide increases[38].

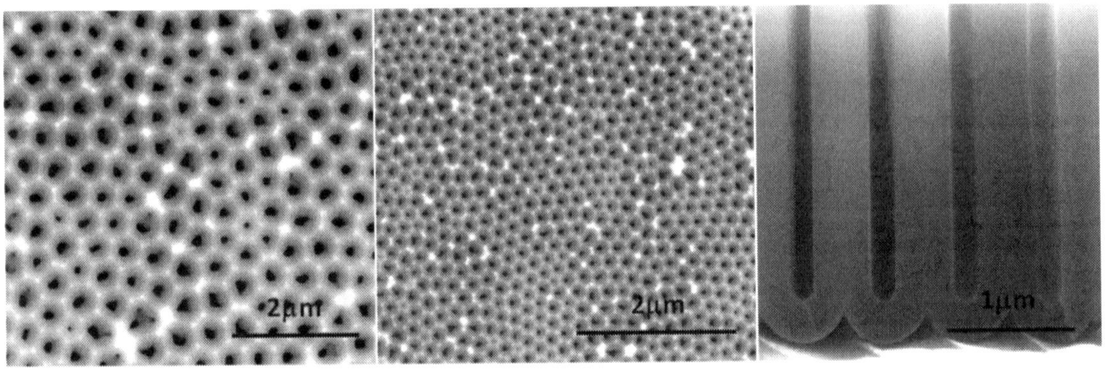

Figure 5. Surface SEM images of PAA grown at 40 V (left), at 195 V (middle) and cross section view of the PAA template grown at 195 V (right).

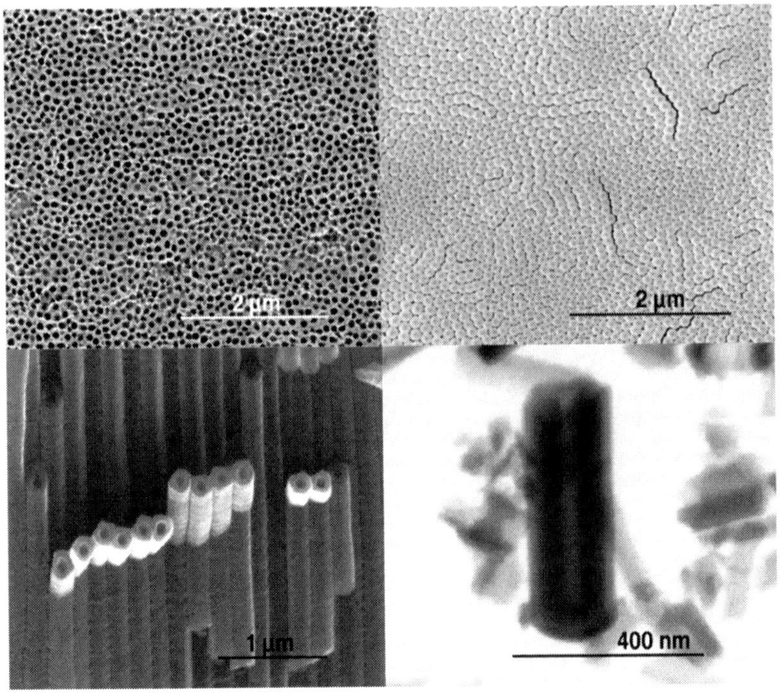

Figure 6. SEM images of PAT: top-view, grown at 50 V (left, top); bottom-view, grown at 60 V (right, top); cross section view of NTs template, grown at 60 V (left, bottom); and STEM image of a single nanotube, grown at 60 V (right, bottom).

Concerning the PAT pore morphology, one also observes the linear dependence between potential and D_p, D_{int} and δ_b, with respective proportionality constants of 0.50 nm/V, 2.41 nm/V and 1.21 nm/V. Our D_p(V) and D_{int}(V) values are similar to those reported in the literature[41]. However, the constant for the barrier layer differs from the one reported (2.2 nm/V)[45], since we used a different organic electrolyte composition, which confirms the dependence of these values on electrolyte composition. Similarly to the PAA templates, the porosity of PAT was also found to decrease with the applied voltage[41, 46].

3.3 Template-assisted growth of Nanowires using PAA

The growth of nickel nanowires inside the PAA pores was performed using DC potentiostaticelectrodeposition (Fig. 7; left) and PED (Fig. 7; right). The PAA oxide barrier layer reduction or removal and the electrodeposition processes were described in detailed in section 2.2. Cross-section SEM micrographs show a uniform deposition with an apparent almost complete filling of the pores independently of the electrodeposition method used.However, top-view images reveled that many pores werenot completely filled. To overcome such non-total filling of the template, the top surface can be consecutively physically etched by ion milling.

The nanowires grown by the DC method present 50 nm in diameter which corresponds to the PAA pore diameter (Fig 6; left). Touse the DC method one requires thick templates (>20 μm) that allow easy handling and are able to endure theexothermic Al removal and barrier layer removal processes. By optimizing the DC electrodeposition potential, nanowires with length larger than 100 μm can be grown in PAA templates [30].

On the other hand, nanowires grown by PED showed a dendritic structure at the bottom (Fig 6;right) due to the tree-like nanopores created during the non-steady-state anodization method used to reduce the bottom of the barrier layer[47]. The PED nanowires also show the original diameter of the PAA pores of 35 nm. In this case one can usually only grow nanowires with lengths up to 20 μm due to the high potentials used which leads to hydrogen evolution. Nevertheless, careful tuning of the PED parameters can lead to high pore filling percentages in thicker templates [20].

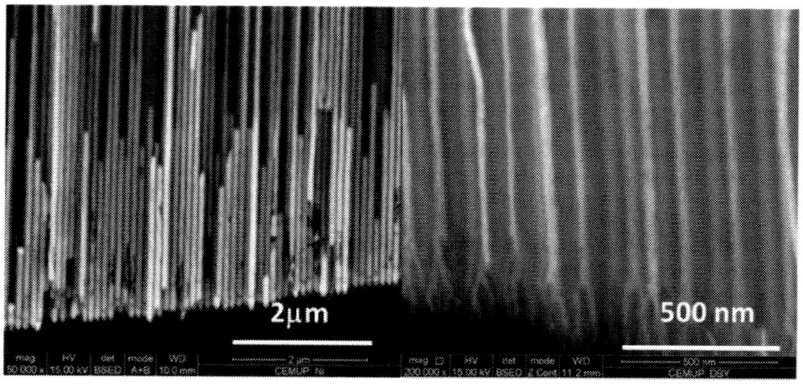

Figure 6. Cross-section SEM images of nickel nanowires grown by DC mode (left) and PED methods (right).

3.4 Template-assisted growth of Nanowires using PAT

For the deposition of materials inside PAT templates, we used the DC electrodeposition technique. Nanotubes bottoms were previously opened, using the step increase potential. Previously to the anodization, we performed a chemical etching pre-treatment onthe Ti foilin aqueous solution of HF (4%) during 15 minutes. A TiO_2 NTs free standing membrane was produced by anodizing at 60 V for 17 hours, immediately followed by prolongation of the anodization at 150 V for 5 minutes to open the NTs bottoms[48]. The use of a large increase of the applied potential at the end of the anodizationprocess, drastically and suddenly increases the NTs diameter at the bottom, resulting in a weaker point for breaking and separate the NTs from the Ti foil. This results in free standing NT membranes with open bottoms [48].

The free-standing membrane with open bottoms was sputtered with gold (to act as an electrical contact for the metal electrodeposition). Finally, Ni was electrodeposited inside TiO_2 NTs, using the potentiostatic mode under a potential of -1.5 V vs Ag/AgCl. Figure 7 shows the 10 μm long Ni NWs electrodeposited inside TiO_2 NTs (formed mainly in the middle length region of the NTs). This unusual and non-uniform Ni deposition along the NTs might be related with the semiconducting properties of TiO_2 and with the fragile nature of the NTs membranes. Additionally, the presence of a double-layer between NTs makes them easily separable from each other[25], which may influence the efficiency of the metal deposition. Compared with PAA, the process of PAT electrodeposition showed lower efficiency, indicating that it requires further optimization.

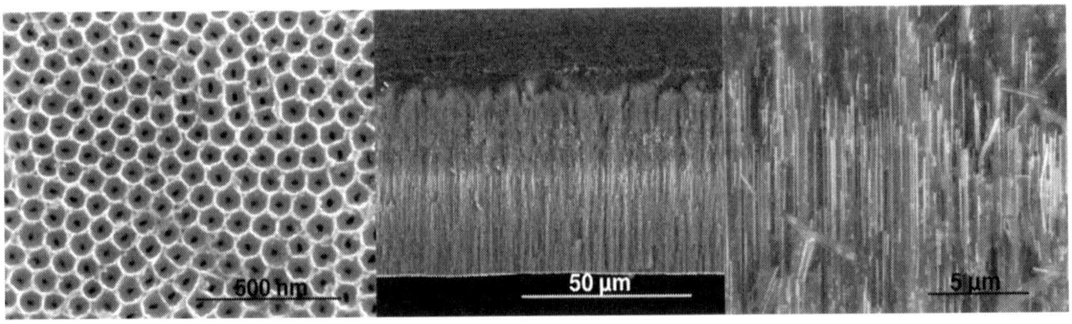

Figure 7. SEM images of PAT with open bottoms (left), cross section view of template after electrodeposition (middle) and electrodeposited nickel inside nanotubes (right).

4. Conclusions

This work was devoted to the detailed description of the design and implementation of a home-made anodization and electrodeposition set-up to fabricate nanoporous oxide templates and ordered arrays of nanowires. Porous anodic templates of aluminium and titanium were synthetized and morphologically characterized. We have also compared the results obtained for nanowires grown in PAA by DC potentiotastic electrodeposition and PED, which present similar morphological properties. The electrodeposition in the PAT nanotubes was also demonstrated using the DC electrodeposition in free-standing PAT membranes with open bottoms. All the fabricated nanostructures have potential applications in a large range of fields from solar cells to biotechnological applications.

Acknowledgements

MPP, CTS and JA are grateful to FCT for grants SFRH/BPD/84948/2012, SFRH/BPD/82010/2011 and SFRH/BD/79207/2011, respectively. JV acknowledges financial support through FSE/POPH.AA acknowledges financial support through PTDC/CTM-MET/118236/2010. The authors also acknowledge funding from FCT through the Associated Laboratory – IN and operation NORTE-070124-FEDER-000070-Multifunctional Nanomaterials funded by FEDER and CCDRN.

References

1. Cordoba, R., et al. 2010 *Microelectron. Eng.*, **87** 1550.
2. Guillamon, I., et al. 2011 *Phys. Rev. Lett.* **106** 077001.
3. Liu, K., et al. 2010 *Science* **329** 197.
4. Proenca, M.P. et al. 2011 *Phys. Chem. Chem. Phys.* **13** 9561.
5. Yue, X.Q., et al. 2011 *Powder Technol.* **211** 95.
6. Teixeira, J.M., et al. 2011 *Phys. Rev. Lett.* **106** 196601.
7. Kakazei, G.N., et al. 2003 *J. Appl. Phys.* **93** 8418.
8. Petruskevicius, R., et al. 2010. *Microelectron. Eng.* **87** 2332.
9. Lee, B.K., et al., 2010 *Microelectron. Eng.*, **87** 2449.
10. Maaz, K., et al., 2009 *J Magn. Magn. Mater.*, **321** 1838.
11. Cave, H.M., S.P. Krumdieck, and M.C. Jermy 2008 *Chem. Eng. J.* **135** 120.
12. Mitchell, D.T., et al. 2002 *J. Am. Chem. Soc.* **124** 11864.
13. Nielsch, K., et al. 2000 *Adv. Mater.*, **12** 582.
14. Plassat, V., et al. 2007 *Int. J. Pharm.* **344** 118.
15. Masuda, H. and K. Fukuda 1995 *Science* **268** 1466.
16. Schoeman, B.J., Sterte J. and Otterstedt J. E.1994 *Zeolites* **14** 110.
17. Hamley, I.W.2009 *Prog. Polym. Sci.* **34** 1161.

18. Yao, J., et al. 2008 *Science* **321** 930.
19. Macak, J., et al. 2007 *Adv. Mater.* **19** 3027.
20. Azevedo, J., et al. 2014 *Materials Research Express* **1** 015028.
21. Yoon, S.L. 2008 *Self-Assembly and Nanotechnology*. (Wiley)
22. Raja, K.S., Misra M. and Paramguru K. 2005 *Electrochim. Acta*, 2005. **51** 154.
23. LaTempa, T.J., et al. 2009 *J. Phys. Chem. C* **113** 16293.
24. Qiu, X.F., et al. 2009 *Nanotechnology* **20** 455601.
25. Su, Z. and Zhou W. 2008 *Adv. Mater.*, 2008. **20**(19): p. 3663-3667.
26. Apolinário, A., et al., *J. Mater. Chem.* (in press).
27. Sousa, C.T., et al. 2012 *J. Mater. Chem.* **22** 6.
28. Marquardt, B., et al. 2008 *Nanotechnology*, 2008. **19** 405607.
29. Schlesinger, M. and Paunovic M. 2000 *Modern Electroplating*. (Wiley).
30. Proenca, M., et al. 2012 *Electrochim. Acta* **72** 215.
31. Sousa, C.T., et al., 2011 *Nanotechnology* **22** 315602.
32. Diggle, J.W., Downie T.C. and Goulding C.W. 1969 *Chem. Rev.* **69** 365.
33. Tajima, S. 1977 *Electrochim. Acta* **22** 995.
34. Shankar, K., et al. 2007 *Nanotechnology* **18** 11.
35. Vega, V., et al., 2008 *J. Non-Crystal. Sol.* **354** 5233.
36. Cai, Q.Y., et al. 2005 *J. Mater. Res.* **20** 230.
37. Guan, D.S. and Wang Y. 2012 *Nanoscale*, **4** 2968.
38. Thompson, G.E. and Wood G.C. 1981 *Nature* **290** 230.
39. Sreekantan, S., et al.2010 *Nanotechnology* **21** 365603.
40. Yuan, X., et al. 2010 *Nanotechnology* **21** 405302.
41. Sulka, G.D., et al. 2010 *Electrochim. Acta* **55** 4359.
42. O'sullivan, J. and Wood G. 1970 *Proc. Royal Soc. London. Series A, Mathematical and Physical Sciences* p511.
43. Ono, S., et al., 2004 *J. Electrochem. Soc.***151** B473.
44. Nielsch, K., et al. 2002 *Nano Lett.* **2** 677.
45. Macak, J.M., et al. 2008 *J. Electroanal. Chem.***621** 254.
46. Su, Z. and Zhou W. 2009 *J. Mater. Chem.* **19** 2301.
47. Sousa, C.T., et al. 2012 *Nanoscale Res. Lett.* (in press).
48. Liao, J., et al. 2012 *Chem. Eng. J.* **211–212** 343.

Journal of Physics: Conference Series **534** (2014) 012002 doi:10.1088/1742-6596/534/1/012002

Mössbauer Spectroscopic Study andMagnetic Investigation of Iron(III) Complexes on a DendrimericBasis

Driss Nariaki[1-2],Faik Lekovic[2], Patrick Homenya[1],Franz Renz[1-2]*

[1] Institut für Anorganische Chemie, Leibniz Universität Hannover, Callinstr. 9, 30167 Hannover, Germany
[2] Institut für Anorganische Chemie und Analytische Chemie, Johannes Gutenberg-Universität Mainz, Duesbergweg 10-14, 55099 Mainz, Germany

E-mail: franz.renz@acd.uni-hannover.de

Abstract. The functionalization of the molecular surface of various dendrimer generations with a phosphorous core and external amine groups is obtained by converting those amine groups into the corresponding imines of salicylaldehyde creating multiple coordination sites for the iron atoms. Treatment with iron(III) chloride yields multinuclear iron(III) complexes on a dendrimeric basis.The obtained multinuclear molecular systems exhibit extremely high total spin values. The influence of the generation growth on this type of coordination compounds is investigated by Mössbauer spectroscopy and SQUIDmagnetometry.

1. Introduction

One current topic in the research field of polymers aredendrimers [1]. These macromolecules have a tunable hyperbranched as well as defined structures and are built generation after generation by repetition of a series of reactions, starting from a central core [2]. Furthermore,a monodisperse nature and high end-group functionality are features of dendrimers. As consequence of these properties, it is possible to get several transition metals into one molecule by creating coordination sites at the end-groups [3, 4]. The dendrimeric backbone has the advantage that every further generation increases the number of metal atoms that can be bound in one molecule.The idea behind synthesizing dendrimers with higher generation numbers and many transition metals,respectively, is the systematic tuning of the magnetic properties in any generation [5].In this way, a large number of complexes can be obtained on small area – a method, which could achieve high densities in possible applications: optoelectronics, catalysis, biology and photophysical processes.

In this contribution, various dendrimer generations with phosphorous core and external amine groups are functionalized to yield iron(III) complexes bridged via a dendrimeric backbone. The products are studied by Mössbauer spectroscopy and SQUID magnetometry [6-8].

2. Experimental Section

Synthesesof the dendrimeric backbone have been carried out byadapting the description in reference [1, 2].

2.1. First generation dendrimeric nitrile

The amine P_0 (3.4 mmol) was dissolved in distilled water and acrylnitrile (122.75 mmol) was added dropwise. After stirring at 80 °C for 12 h the resulting yellow solution was evaporated to dryness.Yield: 83 % (M= 523.67 g/mol).
IR (KBr): ν (cm^{-1}) = 3358 cm^{-1} (N-R, val.),2934 cm^{-1} (CH_2, val.), 2250 cm^{-1} (C≡N,val.).

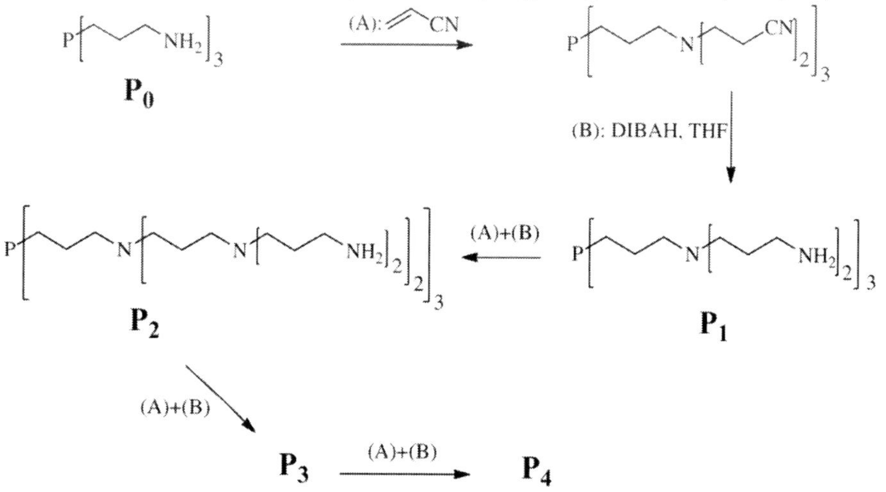

Figure 1. Step by step synthesis of a dendrimer with terminal amines from zeroth to fourth generation

2.2. First generation dendrimeric amine P_1

To a solution of the nitrile P_1 (1.68mmol) in dry THF (300 ml) a 1M solution of DIBAH in heptane (28.22 mmol) was added dropwise under nitrogen atmosphere. The mixture was heated to 80 °C for 48 h. After adding methanol (400 ml) the precipitated aluminium salt was filtered off and the resulting yellow solution was evaporated to dryness.Yield: 86 % (M= 574.86 g/mol).
IR (KBr): ν (cm^{-1}) = 3354 cm^{-1} (N-R, val.), 2934 cm^{-1} (CH_2, val.), 1664 cm^{-1} (NH_2, val.).

2.3. Higher generation dendrimeric nitriles and amines

Thedendrimeric nitrilesanddendrimeric amines of the second (P_2), third (P_3) and fourth (P_4) generation,respectively,were synthesized in analogy to the first generation (P_1) starting from the respective previous generation.(See Fig. 1)

Table 1.Reaction parameters for the synthesis of the corresponding nitriles.

	Educt [mmol]	Acrylnitrile [mmol]	Yield [%]	ν(C≡N) [cm^{-1}]
nitrile P_2	1.73	125	60	2248
nitrile P_3	1.2	172	69	2247.3
nitrile P_4	0.33	95	96	2246.8

Table 2.Reaction parameters for thesynthesis of the corresponding amines.

	Educt [mmol]	DIBAH [mmol]	Yield [%]	$\nu(NH_2)$ [cm^{-1}]
amine P_2	0.50	16	75	1660.3
amine P_3	0.35	23	81	1650.6
amine P_4	0.23	31	63	1655

2.4. Functionalization and complexation of dendrimers
Syntheses of the coordination compounds on the dendrimeric surface have been carried out by adapting the description in [3, 4].

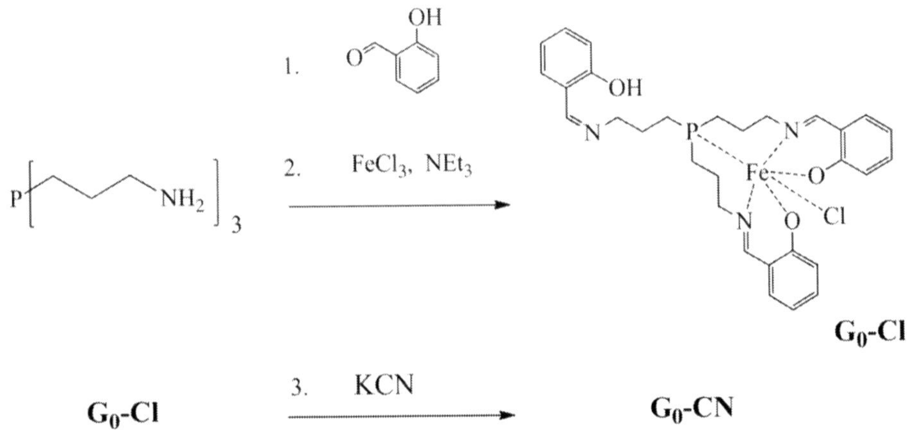

Figure 2. Synthesis of an iron(III) dendrimer compound, starting from a zeroth generation dendrimer and subsequent monodentate ligand-exchange with potassium cyanide.

The corresponding amine P_0 (3.16 mmol) was dissolved in methanol. The mixture was heated under reflux for 10 min and salicylaldehyde (10.42 mmol) was added dropwise. After 30 min of stirring at room temperature a solution of anhydrous iron(III) chloride (3.16 mmol) was added slowly. The resulting solution was heated to 50 °C for 5 min andtriethylamine (6.32 mmol) was added. Reducing the solvent and subsequent cooling precipitated the product. A solid G_0-Clwascollected and dried under reduced pressure. Yields: 60 %, (M= 606.89 g/mol).
IR (KBr): ν (cm^{-1}) = 3418 cm^{-1} (N-R, val.), 2925 cm^{-1} (CH_2, val.), 1612 cm^{-1} (C=N, val.).

The synthesis of the dendrimeric compounds of the first, second, third and fourth generations were performed in analogy to the zeroth generation (see Fig. 2, step 1&2 and Table 3) starting from the respectivedendrimer generation P_1, P_2, P_3 and P_4.

Table 3. Reaction parameters for the synthesis of the corresponding complexes G_n-Cl (n: generation number).

	Salicyl. [mmol]	$FeCl_3$ [mmol]	NEt_3 [mmol]	Yield [%]
G_1-Cl	9.91	4.19	8.38	61
G_2-Cl	5.88	4	4.8	58
G_3-Cl	6.91	2.5	5	50
G_4-Cl	6.28	2.5	5	45

2.5. Ligand exchange with cyanide

The sixth coordination site, a chloride ion, was exchanged by a cyanide ion via reaction with potassium cyanide (see Fig. 2, step 3): The dendrimeric complex G_0-Clwas dissolved in methanol and stirred for 4 h at 50 °C. Then, KCN as powder was added carefully in the reagent flask and the solution was stirred 1 h at room temperature. Reducing the solvent and subsequent cooling precipitated the product. The productG_0-CN was collected and dried under reduced pressure.
Yield: 75%, (M= 597.94 g/mol).
IR (KBr): ν (cm^{-1}) = 3407 cm^{-1} (N-R, val.),2044 cm^{-1} (C≡N, val.), 1618.7 cm^{-1} (C=N, val.).
The procedure for higher generation cyanides was performed in analogy.

Table 4. Reaction parameters for the synthesis of the corresponding complexes G_n-CN.

	KCN [mmol]	Yield [%]		KCN [mmol]	Yield [%]
G_1-CN	1	75	**G_3-CN**	1	75
G_2-CN	1	79	**G_4-CN**	1	63

3. Analytical Data for Dendrimer Compounds

All compounds were characterized by IR spectroscopy and ESI-MS. The IR spectra were measured in KBr pellets (Bruker Optics Tensor) in the 4000-400 cm^{-1} region at ambient temperatures. ESI mass spectra were recorded by "Finnigan MAT LCQ" in methanol (SupraSolv).

Infrared spectroscopy has proved to be a very useful toolto monitor the dendrimeric build-up of certain hyperbranched structures[9-12]. For the synthesizedamines, the IR spectra exhibita band in the range of 1664-1650 cm^{-1} characteristic for NH stretching vibrations. For synthesized nitriles(see Fig. 1), one band appears in the range of 2250-2246 cm^{-1} characteristic of C≡N stretching vibrations, whereas the NH stretching vibrations practically completely disappear. Vice versa, the C≡N stretching vibration band disappears after the reduction with DIBAL-H and NH band appears again. For synthesized complexes(see Fig. 2) bands in the range 1622-1611$cm^{-1}$are corresponding to the stretching vibration of C=N of the imine group, which show a significant shift under the influence of generation growth and ligand-exchange. (See Tab. 5).

Selected examples of ESI-MS signalsof different dendrimeric compounds are given belowin tabular form (see Tab.5). Due to the hyperbranched structure of the dendrimer compounds molecular peaks have been detected seldom in the ESI-MS despite its gentle ionisation method. Nevertheless, a tendency for a relative increase in the number of fragments with high mass-to-charge (m/z) ratioscan be observed for higher generation numbers.
Another method suggesting the successful generation built-up of the compounds is the SQUID magnetometry. Those results will be presented below (see Paragraph 5).

Table 5. Selected examples of infrared spectroscopy bands and electrospray ionisation mass spectrometry (ESI-MS) signals of different dendrimeric complexes.

	M [g/mol]	ν(C=N) [cm^{-1}]	ESI-MS m/z		M [g/mol]	ν(C=N) [cm^{-1}]	ESI-MS m/z
G_0-Cl	606.89	1612	304.2(z =2)	**G_0-CN**	597.94	1618.7	599.4(z =1)
G_1-Cl	1440.36	1617	441.1(z =3)	**G_1-CN**	1430.92	1620.7	441.8(z =3)
G_2-Cl	3017.99	1617.7	613(z =5)	**G_2-CN**	2961.38	1621.4	588(z=5)
G_3-Cl	6173.28	1611.5	996(z =6)	**G_3-CN**	6060.07	1621.8	865.6(z=7)
G_4-Cl	12438.8	1611.3	1234,6(z =10)	G_4-CN	12212.42	1621.5	1745(z=7)

4. Mössbauer Spectroscopy: Results and Discussion

Mössbauer spectra were recorded using a conventional transmission spectrometer at T = 20 K and T = 300 K. ^{57}Co/Rh was used as the source of the radiation and the isomeric shift data are given relative to Fe in Rh at room temperature.

Fig. 3 and Table 6 show the Mössbauer spectra and parameter of the dendrimericcomplexes from zeroth to fourth generation. Mössbauer spectra of all dendrimeric compounds at 300K and at 20 K consist of a signal with the shapeof a simple line doublet indicative for Fe(III) in the high-spin state (S=5/2).

In case of the G_n-Cl dendrimers, the spectra could be described best by a fit with 2 Lorentzian doublets. Since one would not expect discrete positions for the iron centres, for isomeric shift (IS) values the centre of gravity was determined for the complete signal in order to achieve better comparability within this discussion. Respective mean values for the quadrupole splitting (QS) were determined as well.

At 300 K all spectra show typical values for the IS in the range 0.2440(36)-0.2936(85) mm/s, showing the maximum value for the third generation. A maximum was found at low temperature as well. At 20 K the values of the IS are in the range of 0.3459(40)-0.4066(38) mm/s and are therefore well in accordance with the "Second-Order Doppler Shift".

The QS values exhibit a slight temperature dependence. Values in a range of 0.599(36)-1,132(93) mm/s indicate a deviation from an ideal octahedral coordination sphere. The differences occurring on generation growth could be explained by influences of different conformations and agglomeration degrees on the symmetry of the coordination sphere and therefore the electric field gradient.

Figure 4 displays the mean values for the isomeric shift and the quadrupole splitting for the G_n-Cl (n: 1-4) series for 20 and 300 K - visualizing the "Second-Order Doppler Shift" and the generation effect on the electric field gradient.

In case of the G_n-CN dendrimers, a strong asymmetry of the Mössbauer doublet is observed. This asymmetry is particularly pronounced for the lower generations and seems to be temperature dependent. In general, an asymmetric absorption signal in the Mössbauer spectroscopy can be caused by a texture, Goldanski-Karyagin or relaxation effects. The temperature dependence as well as the similar areas of the two resonant lines lead to the assumption that this asymmetry is caused by (spin-spin) relaxation.

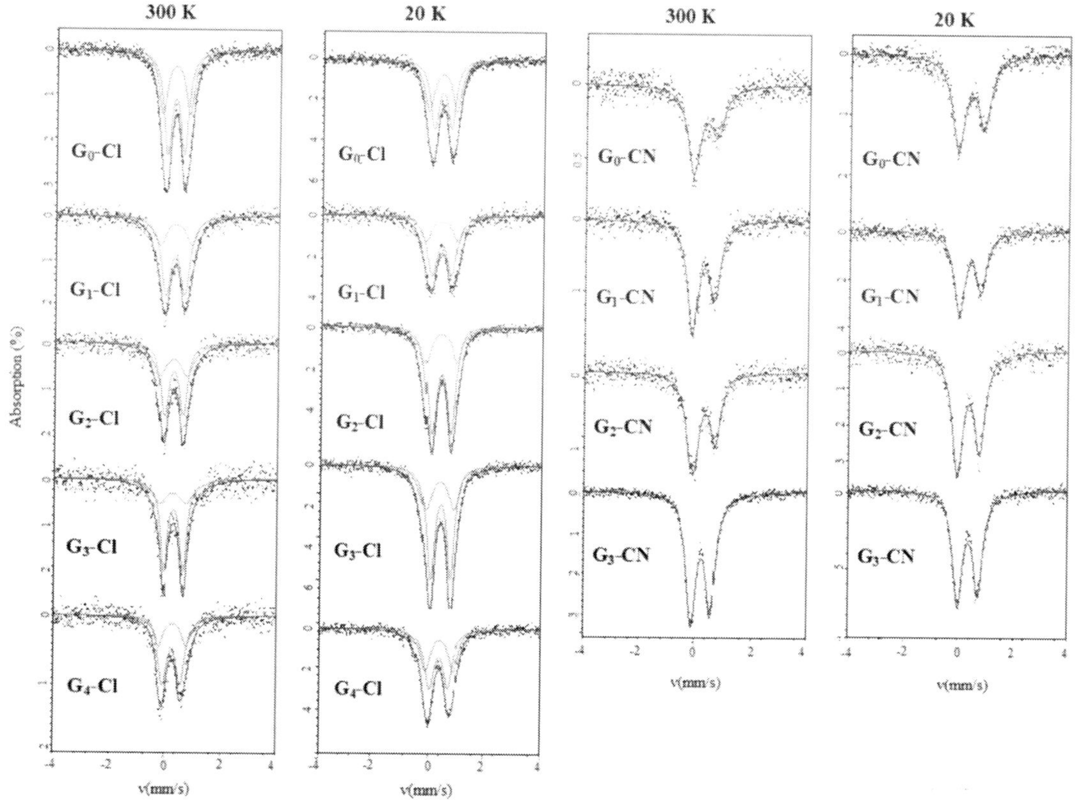

Figure3.Mössbauer spectra at 300 K und at 20 K of Iron(III) dendrimer compound from zeroth to fourth generation.

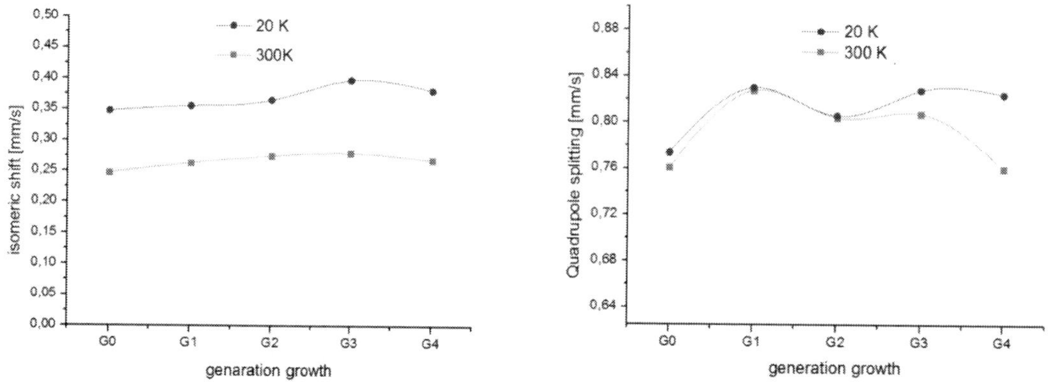

Figure 4. Dendrimeric complexes G_n-Cl (n: 1-4): temperature-dependence of the mean isomeric shift (left), and the mean quadrupole splitting (right).

5. Magnetometry: Results and Discussions

Magnetic susceptibility and magnetisation measurements were performed using a SQUID magnetometer (Quantum Design MPMS-XL)within a temperature range of T = 2 - 300 K and an applied external magnetic field of B = 0.1 T.Field-depended magnetization data were taken at T = 5 K. The effective magnetic moment has been calculated using $\mu_{eff} / \mu_B = 798 (\chi T)^{1/2}$ when SI units are employed. The results were corrected for the magnetisation of the sample holder and for the diamagnetic contributions, which were estimated from Pascal constants.

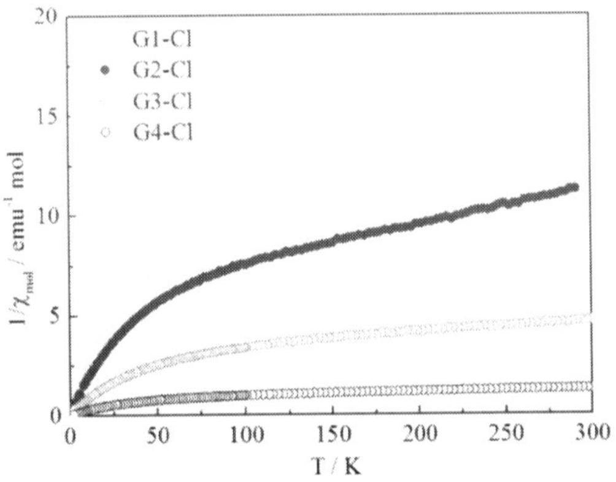

Figure 5. Temperature dependence of a pot of inverse susceptibility for dendrimeric complexes Gn-Cl (n: generations number) in different generations.

All $\chi^{-1}(T)$ curves (see Fig. 5) deviate strongly from a linear behaviourand only at higher temperatures the susceptibilities were found to follow the Curie-Weiss relationship.

By extrapolation to the x-axis negative values ofWeiss constantθwere found which indicate an antiferromagnetic behavior of the compounds. For all four dendrimer generations thisantiferromagnetismappears to be temperature dependent: It decreases with temperature reduction and increases with a temperature rise. The increase due to heating can be explained by the occupation of thermally populated states or by cooperative effectsunderlying the model of superexchange. It is apparent that the critical temperature abovewhich linear Curie-Weiss behavior occurs strongly depends on the dendrimer generation number.Furthermore, it can be observed that the smaller the generation number is the more negative appears the extrapolated Weiss constant, thus, indicating a weaker antiferromagnetic coupling between the spin-bearing iron centres for higher generation numbers which is probably due to inferior magnetic exchange paths within the large molecules.

The temperature dependence of the effective magnetic moment μ_{eff} seems not to have reached saturation at room temperature, whereas the slope of the curve varies heavily with the generation number (see Fig. 6, left). This observation suggests significantly higher values of the total spin for higher generation numbers.

The field-dependent magnetization M exhibits a similar tendency (Fig. 6, right): Saturation of the sample seems to be reached at lower external field-strengths for lower generation numbers.

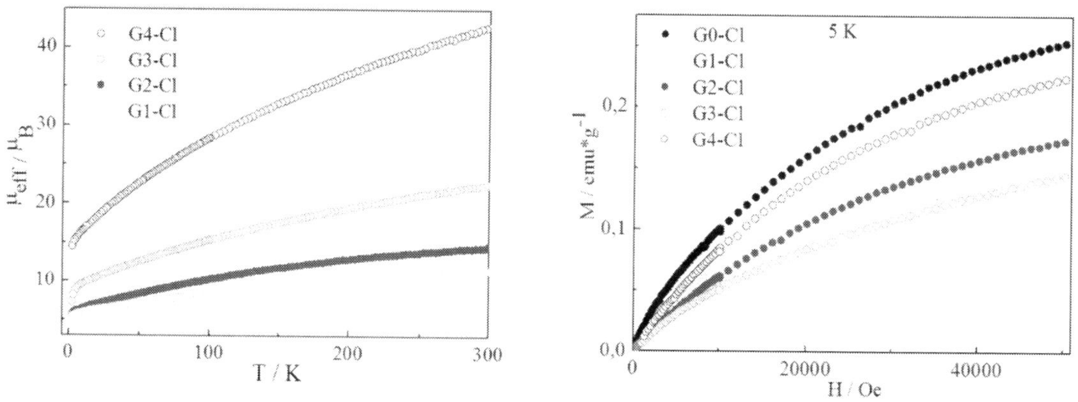

Figure 6. Temperature dependence of effective magnetic moment (left) and field dependence of the magnetization at 5 K (right) for dendrimeric complexes Gn-Cl in different generations.

6. Conclusions

In this contribution, multinuclear iron(III) compounds with a dendrimeric backbone were synthesized and characterized via IR spectroscopy, Mössbauer spectroscopy and SQUID magnetometry.

Even though a pantadente ligand providing a N3O2 coordination surrounding for the iron(III) was chosen which enabled spin crossover in previous studies [4, 6-7], a spin transition could not be detected. Rather, Mössbauer spectroscopic measurements indicate all iron(III)centresto be in their high spin state in the temperature range investigated (20-300 K). The isomeric shiftsare slightly influenced by the generation growth whereasquadrupolesplittingswere found to bestrongly dependent on the generation number. A ligand exchange in the position of themonodentateligand from chloride to cyanide led to a strong asymmetry of the Mössbauer doublet. This asymmetry is particularly pronounced for the lower generations and can probably be explainedbyrelaxations processes or texture effects.

The magnetic investigations revealed antiferromagnetically coupled molecular systems with higher magnetic moments for higher generation numbers. Even thecooperative interactions of the spin bearing centres exhibit aremarkable generation effect: The antiferromagnetic interaction decreases with an increasing generation number which is probably due to inferior magnetic exchange paths.

All results outline the possibility to create molecular systems with extremely high total spin values using the dendrimeric approach.

Table 6.Mössbauer parameters of dendrimer compounds in different generations:Before and after a monodentatelingand exchange.

		Fe(III) HS-state			Fe(III) HS-state		
Compound	T(K)	δ [mm/s]	Δ [mm/s]	Fraction [%]	δ [mm/s]	Δ [mm/s]	Fraction [%]
G_0-Cl	300	0.2440(36)	0.608(30)	59	0.2529(57)	0.980(55)	41
	20	0.3459(40)	0.599(36)	53	0.3508(53)	0.970(59)	47
G_0-CN	300	0.278(16)	0.860(28)	100	---	---	---
	20	0.3636(86)	0.932(15)	100			
G_1-Cl	300	0.2545(64)	0.690(40)	69	0.280(18)	1.132(93)	31
	20	0.364(16)	1.119(74)	32.6	0.355(15)	0.690(11)	67.4
G_1-CN	300	0.2496(78)	0.7591(14)	100	---	---	---
	20	0.3528(70)	0.781(13)	100			
G_2-Cl	300	0.283(12)	0.672(61)	57	0.265(34)	0.98(47)	43
	20	0.3653(24)	0.6772(19)	69.1	0.3657(72)	1.093(69)	30.9
G_2-CN	300	0.2769(90)	0.793(16)	100	---	---	---
	20	0.3672(63)	0.800(11)	100			
G_3-Cl	300	0.2936(85)	0.720(30)	66	0.254(60)	0.977(65)	34
	20	0.4066(38)	0.7142(18)	63	0.383(12)	1.021(13)	37
G_3-CN	300	0.2557(22)	0.7148(38)	100	---	---	---
	20	0.3658(34)	0.7364(60)	100			
G_4-Cl	300	0.258(11)	0.650(35)	68	0.292(23)	0.994(68)	32
	20	0.3676(55)	0.677(18)	56.2	0.3990(78)	1.013(18)	43.8

7. References

[1] Moors R andVögtle F1993*Chem. Ber.* **126** 2133-35
[2] De Brabander-van den Berg E. M. M and Meijer E. W*1993Angew. Chem.* **105**1370-72
[3] Renz F, Gembický M and Boca R2000*Inorganic Chemistry Communications***11** 662-665
[4] Renz FandKerep P 2004*Hyperfine Interactions***156** 371-377
[5] Matsumoto M, Ohta S, Yoshimura C, Ohyoshi A,Kohata S,OkawaH andMaeda Y 1985*J. Chem. Soc. Dalton Trans.* **12** 2575-84
[6] Renz F, Martinez F, Klein M 2008*Hyperfine Interact***184**251–257
[7] Renz F, Hill D, Klein M andHefner J 2007*Polyhedron***26** 2325–29
[8] Piñeroa D, Barana P, Boca B, Herchel R, Klein M,Raptis R G, Renz F andSanakis Y 2007 *Inorg. Chem.* **46**(26) 10981-89
[9] Furer V L, Vandyukov AE,Vandyukova I I, Majoral J P, Caminade A M andKovalenko V I2009*Vibrational Spectroscopy***51**326–332
[10] Li R and Bu J2004 *KoreanJ. Chem. Eng.***21**(1) 98-103
[11] Tarazona-Vasquez F andBalbuena P B2004*J. Phys. Chem.* B**108**15982-91
[12] Furer V L, Kovalenko V I,Vandyukov A E,Majoral J P andCaminade A M2002*SpectrochimicaActa*PartA**58** 2905-12,2003*Vibrational Spectroscopy***31** 71–79,2013*Chemical Physics***421** 57–67

Iron(III) Complexes on a Dendrimeric Basis and Various Amine Core Investigated by Mössbauer Spectroscopy

D. Nariaki[1-2], F. Lekovic[2], P. Homenya[1], B. F. O. Costa[3], M. Menzel[4], R. Boca[5], M. Blumers[2], G. Klingelhöfer[2], F. Renz[1-2*]

[1] Institut für Anorganische Chemie, Leibniz Universität Hannover, Callinstr. 9, 30167 Hannover, Germany
[2] Institut für Anorganische Chemie und Analytische Chemie, Johannes Gutenberg-Universität Mainz, Duesbergweg 10-14, 55099 Mainz, Germany
[3] CEMDRX, Physics Department, University of Coimbra, RuaLarga, P-3004-516 Coimbra,
Portugal
[4] Federal Institute for Materiels Research and testing (BAM), Richard-Willstaetter Str. 11, D -12489 Berlin, Germany
[5] Institute of Inorganic Chemistry, Slovak University of Technology, 812 37 Bratislava,
Slovakia

E-mail:franz.renz@acd.uni-hannover.de

Abstract. Dendrimers of various generations were synthesized by the divergent method. Starting from various amine cores (G_{0a}, G_{0b}, G_{0c}) the generations were built by reaction of the amine with acrylnitrile followed by hydrogenation with DIBAL-H. Treatment with salicylaldehyde creates a fivefold coordination sphere for iron in the molecular periphery. The resulting multinuclear coordination compounds are investigated by Mössbauer spectroscopy.

1. Introduction

Dendrimers represent a novel type of polymeric materials that have been of increasing interest in recent years revealing applications in supramolecular chemistry, nanoscience, medicine and catalysis [1]. These macromolecules are defined as highly branched, spherical and monodisperse molecules composed of monomers that spread from a central core [2]. The dendrimers are produced in iterative sequences of reactions steps, whereby any further sequence leads to a higher generation dendrimer. (See Fig. 1)

In this study, we focused on divergent approach, where the molecule grows from the core to the periphery and whereas convergent synthesis first builds up the dendrons called arms, wich were connected in a merging step to the core fragment. The extraordinary topology of the dendrimers was used to yield multinuclear iron(III) complexes by functionalizing the terminal amine groups and subsequently to observe the influence of the generation growth on the spin state of the resulting metal complex with Mössbauer spectroscopy [3-7].

Figure 1. Scheme of a fourth generation dendrimer complex G_{4a}.

2. Experimental Section

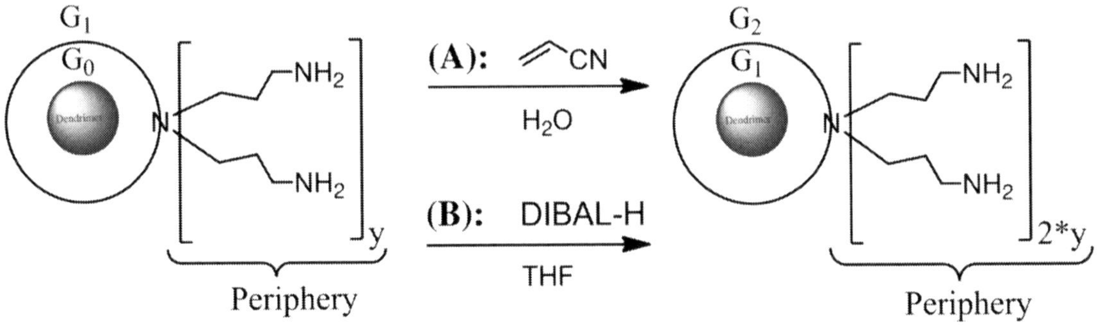

Figure 2. Generation growth of dendrimeric amine; y: indicates number of amines at the molecular periphery.

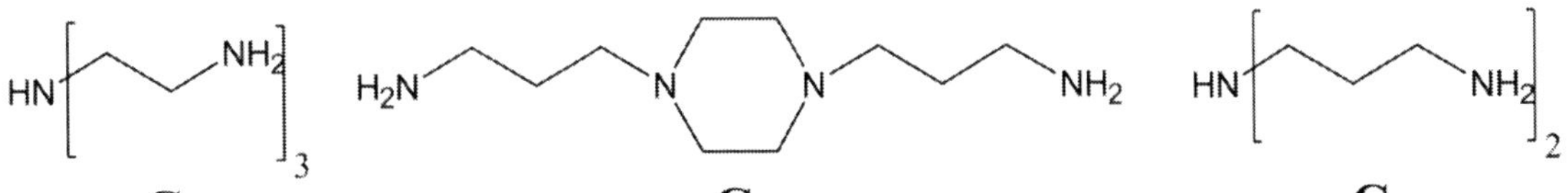

G_{0a} G_{0b} G_{0c}

Figure 3. Scheme of the different amine cores. The subscript numbers indicate the generation; the subscript letters indicate the respective core.

2.1. First generation dendrimeric nitrile G_{1a}

The amine G_{0a}(3 mmol) wasdissolved in distilled water and acrylnitrile (108 mmol) was added dropwise. After stirring at 80 °C for 12 h the resulting yellow solution was evaporated to dryness. Yield: 87 %, (M= 464.31 g/mol).
IR (KBr): ν(cm^{-1}) = 3350 cm^{-1}(N-R, val.), 2954 cm^{-1} (CH_2, val.), 2247cm^{-1} (C≡N, val.).

2.2. First generation dendrimeric amine G_{1a}

To a solution of the nitrile G_{1a} (2 mmol) in dry THF a 1M solution of DIBAL-H in heptane (33.6 mmol) was added dropwise under nitrogen atmosphere. The mixture was heated to 80 °C for 48 h. After adding methanol the precipated aluminium salt was filtered off and the resulting yellow solution was evaporated to dryness. Yield: 81%, (M= 484.47 g/mol).
IR (KBr): ν (cm^{-1}) = 3341cm^{-1}(N-R, val.), 2926 cm^{-1} (CH_2, val.), 1643 cm^{-1} (NH_2, val.).

2.3. Higher generation dendrimeric nitrile and amine

The dendrimeric nitriles and dendrimeric amines of the second (G_2), third (G_3) and fourth (G_4) generation were synthesized in analogy to first generation dendrimeric nitriles and amines starting from the respective amines and niriles of the previous generation. (See Fig. 2)

Table 1. Reaction parameters for the synthesis of the corresponding amines of dendrimers a, b and c.

	Educt[mmol]	DIBAL-H[mmol]	Yield[%]
amine G_{1a}	0.86	12.38	81
amine G_{2a}	0.44	11.73	95
amine G_{3a}	0.16	8.63	91

	Educt [mmol]	DIBAL-H[mmol]	Yield[%]
amine G_{1b}	1.30	13.2	87
amine G_{2b}	0.43	10.5	72
amine G_{3b}	0.25	12.0	60
amine G_{4b}	0.01	9.46	97

	Educt [mmol]	DIBAL-H[mmol]	Yield [%]
amine G_{1c}	1.46	12.85	91
amine G_{2c}	0.74	13.02	57
amine G_{3c}	0.37	13.18	86
amine G_{4c}	0.18	13.73	87

2.4. Functionalization and complexation of dendrimers

The amine G_{0a} (2 mmol) was dissolved in methanol. The mixture was heated under reflux for 10 min and salicylaldehyde (6.6 mmol) was added dropwise. After 30 min of stirring at room temperature a solution of anhydrous iron(III) chloride (2 mmol) in methaol was added slowly. The resulting solution was heated to 50 °C for 5 min and triethylamine (4 mmol) was added. Reducing the solvent and cooling precipitates the product. A solid G_{0a}-Cl was collected and dried under reduced pressure. Yield: 55 %, (M= 546.87 g/mol).
IR (KBr): ν (cm^{-1}) = 3347 cm^{-1} (N-R, val.), 2935 cm^{-1} (CH_2, val.), 1619 cm^{-1} (C=N, val.).

The synthesis of the dendrimeric compounds of the first (G_1), second (G_2), third (G_3) and fourth (G_4) generations were performed in analogy to the zeroth generation. The synthesis of dendremeric compounds with a different coreG_b and G_c were performed in analogy to G_a. (See Fig. 3 & 4)

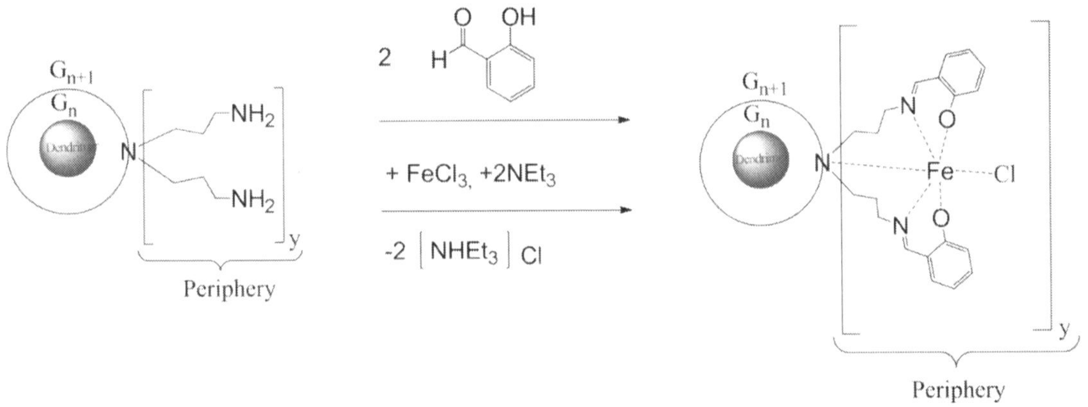

Figure 4. Scheme of the functionalization and the complexation of dendrimers; n: generation number.

Table 2. Reaction parameters for the synthesis of the corresponding complexes G_{na}-Cl, G_{nb}-Cl and G_{nc}-Cl(n: generation number).

	Salicyl.[mmol]	$FeCl_3$[mmol]	NEt_3 [mmol]	Yield [%]
G_{0a}-Cl	15.41	7.04	15.41	55
G_{1a}-Cl	6.12	3.06	6.12	28
G_{2a}Cl	5.00	2.50	5.00	48
G_{a3}-Cl	3.57	1.79	3.57	29

	Salicyl.[mmol]	$FeCl_3$[mmol]	NEt_3 [mmol]	Yield [%]
G_{1b}-Cl	0.48	0.24	0.48	20
G_{2b}-Cl	1.01	0.51	1.01	66.5
G_{3b}-Cl	2.80	1.40	2.88	63.9
G_{4b}-Cl	5.89	2.94	5.89	15

	Salicyl.[mmol]	$FeCl_3$[mmol]	NEt_3 [mmol]	Yield [%]
G_{1c}-Cl	10.44	5.22	10.44	45
G_{2c}-Cl	3.36	1.68	3.36	71
G_{3c}-Cl	5.12	2.56	5.12	18
G_{4c}-Cl	7.04	3.52	7.04	13

3. Structure Investigation of the Dendrimeric Compounds

All compounds were characterized by IR spectroscopy and ESI-MS. The IR spectra were measured in KBr matrix on a Brucker Optics Tensor in the 4000-400 cm^{-1} region at ambient temperatures. Infrared spectroscopy has proved to be a very useful tool in the structural analysis and identification of different dendrimeric compounds[8, 9]. For the synthesized amines, the IR spectra show a band in the range of 1654-1645 cm^{-1} characteristic of NH stretching vibrations.For synthesized nitriles, one band appears in the range of 2250-2240 cm^{-1} characteristic of C≡N stretching vibrations, whereas the NH stretching vibrations disappear completely. Vice versa, the C≡N stretching vibration band disappears after the reduction with DIBAL-H.

For synthesized dendrimeric complexes, bands in the range of1618-1621 cm^{-1} are corresponding to the stretching vibration of C=N of the imine group. (See Tab. 3)Selected examples of ESI-MS signals of different dendrimeric compounds are given below in tabular form (see Tab. 3). Due to thehyperbranched structure of the dendrimer compounds molecular peaks have been detected seldom in the ESI-MS despite its gentle ionisation method. Nevertheless, a tendency for a relative increase in the number of fragments with high mass-to-charge (m/z) ratios can be observed for higher generation numbers.

Table 3. Selected examples of infrared spectroscopy bands and electrospray ionisation mass spectrometry (ESI MS) signals of different dendrimeric complexes

	M [g/mol]	υ(C=N) [cm^{-1}]	ESI-MS m/z		M [g/mol]	υ(C=N) [cm^{-1}]	ESI-MS m/z
G_{0a}-Cl	546.87	1619.3	547.1(z =1)	**G_{1b}-Cl**	1115.79	1618.7	1061(z =1)
G_{1a}-Cl	1380.39	1621.5	1337(z =2)	**G_{2b}-Cl**	2472.45	1619.1	1236(z =2)
G_{2a}-Cl	2959.17	1620.7	1479.1(z =2)	**G_{3b}-Cl**	5102.85	1620.2	1595.6(z=3)
G_{3a}-Cl	6124.62	1619.6	1524(z =4)	**G_{4b}-Cl**	10364.02	1620.0	1727.3(z=6)

	M [g/mol]	υ(C=N) [cm^{-1}]	ESI-MS m/z
G_{1c}-Cl	1023.69	1618.7	476.5(z = 2)
G_{2c}-Cl	2037.68	1619.8	1019.3(z = 2)
G_{3c}-Cl	4179.20	1620.2	10358.3(z = 3)
G_{4c}-Cl	8413.85	1620.1	1542.1(z = 5)

4. Mössbauer Results and Discussion

The Mössbauer spectra were recorded using a conventional transmission spectrometer at T = 20 K and T = 300 K. ^{57}Co/Rh was used as the source of the radiation and the isomeric shift data are given relative to Fe in Rh at room temperature.

Fig. 6 and Table 4 show the Mössbauer spectra and the parameter of the dendrimeric complexes from zeroth to fourth generation. Mössbauer spectra of all dendrimeric compounds at 300K and at 20 K exhibit the shape of a simple line doublet indicating Fe(III) in high-spin state (S=5/2).

Some spectra could be described best by a fit with 2 or 3 lorentzian doublets. Since one would not expect discrete positions for the iron centres, for both isomeric shift (IS) and quadrupole splitting values the centre of gravity was determined for the complete signal in order to achieve better comparability within this discussion.

At 300 K all spectra exhibit typical values for the IS in the range of 0.237(22)-0.341(13) mm/s. The observed higher values at 20 K are well in accordance with the "Second-Order Doppler Shift" (in the range of0.295(37)-0.440(11) mm/s). There seems to be no consistent trend of an influence of the generation growth on the IS. Deviations of individual values could be caused by agglomeration effects. (See Tab. 4)

The QS values exhibit a slight temperature dependence. Values in a range of 0.554(21)-1.300(30) mm/s indicate a deviation from an ideal octahedral coordination sphere. The differences occurring on generation growth could be explained by influences of different conformations and agglomeration degrees on the symmetry of the coordination sphere and therefore the electric field gradient. (See Fig. 5& Tab. 4)

In case of the G_c-Dendrimers strong asymmetry of the Mössbauer doublet is observed (see fig. 6). This asymmetry is particularly pronounced for the first and third generation and seems to be temperature dependent. In general, an asymmetric absorption signal in the Mössbauer spectroscopy can be caused by a texture, Goldanski-Karyagin or relaxation effect. The temperature dependence as well as the similar areas of the two resonant lines lead to the assumption that this asymmetry is caused by (spin-spin) relaxation.

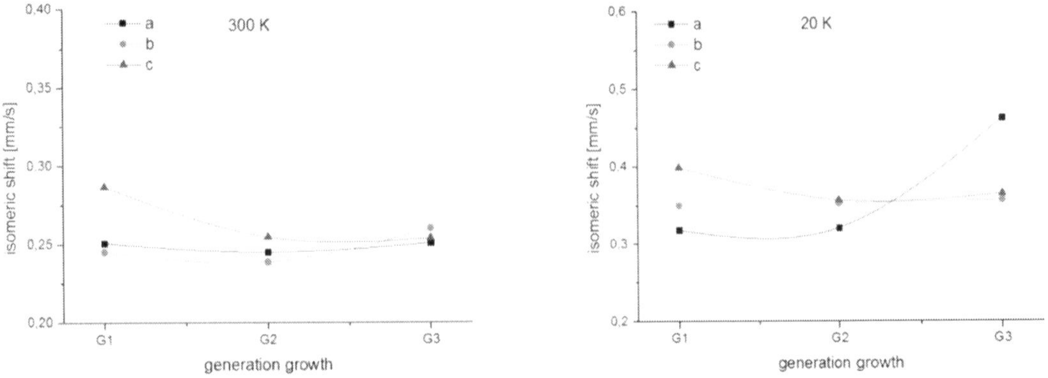

Figure 5. Dependence of the Isomeric shift on the dendremeric cores a, b and c and the generation at 20 K and 300 K..

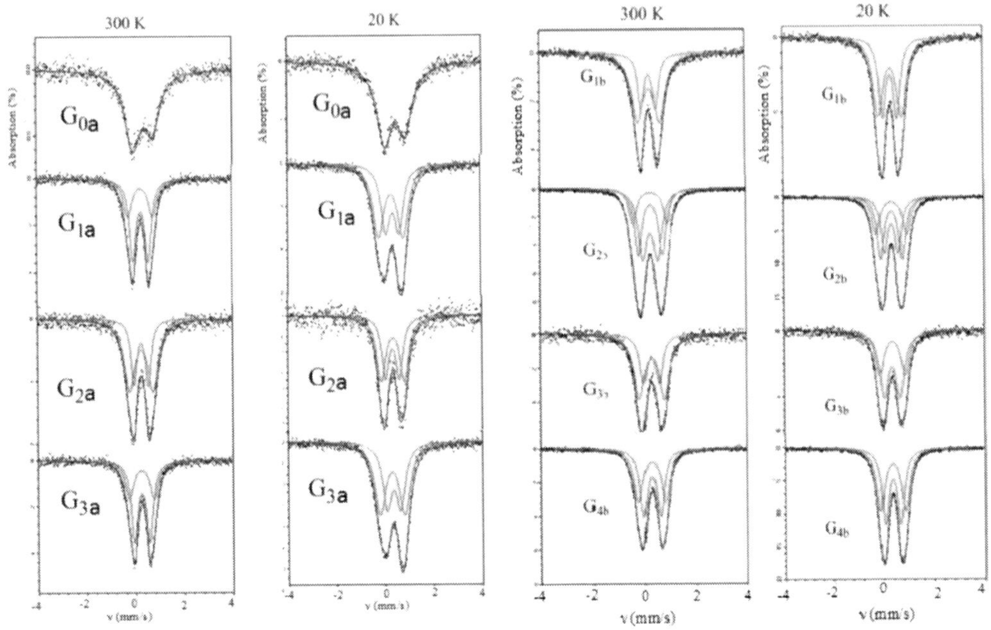

Figure 6a.Mössbauer spectra at 300 K and at 20 K of dendrimeric Iron(III) compound G_{na}, and G_{nb}(n: generation number)

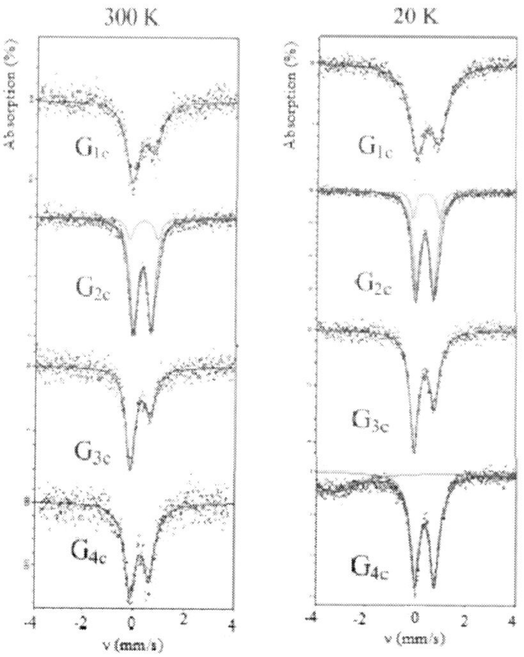

Figure 6b. Mössbauer spectra at 300 K and at 20 K of dendrimeric Iron(III) compound G_{nc}(n: generation number)

Table 4: Mössbauer parameters of dendrimeric Iron(III) compounds G_{na}, G_{nb} and G_{nc} (n: generation number)

		Fe(III) HS-state			Fe(III) HS-state		
Complexes core a	T(K)	δ [mm/s]	Δ [mm/s]	Fraction [%]	δ [mm/s]	Δ [mm/s]	Fraction[%]
G_{0a}-Cl	300	0.341(13)	0.799(18)	100	---	---	---
	20	0.440(11)	0.845(16)	100	---	---	---
G_{1a}-Cl	300	0.2457(67)	1.018(22)	39	0.2558(30)	0.620(19)	61
	20	0.358(17)	0.563(11)	43	0.295(19)	1.024(11)	57
G_{2a}-Cl	300	0.247(13)	0.963(19)	66	0.242(27)	0.587(13)	34
	20	0.341(20)	0.960(24)	57	0.295(14)	1.024(32)	43
G_{3a}-Cl	300	0.245(13)	01.018(17)	39	0.255(12)	0.620(17)	61
	20	0.358(11)	0.563(30)	42	0.547(25)	0.848(15)	57

		Fe(III) HS-state			Fe(III) HS-state		
Complexes core b	T(K)	δ [mm/s]	Δ [mm/s]	Fraction[%]	δ [mm/s]	Δ [mm/s]	Fraction[%]
G_{1b}-Cl	300	0.2423(32)	0.554(21)	31	0.2472(37)	0.837(65)	69
	20	0.3402(29)	0.803(43)	59	0.360(25)	0.340(21)	41
G_{2b}-Cl	300	0.2373(22)	1.300(30)	20	0.242(12)	0.909(14)	37
					0.2378(14)	0.555(27)	43
	20	0.3552(11)	0.848(15)	39	0.3500(21)	1.215(24)	27
					0.3528(13)	0.535(24)	34
G_{3b}-Cl	300	0.2654(59)	0.920(69)	65	0.2524(37)	0.569(52)	35
	20	0.367(12)	1.003(41)	40	0.350(13)	0.626(31)	60
G_{4b}-Cl	300	0.2563(15)	0.621(11)	50	0.2545(19)	1.007(20)	49
	20	0.3670(11)	0.6150(19)	50	0.3672(14)	0.982(15)	50

		Fe(III) HS-state			Fe(III) HS-state		
Complexes core c	T(K)	δ [mm/s]	Δ [mm/s]	Fraction [%]	δ [mm/s]	Δ [mm/s]	Fraction [%]
G_{1c}-Cl	300	0.287(20)	0.864(33)	100	---	---	---
	20	0.3990(83)	0.921(13)	100			
G_{2c}-Cl	300	0.2673(47)	0.696(11)	72	0.315(33)	1.100(47)	28
	20	0.4073(31)	1.100(17)	17	0.3746(24)	0.608(57)	83
G_{3c}-Cl	300	0.250(16)	0.755(28)	100	---	---	-----
	20	0.3654(43)	0.8269(78)	100			

5. References

[1] Moors R andVögtleF 1993*Chem. Ber.* **126** 2133-35
[2] De Brabander-van den BergE MMandMeijer E W1993 *Angew. Chem.***105** 9 1370-72
[3] Renz F, Gembický M andBoca R 2000*Inorganic Chemistry Communications***3**, 11, 662-665
[4] Frey H, Lorenz K and Lach C 1996 *Chemie in unserer Zeit***30** 2 75-85
[5] RenzF andKerep P 2004*Hyperfine Interactions***156**, 371-377
[6] Renz F, Hill D, Kerep P, Klein M, Müller-Seipel R and Werner F2006*Hyperfine Interactions***168**, 1051-56
[7] Tahir M N, Messerschmidt M, Klein M, Martinez V, Theato P, Metz N, Hartmann S, Kolb U, Ksenofontov V, Renz F andTremel W2009*Polyhedron***28**, 1728-1733
[8] Li R and Bu J 2004*Korean J. Chem. Eng.***21**(1) 98-103
[9] Furer V L, Kovalenko V I,Vandyukov A E,Majoral J P andCaminade A M2002*SpectrochimicaActa*Part A**58** 2905-12.2003*Vibrational Spectroscopy***31**, 71–79.2013*Chemical Physics***421**, 57–67

Extended X–ray absorption fine structural studies of copper and nickel ferrites

P. K. Malviya[1], P. Sharma[2], Ashutosh Mishra[3] and D. Bhalse[4]
[1]Govt. College, Nagda, M P, India
[2]Govt. Holkar Science College, Indore, M P, India
[3]School of Physics, Devi Ahilya University Indore, M P, India
[4]Govt. P. G. College, Khargone, M P, India

Email-malviyapramod35@yahoo.co.in

Abstract:-The Cu-Ni ferrites with general formula $Cu_{1-x}Ni_x Fe_2O_4$ (where x=0.0, 0.05, 0.10, 0.15, 0.20) were prepared by solid state root method. X-ray, K- absorption fine structural measurements were carried out. EXAFS spectra have been recorded at the K-edge of Fe using the dispersive EXAFS (DEXAFS) beam line at 2.5GeV Indus-2 synchrotron radiation source RRCAT, Indore, India. The EXAFS data have been analysed using the computer software Athena. These have been used to determine the bond lengths in these ferrites with the help of four different methods, namely, Levy's, Lytle's and Lytle, Sayers and Stern's (LSS) methods.

1. Introduction

Ferrites have gained technological important due to their high electrical resistivity ,low eddy current and low dielectric loss These material are extensively used in microwave device , computer memory chips , magnetic recording media ,transformer cores, rod antennas etc. Usually ferrites have spinel structure (AB_2O_4), where A is a divalent and B is a trivalent metal ion. Additives play an important role in controlling the magnetic and electrical properties of ferrites. Cu-Ni ferrites belong to the group of soft ferrite materials characterized by high magnetic permeability's and low losses and have numerous electronic applications [1, 2].In Cu-Ni ferrites, the minor additions of bivalent ions decrease the electrical conductivity .However; higher valent additive ions received very little attention [3, 4].Co –Ni ferrites are mixed ferrites and the co-ordination of the Fe+3 ion in the system is very little affected by the changes in compositional parameters [5].It is reported that the vacancy concentration of Oxygen is important parameter in the sintering process of spinel ferrites. It is also reported that electrical resistance of Cu ferrites decreases whit increasing quenching temperature, which is mainly attributed to the decrease of grain boundary resistance in Fe-excess Co- ferrites [6, 7].Extended X- ray absorption fine structure, which start approximately from 50 eV to continue up to 1000 eV above the edge. XANES is strongly sensitive to formed oxidation state and coordination chemistry of the absorbing atom while the EXAFS is used to determine the bond distance, Coordination number and species of the neighbours of the absorbing atom [8].

2. Experimental

The low cost parent oxides (Fe_2O_3, CuO, NiO) were used are raw materials. The ferrites of different composition [$Fe_2Cu_{(1-x)}Ni_{(x)}O_4$,x=0.00,0.05,0.10,0.15,0.20] were prepared using conventional solid state method. Powder was thoroughly mild and mixed by hand grinding tool. The sample heated for 10 -hours in 1000^0Cby furnace. The X-ray absorption spectra have been recorded using synchrotron radiation source .The X- ray spectroscopy setup is available at Raja Ramanna Centre for Advanced Technology (RRCAT) and is called beam line .This beam line BL-8 has been recently commissioned at the 2.5 GeV Indus-2 synchrotron radiation source.

3. Results and discussion

We have determined the bond lengths for the Cu-Ni ferrite complexes with the help of Levy's [2] method and graphical methods [3]. We have determined bond lengths using the slope of n Vs k Plots, which gives the value of (R – α) where R is the bond length. The parameter α depends to a large extent on the central absorbing atom. It is found that for chemically similar system, the value of α remains more or less the same. The values obtained for R are given in Table 1. We have also calculated the bond lengths by Levy's method and these are also included in Table 1. It is important to note here that the distance R-α should be equal to the distance found from the L.S.S. graphical method outlined above. Hence, both the L.S.S. method and the Fourier transformation method give the value R-α, i.e., both the methods give the value of bond lengths which have not been corrected for the phase shifts. We have called this distance as the phase uncorrected bond length. It is seen from this table 1 that the value of R-α as determined from L.S.S. method and that determined from the Fourier transformation method are in good agreement with each other, i.e., both the L.S.S. method and Fourier transformation method give nearly the same value of the phase uncorrected bond length, i.e., R-α.

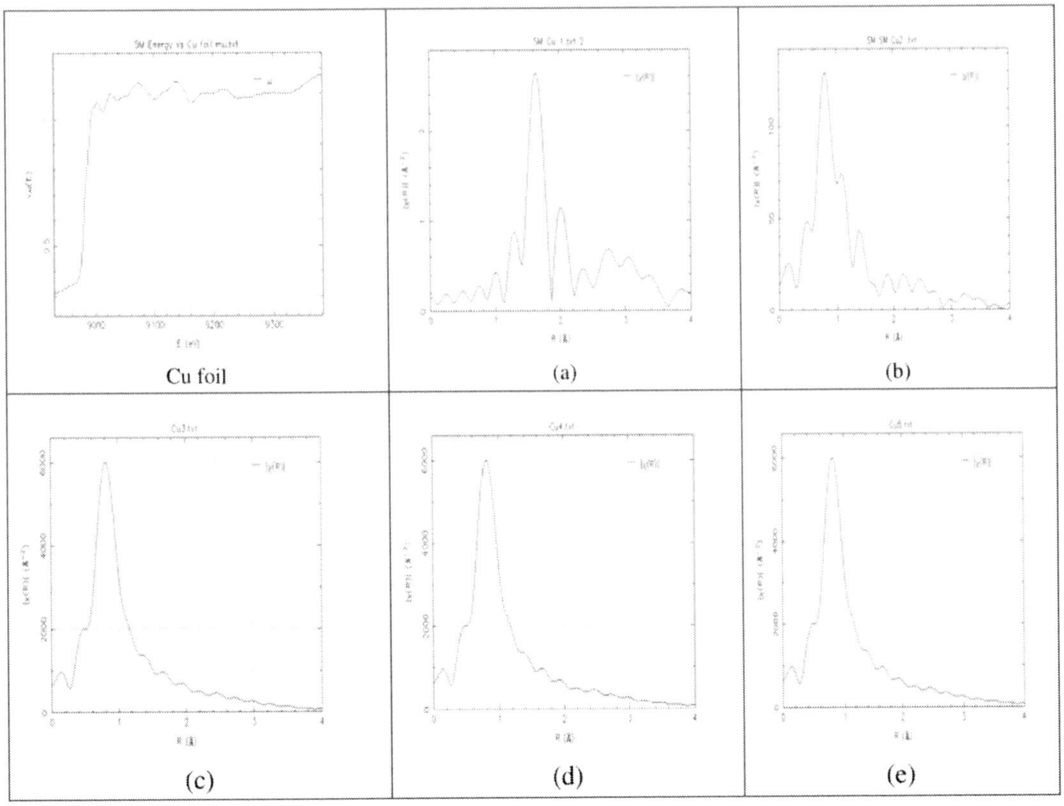

Figure 1. EXAFS spectrum of Cu foil. Fourier transform of EXAFS of Cu-Ni ferrites (a, b, c, d and e)

Table 1. Values of first shell bond length (in Å).

S.No.	Complex	Phase corrected		Phase uncorrected	
		Levy's method	Lytle's method	LSS method	FT method
		R_1	R_s	$R_1 - \alpha_1$	R
1.	Fe_2O_4Cu	1.41	1.41	1.40	1.26
2.	$Fe_2O_4Cu_{(.05)}Ni_{(.05)}$	1.50	1.51	1.51	1.24
3.	$Fe_2O_4Cu_{(.09)}Ni_{(.10)}$	1.49	1.50	1.51	1.38
4	$Fe_2O_4Cu_{(.85)}Ni_{(.15)}$	1.43	1.44	1.42	1.37
5	$Fe_2O_4Cu_{(.8)}Ni_{(.20)}$	1.40	1.41	1.44	1.38

4. Conclusions

It has been observed that the value of the phase uncorrected bond length, i.e., R-α as determined from L.S.S. method and that determined from the Fourier Transformation methods are in good agreement with each other, i.e., both the L.S.S. method and Fourier transformation method give nearly the same value of the phase uncorrected bond length. From the positions of the EXAFS maxima and minima, the bond lengths in the complexes have been determined by three different methods viz. Levy's, (1.41 to 1.43Å)Lytle's(1.41 to 1.51Å) and Lytle, Sayers and Stern's (L.S.S.) (1.40 to 1.51Å) methods. The normalized spectra, i.e., μ (E) versus E curves have been obtained. From these curves, χ(k) versus k curves have been obtained, which has then been Fourier transformed using the software Athena. From the Fourier transforms (1.24 to 1.38Å) of the EXAFS spectra the bond lengths (uncorrected for phase shift) have been determined.

References

[1] Malik F. A, Zahid M and Hassan, M. A 1995 *Proceeding 4th international symposium on advanced materials* 618.
[2] Rana M. U, Islam M. U, Ahmad I and Abbas I 1997, *Proceeding 5th international symposium on advanced materials* 283.
[3] Abbas T., Islam M. U. 1995 *Mod. Phys. Lett. B* **9** 1419.
[4] Cullity B. D. 1956 *Elements of X-ray diffraction* 1956, (Addison-Wesley)
[5] Scherrer P 1990 *Go¨ttinger Nachrichten.* **2** 98.
[6] Agrawal B. K. and Verma L.P. 1970 *J. Phys. C* **3** 535
[7] Teo B.K. and Lee P. A. 1979 *J. Am. Chem. Soc.* **101** 2815.
[8] Lytle F W, Sayers D E and Lytle F W 1975 *Phys. Rev.* B **11** 4836.

Journal of Physics: Conference Series **534** (2014) 012005 doi:10.1088/1742-6596/534/1/012005

X-ray absorption fine structure (XAFS) studies of copper (II) mixed ligand complexes having tetramethylethylenediamine as one of the ligands

Sharad Sharma[1], S K Joshi[2], B D Shrivastava[1], V K Hinge[3], J Prasad[4] and K Srivastava[4]

[1] School of Studies in Physics, Vikram University, Ujjain 456001, India
[2] S. V. Govt. Commerce College, Ratlam 457001, India
[3] Physics Department , Ujjain Engineering College, Ujjain 456001, India
[4] Department of Chemistry, University of Allahabad, Allahabad 211002, India

Email: androw_and@yahoo.com

Abstract. X-ray absorption fine structure (XAFS) has been studied at the K-edge of copper in copper(II) mixed ligand complexes, having tetramethylethylenediamine (tmen) as one of the ligands, viz., $Cu(tmen)(gly)ClO_4$, $Cu(tmen)(bipy)(ClO_4)_2$ and $Cu(tmen)(phen)(ClO_4)_2$. The spectra have been recorded at the dispersive XAFS beamline (BL-8) at the 2.5 GeV INDUS-2 synchrotron, RRCAT, Indore, India. The data obtained has been processed and analyzed using the computer program *Athena*. It has been observed that K-edge has been found to split in two edges, K and K′, in each of the complex. The chemical shift has been utilized to determine the oxidation state of copper in the complexes and also the effective nuclear charge (ENC). The EXAFS data has been analyzed to obtain the bond lengths in the complexes using Levy's, Lytle's and Lytle, Sayers and Stern's (LSS) methods. The first peak in the Fourier transform of the spectra gives the value of first shell phase uncorrected bond length. The results obtained from the Fourier transformation and LSS methods are in good agreement.

1. Introduction

The X-ray absorption spectroscopy provides information regarding the nature of bonding between the central metal ion and the nearest neighbours. The extended X-ray absorption fine structure (EXAFS) yields information about bond lengths. In the present work, the X-ray K absorption spectra of copper have been studied in the mixed ligand copper(II) complexes having tetramethylethylenediamine as primary ligand, other ligands being – glycine, 2 2'-bipyridine and 1,10-phenanthroline.

2. Experimental

The X-ray absorption spectra at the K-edge of copper of the complexes listed in table 1 have been recorded at BL-8 EXAFS beamline at the Indus-2 Synchrotron Source, Indore, India. The various aspects of the calibration and the procedure to be followed on this beamline were similar to those described by Gaur et al [1].The experimental data have been analyzed using the available computer software package *Athena*.

3. Results

The shapes of the copper K-edge and the near edge structure (XANES) for all the complexes are shown in figures1(a-d). The energies of the various absorption features are given in table 1.

Table 1. XANES data for the K-absorption edge of copper in the complexes .

Complex	E_K (eV)	$E_{K'}$ (eV)	E_A (eV)	Chemical shift $\Delta E_K = (E_{complex} - E_{metal})$ (eV)	ENC	Edge-width $(E_A - E_K)$ (eV)
Cu metal	8980.5	8991.8	8995.0	-	-	14.5
$Cu(tmen)(gly)ClO_4$	8988.4	8992.5	8999.7	7.9	0.92	11.3
$Cu(tmen)(bipy)(ClO_4)_2$	8987.6	8994.4	9001.8	7.1	0.82	14.2
$Cu(tmen)(phen)(ClO_4)_2$	8989.2	8994.9	9002.6	8.7	1.00	13.4

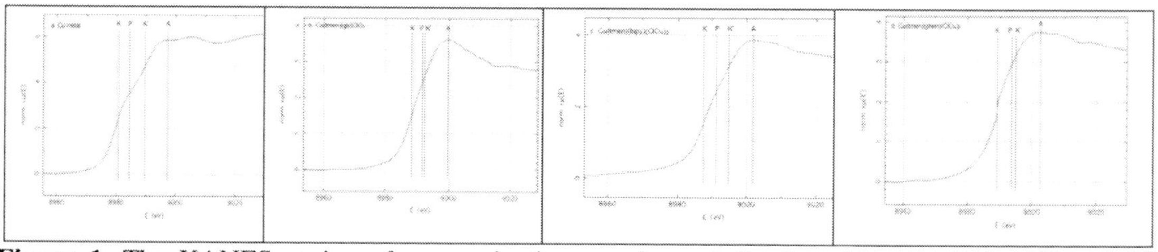

Figure 1. The XANES region of copper in the complexes indicating absorption edges K and K′, shoulder P and the principal absorption maxima A.

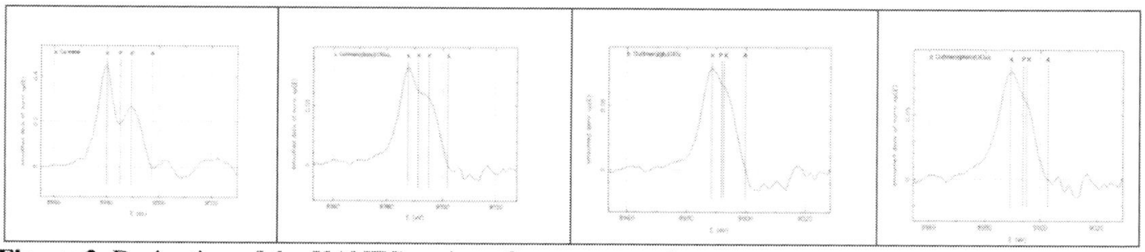

Figure 2. Derivative of the XANES region of copper complexes.

Table 2. Values of first shell bond length (in Å).

S.No.	Complex	Phase corrected		Phase uncorrected	
		Levy's method	Lytle's method	LSS method	FT method
		R_1	R_s	$R_1 - \alpha_1$	R
1.	$Cu(tmen)(gly)ClO_4$	1.73	1.79	1.49	1.48
2.	$Cu(tmen)(bipy)(ClO_4)_2$	1.75	1.58	1.44	1.45
3.	$Cu(tmen)(phen)(ClO_4)_2$	1.73	1.65	1.41	1.42

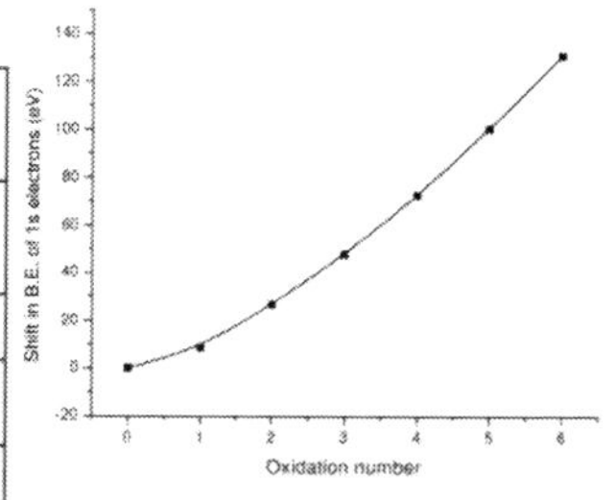

Figure 3. Plot of oxidation number vs binding energy

4. Discussion

4.1 *Splitting of the main edge*

The copper K-edge, as shown in figures 1(a–d), has been found to split into two components, i.e. K and K′. The splitting is seen clearly in the first derivative of these spectra, given in figures 2(a–d).

According to Kau et al.,[2] the characteristic feature of Cu(II) is the maximum A at ~9000 eV, which corresponds to $1s \rightarrow 4p_z$ transition as well as to the transitions to continuum states. In the absorption edge of the complexes, as seen in figures 2(a–d), apart from shoulder P at ~8985 eV, the feature A at ~9000 eV are present, both of which confirm oxidation state +2 in these complexes.

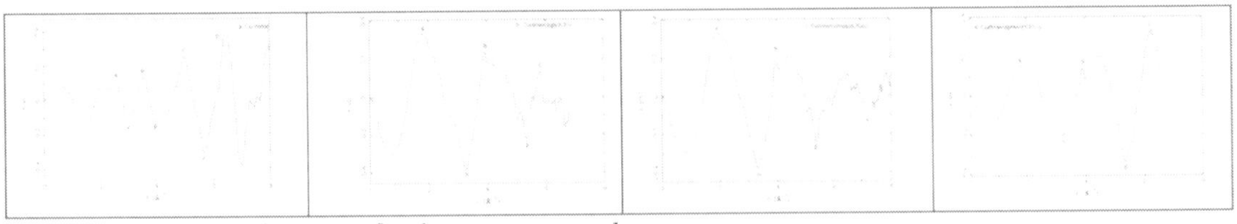

Figure 4. $\chi(k)$ versus k curves for the copper complexes

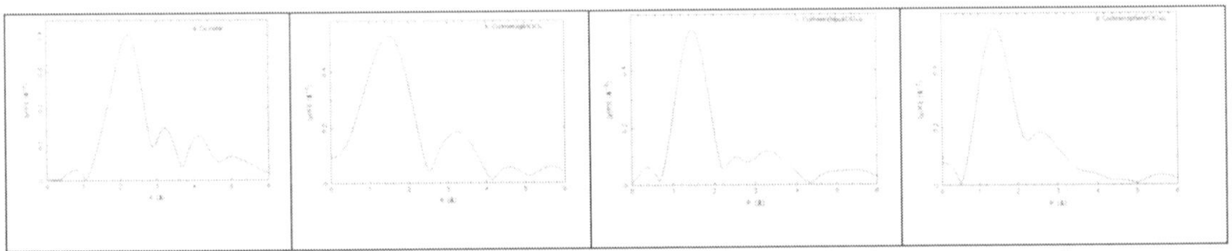

Figure 5. Magnitude of the Fourier transform of the $\chi(k)$ vs k curve for the copper complexes.

4.2 *Chemical shift*

All the three complexes have the values of chemical shifts between 7.1 to 8.7 eV (table 1). Hence, on the basis of values of the chemical shifts, all the complexes are found to have copper in oxidation state +2 [3].

4.3 *Effective nuclear charge (ENC)*

If the binding energies of K-electron of copper in different oxidation states are determined, one can find from the difference in binding energies of the neutral atom and the ionized atom, the so called theoretical shifts in the X-ray absorption edge. In the present work, the theoretical values of binding energies have been taken from the tables of Clementi and Roetti[4]. A plot of oxidation number versus theoretical shift in binding energy of 1s electron is shown in figure 3. The values of ENC so determined have given in table 1.

4.4 *Edge-width*

In general, the edge-width of the K-absorption edge (E_A - E_k) increases with the increase of covalent character of the bonds provided other factors like molecular symmetry etc. remain the same [5]. The edge-width values depend on stereochemistry and on the nature of bonding, i.e., ionic/ covalent. For the pair of complexes $Cu(tmen)(bipy)(ClO_4)_2$ and $Cu(tmen)(phen)(ClO_4)_2$ though the stereochemistry is same, i.e., Cu:N(1:4) but the edge-width values are different, i.e., (bipy) complex(14.2 eV) and (phen) complex(13.4 eV). This may be explained in terms of the higher covalent character in (bipy) complex as compared to (phen) complex as evident from the chemical shift values [6].

4.5 *Determination of bond length*

In Levy's method, the bond lengths are calculated by using the relation $R_1 = [151/\Delta E]^{1/2}$ Å, where ΔE is the difference in eV of the energies of the EXAFS maximum B and minimum β [7]. The bond lengths estimated in this way are tabulated in Table 2.

The $\chi(k)$ versus k curves for copper complexes have been shown in figures 4(a-d). In Lytle's method [8], the energy values (E) of the absorption maxima as obtained from the measurements are plotted, against the Q values for p symmetry, i.e., Q = 2.04, 6.04, 12.00 and 20.00. The slope M of the

E vs Q plots are used to evaluate the radius R_s by using the relation $R_s = [37.60/M]^{1/2}$. The values of R_s calculated with the help of this method are reported in table 2.

Lytle, Sayers and Stern's (L.S.S.) have given a simple method to determine the nearest neighbour distances [9]. To estimate the bond length by this method, n vs k graph for the complexes studied have been plotted. The slope of n vs k plot gives the value of $2(R_1 - \alpha_1)/\pi$ where R_1 is the bond length. The parameter α_1, depends to a large extent on the central absorbing atom. The values of $(R_1 - \alpha_1)$ thus obtained are given in table 2.

4.6 By Fourier transform of EXAFS spectra

The Fourier transform of the $\chi(k)$ vs k spectra peaks at the radial distances of the neighbouring atoms from the absorbing atom. The magnitude of the Fourier transform of the $\chi(k)$ vs k curve for the copper complexes is shown in figure 5. However, the distance found from Fourier transform is shorter than the actual distance as a result of energy dependence of the phase factors in the sine function of the EXAFS equation. The peaks in the Fourier transform are shifted towards the origin by an amount α_j, and hence, the position of the first peak also determines the distance $(R_1\text{-}\alpha_1)$ [10].

5. Conclusions

The chemical shift has been used to determine the effective nuclear charge on the absorbing atom. The values of the chemical shifts suggest that copper is in oxidation state +2 in all of the complexes. The bond lengths have been determined. It has been observed that the values of the phase uncorrected bond length, i.e., $(R_1\text{-}\alpha_1)$, as determined from LSS method and that determined from the Fourier transformation method, are in good agreement with each other. The present work shows that the methods outlined above provide reliable results for the first coordination shell bond lengths.

References

[1] Gaur A, Johari A, Shrivastava B D, Gaur D C, Jha S N, Bhattacharya D, Poswal A and Deb S K 2011 *Sadhana* **36** 339

[2] Kau L S, Spira-Solomon D J, Penner-Hahn J E, Hodgson K O, Solomon E I 1987 *J. Am. Chem. Soc.* **109** 6433

[3] Rockwell M, Melden R A, Copeland K, Hardman C P, Decicco W W and Degrado W F 1996 *J. Am. Chem. Soc.* **18** 10337

[4] Clementi E and Roetti C 1974 *At. Data Nucl. Data Tables* **14** 177

[5] Kumar A, Nigam A N and Shrivastava B D *X-ray Spectrom.* **10** 25

[6] Nigam H L and Srivastava U C 1971 *Chem. Comm.* **14** 761

[7] Levy R M 1965 *J. Chem. Phys.* **43** 1846

[8] Lytel F W 1966 *Adv. X-ray Annl.* **9** 398

[9] Lytle F W, Stern E A and Sayers D E 1975 *Phys. Rev.* B **11** 4825

[10] Lytel F W, Sayers D E and Lytle F W 1975 *Phys. Rev.* B **11** 4836

Journal of Physics: Conference Series **534** (2014) 012006 doi:10.1088/1742-6596/534/1/012006

X-ray absorption spectral studies of copper(II) mixed ligand complexes having ethylenediamine as one of the ligands

Shikha Dwivedi [1], S K Joshi [2], V K Hinge[3], B D Shrivastava[1], J Prasad[4] and K Srivastava[4]

[1] School of Studies in Physics, Vikram University, Ujjain 456001, India
[2] S. V. Govt. Commerce College, Ratlam 457001, India
[3] Physics Department, Ujjain Engineering College, Ujjain 456010, India.
[4] Department of Chemistry, University of Allahabad, Allahabad 211002, India

Email: shi.dwivedi@gmail.com

Abstract. X-ray absorption spectra of copper(II) mixed ligand complexes, having ethylenediamine (en) as one of the ligands, have been recorded at the K-edge of copper at the dispersive extended X- ray absorption fine structure (EXAFS) beamline (BL-8) at the 2.5 GeV INDUS-2 Synchrotron, RRCAT, Indore, India. The samples studied are: $Cu(en)_2(ClO_4)_2$, $Cu(en)_2Br_2.H_2O$ and $Cu(en)_2SO_4$. The data obtained has been processed using EXAFS data analysis program *Athena*. The K-edge has been found to split in two edges K and K' in each of the complex. The energies of the edges K(E_K) and K'($E_{K'}$) and the principal absorption maximum A(E_A) have been determined from the derivative spectra. The chemical shift has been utilized to determine the oxidation state of copper in the complexes and to estimate effective nuclear charge (ENC) on the absorbing atom. The EXAFS data has been used to determine the bond lengths in the complexes using three different graphical methods. The bond lengths, obtained from one of these methods and the Fourier transformation method, are comparable with each other, showing that both of these methods give phase uncorrected bond lengths.

1. Introduction

In the present paper, the X-ray absorption spectra of three mixed ligand complexes of Cu(II) having ethylenediamine (en) as one of the ligands, viz., $Cu(en)_2(ClO_4)_2$, $Cu(en)_2Br_2.H_2O$ and $Cu(en)_2SO_4$ have been investigated. Earlier these samples have been studied using laboratory X-ray spectrometer setup [1]. In the present work the spectra have been studied using synchrotron source.

2. Experimental

The complexes have been prepared and characterized using standard methods [2]. The absorption spectra at the K-edge of copper of these complexes have been recorded at BL-8 EXAFS beamline at INDUS-2 synchrotron source at RRCAT, Indore, India. The experimental data have been analyzed using the available computer software *Athena*.

3. Results

Various XANES features indicating absorption edge K and K' and A are shown in figures 1(a-d). The derivative spectra for all the three complexes are depicted in figures 2(a-d). The values of The energies of the edges K (E_K) and K' (E_K') and the principal absorption maximum A (E_A) have been determined from the derivative spectra and are given in table 1. The magnitudes of Fourier transform of figures 4(a-d) are shown in figures 5(a-d).

Table 1. XANES data for the K-absorption edge of copper in the complexes .

Complex	E_K (eV)	$E_{K'}$ (eV)	E_A (eV)	Chemical shift $\Delta E_K = (E_{complex} - E_{metal})$ (eV)	ENC
Cu metal	8980.5	8991.8	8995.0	-	-
$Cu(en)_2(ClO_4)_2$	8986.4	8991.7	9000.8	5.9	0.62
$Cu(en)_2Br_2.H_2O$	8986.1	8991.2	9000.3	5.6	0.58
$Cu(en)_2SO_4$	8987.6	8991.2	9002.7	7.1	0.82

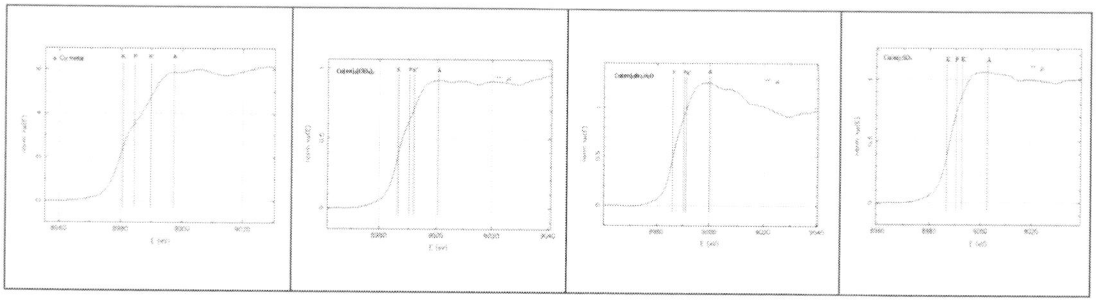

Figure 1. The XANES region of copper in the complexes indicating K and K', shoulder P and A.

4. Discussion

4.1 Splitting of the main edge

The copper K-edge, as shown in figures 1(a-d), has been found to split into two components, i.e. K and K' in all the complexes with a shoulder P in between as shown in figures 2(a-d). According to Kau *et al.*,[3] the characteristic feature of Cu(II) is the maximum A at ~ 9000eV, which also corresponds to $1s \rightarrow 4p_z$ transition as well as to the transitions to continuum states. In the absorption edge of the complexes, apart from shoulder P at ~ 8985eV, the feature A at ~ 9000eV are present, both of which are characteristics of Cu(II). Thus, copper has been found to have oxidation state +2 in these complexes.

4.2 Chemical shift

The compounds having copper in oxidation state +1 show chemical shifts less than 5 eV while those having copper in oxidation state +2 show chemical shifts more than 5eV [3]. In the presently studied three complexes, the values of chemical shifts are found to be between 5.6 to 7.1eV as shown in table 1. Hence, on the basis of values of the chemical shifts, all the complexes are found to have copper in oxidation state +2. The sequence of the chemical shift is: Cu (en) $_2Br_2.H_2O$ < Cu (en) $_2(ClO_4)$ $_2$< Cu (en) $_2SO_4$.

4.3 Effective nuclear charge (ENC)

The value of binding energies have been determined as given by Clementi and Roetti[4]. The theoretical shifts are plotted against the oxidation number. Such a graph for copper atom is shown in figure 3. From this graph, the effective nuclear charges (ENC) on copper in its complexes have been determined and are given in table 1.

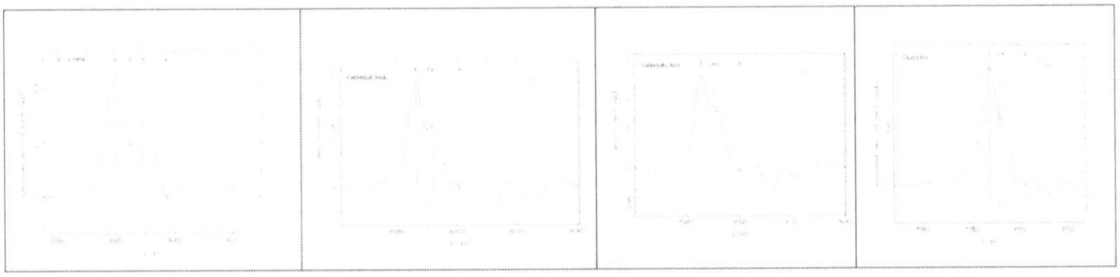

Figure 2. Derivative of the XANES region indicating positions of K and K', shoulder P and A.

Table 2. Values of first shell bond length (in Å)

S.No.	Complex	Phase corrected bond length		Phase uncorrected bond length	
		Levy's method	Lytle's method	LSS method	FT method
		R_1	R_s	$R_1 - \alpha_1$	R
1.	$Cu(en)_2(ClO_4)_2$	2.24	1.5	1.41	1.47
2.	$Cu(en)_2Br_2.H_2O$	1.80	1.6	1.60	1.81
3.	$Cu(en)_2SO_4$	1.92	1.45	1.57	1.51

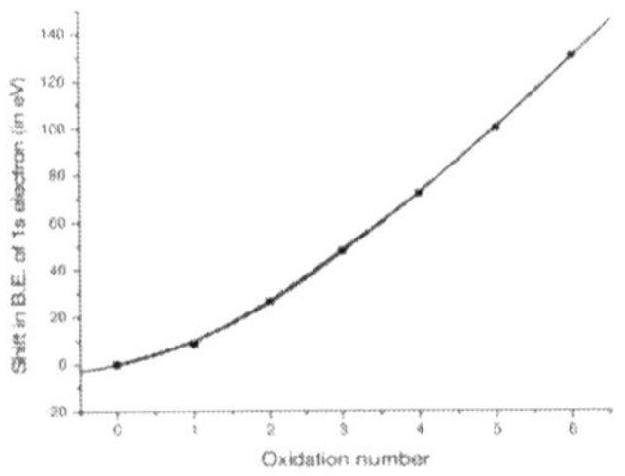

Figure 3. Plot of oxidation number versus theoretical shift in binding energy (BE) of 1s electrons

4.4 Determination of bond length by graphical methods
In Levy's method[5], the bond lengths are calculated by using the relation $R_1 = [151/\Delta E]^{1/2}$ Å, where ΔE is the difference in eV of the energies of the EXAFS maximum B and minimum β, and R_1 is the radius of the first coordination sphere.

For Lytle's method [6], the energy value (E) of the EXAFS maxima are plotted against the Q values for p symmetry, i.e., Q = 2.04, 6.04, 12.0 and 20.0. The (E,Q) plots have been found to be linear (not reproduced here). The slope M of the E versus Q plots have been used to evaluate the radius R_s of equivalent polyhedron, by using the relation, i.e. $R_s = [37.60/M]^{1/2}$.

In the Lytle, Sayers and Stern's (LSS) method [7], the determination of the nearest neighbor distances, n versus k graph is plotted. The plots have been found to be linear for all the complexes (not reproduced here). The slope of n versus k plot gives the value of $2(R_1-\alpha_1)/\pi$ where R_1 is the bond length. The parameter α_1 depends to a large extent on the central absorbing atom [8]. It is found that for chemically similar system, the value of α_1 remains more or less the same.

4.5 Determination of bond length by Fourier transform (FT) method
$\chi(k)$ versus k curves for the copper complexes are given in figure 4. Figure 5 gives the magnitude of the Fourier transform of the $\chi(k)$ versus k curves for the copper complexes. The Fourier transform of the spectra peaks at the radial distance of the neighboring atoms from the absorbing atom. However, the distance found from Fourier transform is about 0.2 Å - 0.5 Å shorter than the actual distance, as a result of energy dependence of the phase factor in the sine function of the EXAFS equation. Hence, the peaks in the Fourier transform are shifted toward the origin by an amount α_j. Therefore, the peaks are at distance $(R_j - \alpha_j)$ [9]. For the first peak j=1 and, hence the position of the first peak determines the distance $(R_1-\alpha_1)$.

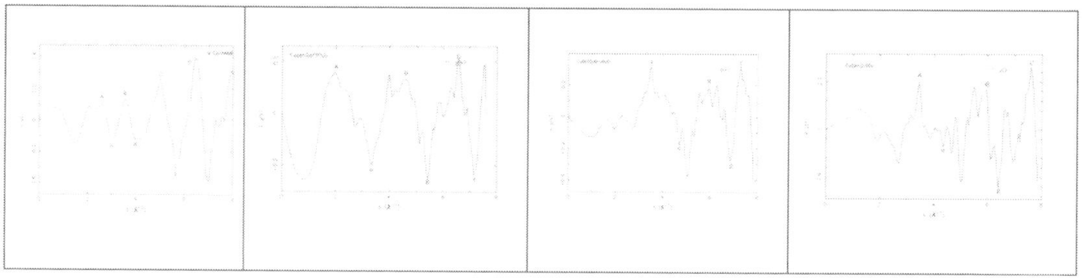

Figure 4. $\chi(k)$ versus k curves for the copper complexes

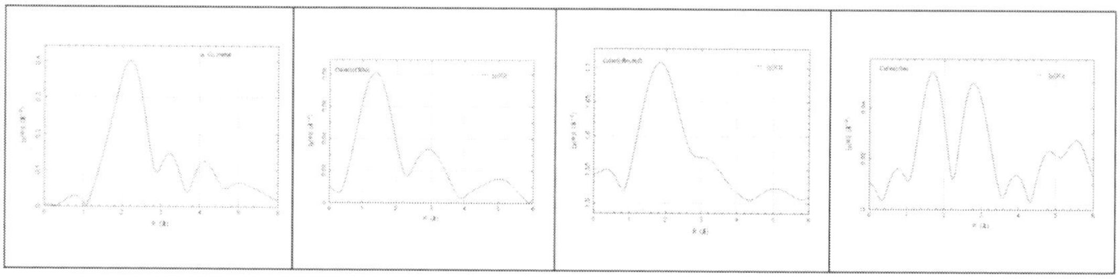

Figure 5. Magnitude of the Fourier transform of the $\chi(k)$ versus k curve for the copper complexes.

5. Conclusions

The chemical shift has been used to determine the effective nuclear charge on the absorbing atom. The values of the chemical shifts suggest that copper is in oxidation state +2 in all three the complexes. The order of the chemical shift may also be taken as representative of the relative ionic character of the bonding in these complexes. The bond lengths in the complexes have been determined by three different methods viz., Levy's, Lytle's and Lytle, Sayers and Stern's (LSS) methods. From the Fourier transforms of the EXAFS spectra, the bond lengths (uncorrected for phase shift) have been determined. It has been observed that the values of the phase uncorrected bond length, i.e. R_1-α_1, as determined from LSS method and that determined from the Fourier transformation method, are in good agreement with each other. This shows that the LSS method also provides reliable results for the bond length of the first co-ordination shell in the complexes.

References

[1] Hinge V K 2010 *PhD Thesis, Vikram University, Ujjain*
[2] Goodgame M and Haines L I B 1966 *J.Chem. Soc.* A 174
[3] Kau L S, Spira-Solomon D J, Penner-Hahn J E, Solomon E I and Hodgson K O 1987 *J. Am. Chem. Soc.* **109** 6433
[4] Clementi E and Roetti C 1974 *At. Data Nucl. Data Tables* **14** 177
[5] Levy R M 1965 *J. Chem. Phys.* **43** 1846
[6] Lytle F W 1966 *Adv. X-ray Annl.* **9** 398
[7] Lytle F W, Stern E A and Sayers D E 1975 *Phys. Rev.* B **11** 4825
[8] Nigam A K and Gupta M K 1974 *J. Phys. Metal Phys.* F **4** 1084
[9] Stern E A, Sayers D E and Lytle F W 1975 *Phys.Rev.* B **11** 4836

Elemental analysis of $BaTi_{0.96}Fe_{0.04}O_3$ & $BaTi_{0.96}Co_{0.04}O_3$

Ashutosh Mishra, Niyati Mishra, Kanaka Mahalakshmi, Supriya Bisen
School of Physics, Devi Ahilya University, Indore-452001, India

E-mail: jkanaka.hvp@gmail.com

Abstract. A finest possible sample of 4% Iron doped & cobalt doped $BaTiO_3$ (BTO) with possible tetragonal & cubic structure via solid state route was prepared. Prepared samples were characterized by X-ray diffraction (XRD) using by Bruker D8 advance XRD instrument, the value of 2Θ is in between 20^0 to 80^0. X-ray photoelectron spectroscopy (XPS) spectra were measured using a VSW Scientific instruments at UGC-DAE CSR.

1. Introduction
Barium titanate ($BaTiO_3$) is one of the best known Perovskite ferroelectric compounds (A2+B4+O3) that have been extensively studied [1, 2] due to the simplicity of its crystal structure, which can accommodate different types of dopant. Barium titanate ($BaTiO_3$) is another type of ABO_3 perovskite, which has not only large ferroelectric response, but also very large nonlinear optical and electro-optic coefficients, which is attractive to designing nonlinear optical devices [3]. These properties can be dramatically enhanced when they are doped with transition metals. It is well known that $BaTiO_3$ has a paraelectric to ferroelectric transition at about 120 ^{0}C with a very high dielectric constant. Thus, considerable attentions were paid to the doped $BaTiO_3$ [4-5] to modify the performance of the materials.

2. Synthesis of Iron & cobalt doped barium titanate
Both samples of Fe and Co doped barium titanate were synthesized via solid state route. We started from highly pure fine powdered samples of $BaCO_3$, TiO_2, Co_3O_4 and Fe_2O_3 for bulk sample of doped $BaTiO_3$. All these were mixed in the calculated percentage ratio. Samples were grinded for 8-hours and calcinations were done at 900 ^{0}C and 1100 ^{0}C than sintered at 1200^{0}C for 24 hours.

3. Experimental technique
3.1. X-ray diffraction
The synthesized powders were characterized by X-ray diffraction using a D8 Advanced X-ray diffractometer. XRD pattern for Fe-BTO & Co-BTO was carried out using CuKα radiation in the 2Θ range of 20^0 to 80^0 shown in figure 1(a) & 1(b).In Figure 1(a) Fitted XRD patterns reports the single phase tetragonal crystal system of the space group P4mm and pattern matched with the standard pattern JCPDS, no. 81-2203. In Figure 1(b) Fitted XRD patterns reports the single phase cubic crystal system of the space group Pm3m and pattern matched with the standard pattern JCPDS, no. 75-0461.

3.2. X-ray photoelectron spectroscopy
X-ray Photoelectron Spectroscopy (XPS) is also known as the Electron Spectroscopy for Chemical Analysis (ESCA).This technique is basically used for detection of chemical information of the surface

of the solid materials. X-ray Photoelectron spectroscopy (XPS) was used to check the Fe & Co composition and chemical nature for the sample. XPS spectra were measured using a VSW Scientific instruments at UGC-DAE CSR. The Fe-2P peaks are observed at 725.87 eV and 711.48 eV. Furthermore the O-1s peak is observed at 529.73 eV and Ba-3d5/2 peaks are observed at 794.53 eV and 779.23 eV and Ti-2p peak is observed at 463.41 eV and 457.60 eV respectively as was also shown in an earlier report [6].

Table 1: Elemental peak parameter of $BaTi_{0.96}Fe_{0.04}O_3$ modified by XPS technique.

Elements	Peak	B. E. (eV)	Area	FWHM	%GL (%)
Ba-3d	0	794.539	17116.510	2.998	80
	1	779.236	21234.470	2.660	80
O-1s	0	529.730	5946.991	2.904	80
Fe-2p	0	725.876	3300.913	14.453	80
	1	711.482	2098.359	7.585	80
Ti-2p	0	463.416	1776.646	3.742	80
	1	457.609	2048.316	1.771	80

The Co peak is observed at 770.178 eV. Furthermore the O-1s peak is observed at 531.959 eV and Ba-3d$_{5/2}$ peaks are observed at 795.644 eV and 780.243 eV and Ba-4d$_{5/2}$ peak is observed at 88.890 eV and Ti-2p peaks are observed at 461.015 eV, 457.669eV, 399.228 eV, 446.000 eV and 439.000 eV respectively as was also shown in an earlier report [6].

Table 2: Elemental peak parameter of $BaTi_{0.96}Co_{0.04}O_3$ modified by XPS technique.

Elements	Peak	B. E. (eV)	Area	FWHM	%GL (%)
Ba-3d	0	795.644	10535.280	3.732	80
	1	780.243	8928.859	2.702	80
Ba-4d	0	88.890	2348.184	3.260	80
Co	0	770.178	962.975	4.937	80
O-1s	0	531.959	10227.890	2.674	80
Ti-2p	0	461.015	1753.040	10.053	80
	1	457.669	268.304	1.192	80
	2	399.228	160.658	0.983	80
	3	446.000	0.100	1.000	80
	4	439.000	31.972	1.202	80

4. Conclusion

The composition of Iron & Cobalt doped Barium Titanate synthesized by solid state route are in single phase. XRD pattern of Fe doped BTO reports the single phase tetragonal crystal system of the space group P4mm and pattern matched with the standard pattern JCPDS, no. 81-2203 and Co doped BTO reports the single phase cubic crystal system of the space group Pm3m and pattern matched with the standard pattern JCPDS no.75-0461. X-ray Photoelectron Spectroscopy (XPS) was used to check the Fe composition and chemical nature for the sample. Shows the Fe-2p & Co-2p XPS signal of the Fe-doped & Co-doped $BaTiO_3$. The binding energy values of the Fe-2p$_{3/2}$ & Co-2p peaks are 711.48 eV & 770.178 eV, indicating the presence of iron & cobalt oxide. It is also convenient to quantify the Fe & Co composition by XPS measurement.

Journal of Physics: Conference Series **534** (2014) 012007 doi:10.1088/1742-6596/534/1/012007

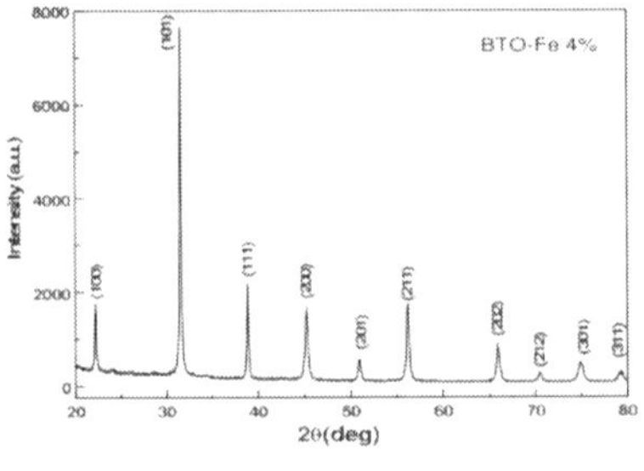

Figure 1(a). XRD pattern of $BaTi_{0.96}Fe_{0.04}O_3$ doped BTO at 900^0C.

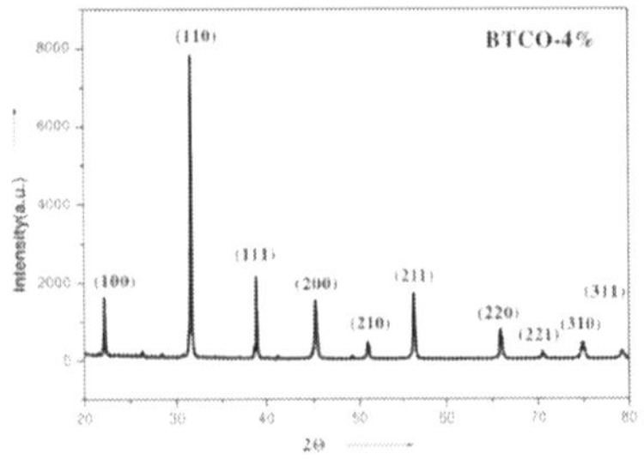

Figure 1(b). XRD pattern of $BaTi_{0.96}Co_{0.04}O_3$ doped BTO at 900^0C.

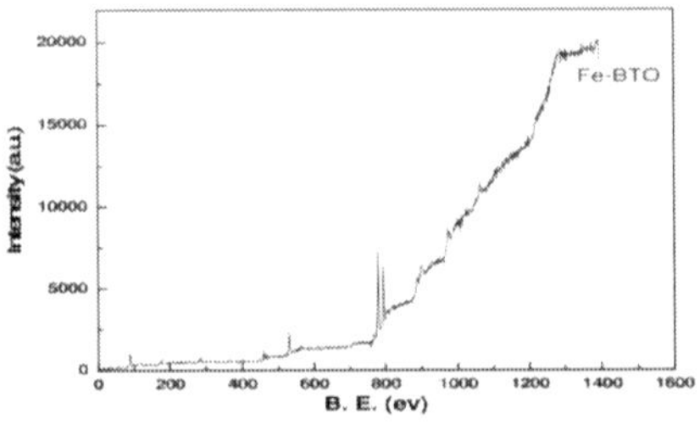

Figure 2. XPS spectrum of $BaTi_{0.96}Fe_{0.04}O_3$

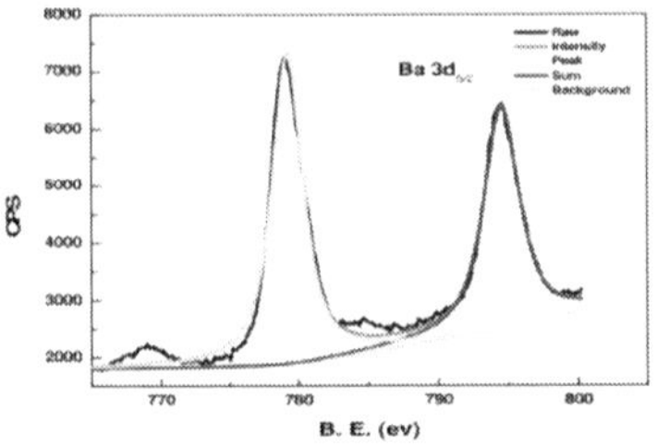

Figure 2(a). Ba-$3d_{5/2}$ peak by XPS

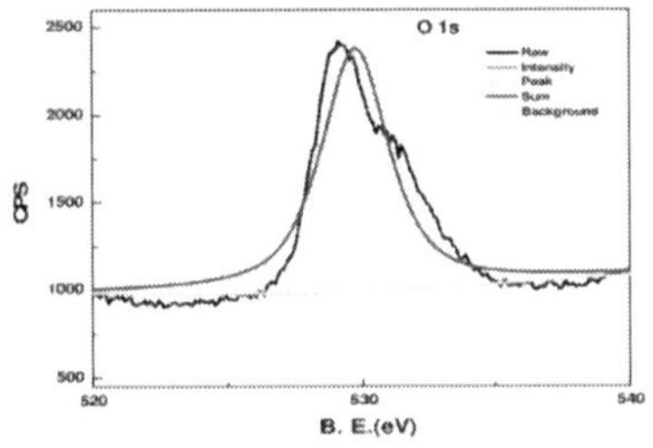

Figure 2(b). O-1s peak by XPS

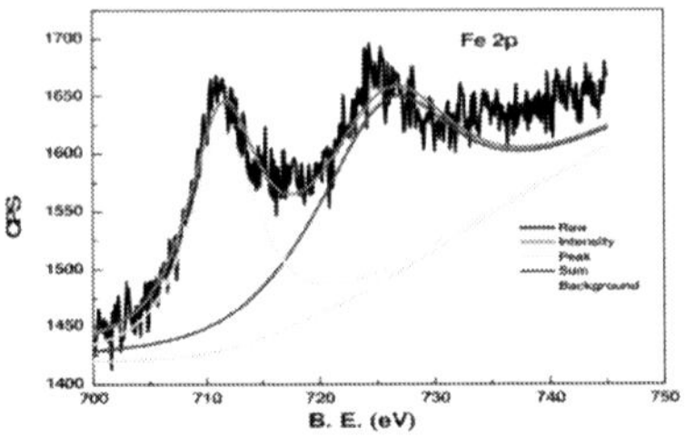

Figure 2(c). Fe-2p peak by XPS

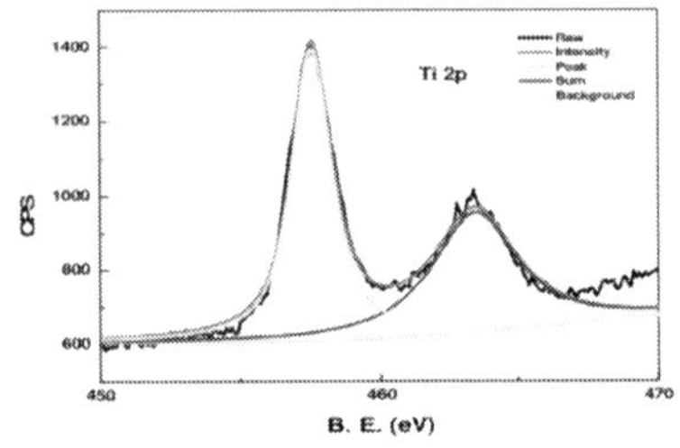

Figure 2(d). Ti-2p peak by XPS

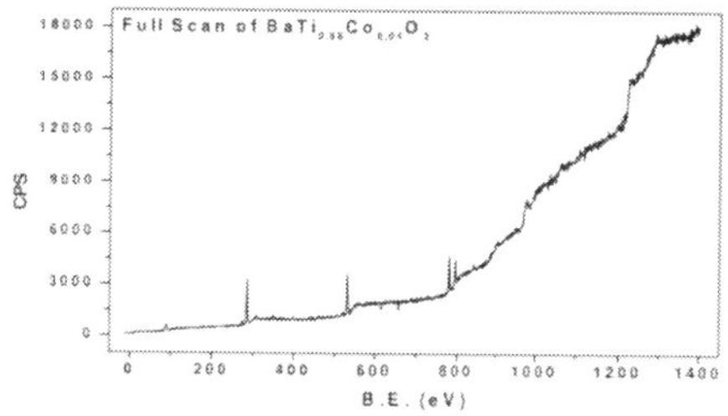

Figure 3. XPS spectrum of $BaTi_{0.96}Co_{0.04}O_3$

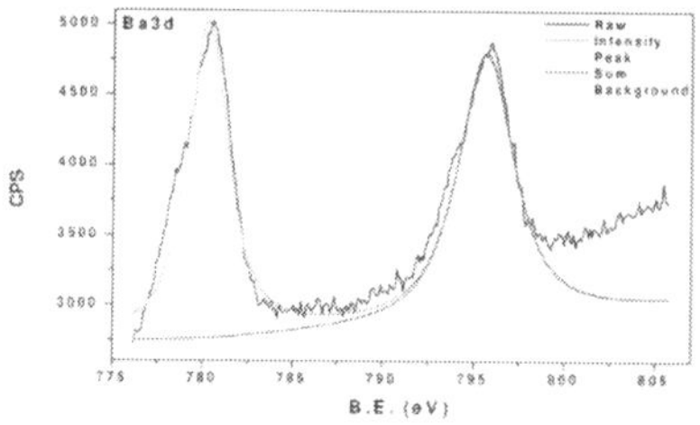

Figure 3(a). Ba-3d$_{5/2}$ peak by XPS

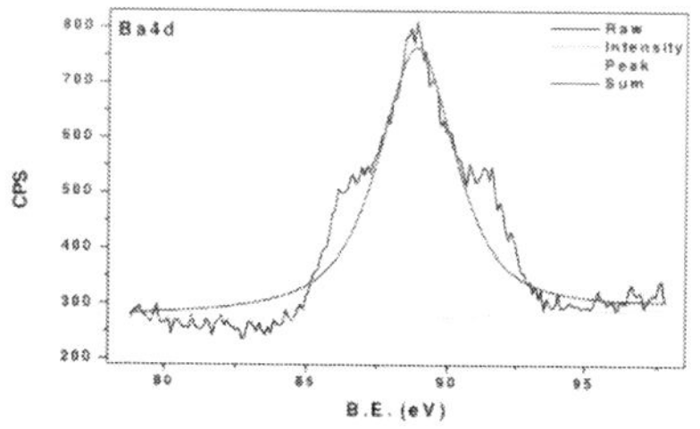

Figure 3(b). Ba-4d$_{5/2}$ peak by XPS

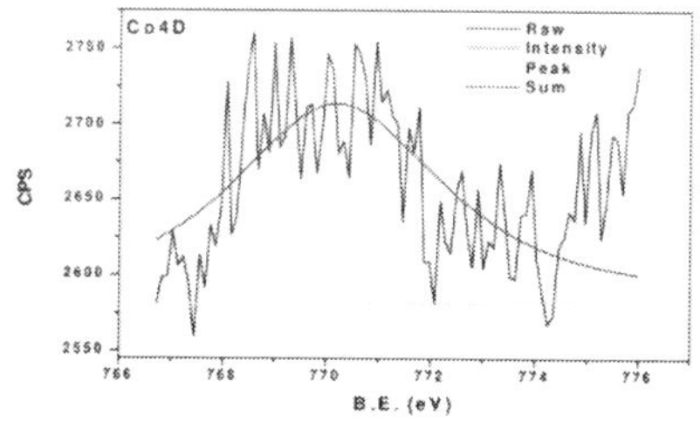

Figure 3(c). Co-2p peak by XPS

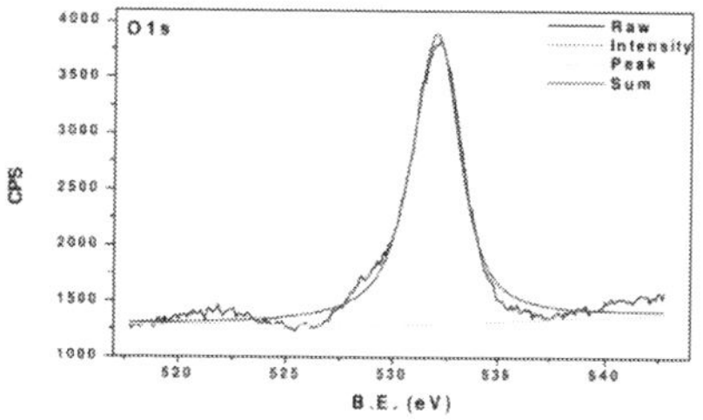

Figure 3(d). O-1s peak by XPS

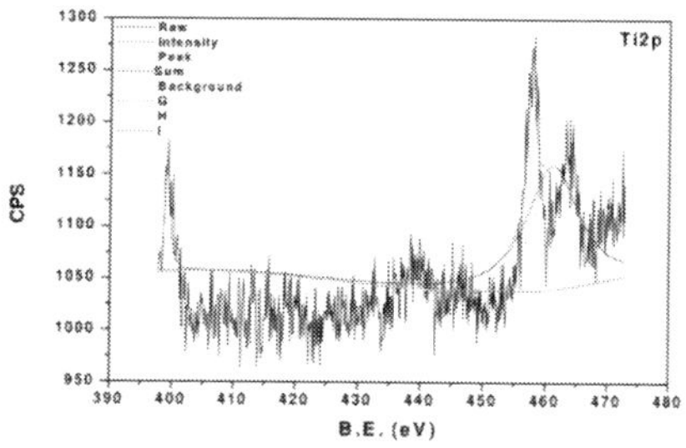

Figure 3(e). Ti-2p peak by XPS

Acknowledgement
The authors are thankful to Prof. Ajay Gupta, Dr. Mukul Gupta, UGC-DAE CSR for XRD, Dr. T. Shripati and Dr. Uday Deshpandey for XPS measurements, Indore for providing the experimental facilities and for the precious guidance. One of the authors (Niyati Mishra) is thankful to UGC-DAE CSR Indore for providing the research fellowship.

References
[1] Makovec D, Samadmija Z and Drofenik M, 2004 *J. Am. Ceram. Soc.* **87** 1324.
[2] Maga D, Igor P and Sergei M 2000 *J. Mater. Chem.* **10** 941
[3] Jana A, Kundu T.K , Pradhan S.K and Chakravorty D 2005 *J. Appl. Phys.* **97** 044311
[4] Smolenski G.A, Isupov A.V 1954 *Dokl. Akad. Nauk. SSSR* **97** 653.
[5] Ismailzade I.H, Ismailov R.M 1980 *Phys. Stat. Sol.* (a) **59** K191.
[6] Shi W S, Chen Z H, Liu N N, Lu H B, Zhou Y L, Cui D F and Chen Z H 1999 *Appl. Phys. Lett.* **75** 1547.

Journal of Physics: Conference Series **534** (2014) 012008 doi:10.1088/1742-6596/534/1/012008

X-ray absorption spectral studies of copper (II) mixed ligand complexes

B Soni[1], Davood Ah Dar[1], B D Shrivastava[1]*, J Prasad[2], K Srivastava[2]
[1]School of Studies in Physics, Vikram University, Ujjain-452001, India
[2]Department of Chemistry, University of Allahabad, Allahabad 211002, India.

email: rashmibasant@gmail.com

Abstract. X-ray absorption spectra at the K-edge of copper have been studied in two copper mixed ligand complexes, one having tetramethyethylenediamine (tmen) and the other having tetraethyethylenediamine (teen) as one of the ligands. The spectra have been recorded at BL-8 dispersive extended X-ray absorption fine structure (EXAFS) beamline at the 2.5 GeV INDUS-2 synchrotron, RRCAT, Indore, India. The data obtained has been processed using the data analysis program *Athena*. The energy of the K-absorption edge, chemical shift, edge-width and shift of the principal absorption maximum in the complexes have been determined and discussed. The values of these parameters have been found to be approximately the same in both the complexes indicating that the two complexes possess similar chemical environment around the copper metal atom. The chemical shift has been utilized to estimate effective nuclear charge on the absorbing atom. The normalized EXAFS spectra have been Fourier transformed. The position of the first peak in the Fourier transform gives the value of first shell bond length, which is shorter than the actual bond length because of energy dependence of the phase factors in the sine function of the EXAFS equation. This distance is thus the phase-uncorrected bond length. Bond length has also been determined by Levy's, Lytle's and Lytle, Sayers and Stern's (LSS) methods. The results obtained from LSS and the Fourier transformation methods are comparable with each other, since both are phase uncorrected bond lengths.

1. Introduction

Chelating diamine ligands form a variety of complexes with transition metal ions. Biological as well as catalytic activity of copper(II) complexes having ethylenediamine as a ligand has also been studied [1]. Tetramethylethylenediamine (tmen) can be classified as an organic base like amines and amidines that are regarded as the strong base due to the resonance stability of their conjugate acid. Nishiyama *et al.* [2] have shown that tmen acts as a powerful ligand for the hydrosilylation of ketones. Sano *et al.* [3] have demonstrated that tmen could induce the acylation of alcohol with benzoyl chloride.

X-ray absorption spectra at the K-edge of copper have been studied in two copper mixed ligand complexes, $Cu_2(teen)_4(OH)_2(ClO_4)_2$ **(1)** and $Cu_2(tmen)_4(OH)_2(ClO_4)_2$ **(2)** (tmen = tetramethylethylenediamine and teen = tetraethylethylenediamine).

2. Experimental details

The X-ray absorption spectra have been recorded at the BL-8 dispersive extended X-ray absorption fine structure (EXAFS) beamline at the 2.5 GeV Indus-2 synchrotron at RRCAT, Indore, India [4]. The energy calibration of the spectra has been done following the procedure outlined in reference [5].

3. Results and discussion

The spectral data has been analyzed using the computer program *Athena* [6]. Normalized $\mu(E)x$ (absorption) vs. E (energy) curves and $\chi(k)$ (oscillations) vs. k (wave vector) curves for the complexes are shown in figures 1 and 2 respectively. From the first derivative of figure 1, shown in figure 3, it has been observed that K-edge has been found to split in two edges, K and K′, with a shoulder P in between, in each complex. The spectra in figures 1,2 and 3 have been shifted vertically by suitable values so that the spectra may be easily compared with each other. The values of the energies of the K-absorption edges (E_K and $E_{K'}$), shoulder P and principal absorption maximum A (E_A) are presented in table 1.

3.1. *Splitting of the edge.* The maximum A at ~9000 eV corresponds to $1s \rightarrow 4p_z$ transition as well as to the transitions to continuum states. Both of the features, i.e., shoulder P at ~8988 eV and maximum A at ~9000 eV are characteristics of Cu(II). Thus, copper has been found to have oxidation state +2 in these complexes.

3.2. *Chemical shift.* Taking the energy of copper metal K-edge as 8980.5 eV, the chemical shift, i.e., the shift of the K-edge in the complex with respect to that in the metal has been found to be the same in both the complexes, i.e., 5.7 eV. Since +2 oxidation state of copper is known to give chemical shift of the K-absorption edge of copper of the order of 5 eV or more, hence the obtained values of the chemical shift suggest that copper is in +2 oxidation state in both the complexes [7].

3.3. *Principal absorption maximum.* The principal absorption maximum (A) has been found to be shifted towards the higher energy side with respect to copper metal by 3.4 eV in both the complexes. Hence, on the basis of this shift also, it can be inferred that copper is in +2 oxidation state in these complexes. Thus, the relative ionic character has been found to be the same in both the complexes.

3.4. *Edge-width.* In general, edge-width (E_A-E_K) of the K absorption edge increases with the increase of covalent character of the bonds provided other factors like molecular symmetry etc., remain the same [8]. In the present study, the edge-width is found to be the same in both the complexes showing that both the complexes have similar molecular symmetry.

3.5. *Effective nuclear charge (Z_{eff}).* The effective nuclear charge (Z_{eff}) has been obtained from the measured chemical shift by using the method outlined by Nigam and Gupta [9] and Parsai *et al.* [10]. The effective nuclear charge on the copper in both the complexes has been found to be 0.57 electrons/atom.

3.6. *Determination of bond lengths.* The bond length has been determined from the extended X-ray absorption fine structure (EXAFS) data by Levy's [11], Lytle's [12] and Lytle, Sayers and Stern's (LSS) [13] methods. These methods are well known methods and have been described recently by Parsai *et al.* [10]. In fact, before the Fourier transformation followed by the fitting of theoretical spectrum to the experimental spectrum became a standard and routine technique, the bond length used to be extracted from the EXAFS data by these three methods. The results obtained for the complexes by these methods are given in table 3. Levy's method gives the radius of coordination sphere (R_1) directly, where as Lytle's method does not. In the latter method, the interatomic spacing is obtained by multiplying the obtained bond length (R_s) with a factor appropriate to the geometry of the system. The LSS method gives the value of (R_1-α_1) where R_1 is the bond length and the parameter α_1 depends to a large extent on the central absorbing atom [9]. For chemically similar system, the value of α_1 remains more or less the same.

The normalized EXAFS spectra of figure 3 have been Fourier transformed and are given in figure 4. The position of the first peak in the Fourier transform gives the value of first shell bond length, which is shorter than the actual bond length because of energy dependence of the phase factors in the sine function of the EXAFS equation. This distance is thus the phase-uncorrected bond length. The peaks in the Fourier transform are shifted towards the origin by an amount α_j, and hence, the peaks are at distances R_j-α_j [14]. For the first peak, j = 1 and, hence, the position of the first peak determines the distance R_1-α_1. The distance R_1-α_1 should be equal to the distance found from the LSS graphical method. Hence, both the LSS method and the Fourier transformation (FT) method give the value R_1-α_1, i.e. both the methods give the value of bond lengths, which have not been corrected for the phase shifts. This distance is called as the phase-uncorrected bond length. The position of the first peak in the Fourier transform which gives the value of R_1-α_1, are given in table 3 for both the complexes. It is seen from table 3 that the values of R_1-α_1 as determined from LSS method and those determined from the Fourier transformation method are in good agreement with each other, i.e. both the LSS method and Fourier transformation method give nearly the same value of the phase-uncorrected bond length, i.e. R_1-α_1.

4. Conclusion

The edge features, chemical shift, edge-width and shift of the principal absorption maximum in the complexes have been found to be approximately same in both the complexes. Copper has been found to be in oxidation state +2 in both the complexes. The effective nuclear charge on the copper in the complexes has been reported. The present study indicates that the two complexes possess similar chemical environment around the copper metal atom. Bond length has also been determined by Levy's, Lytle's and Lytle, Sayers and Stern's (LSS) methods. The results obtained from LSS and the Fourier transformation methods are comparable with each other, since both are phase uncorrected bond lengths.

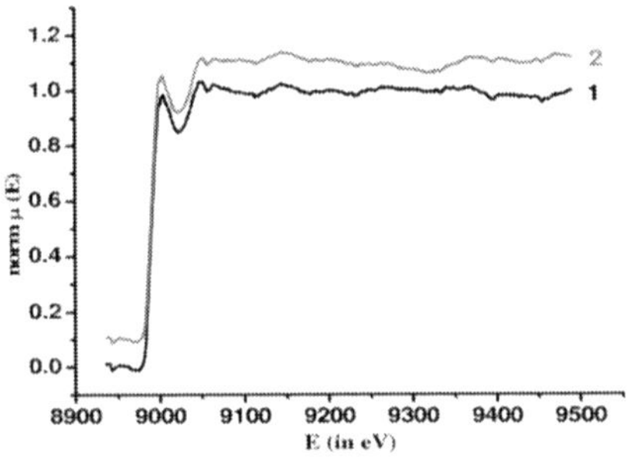

Figure 1. Normalized μ(E) vs. E curves for complexes **1** and **2**

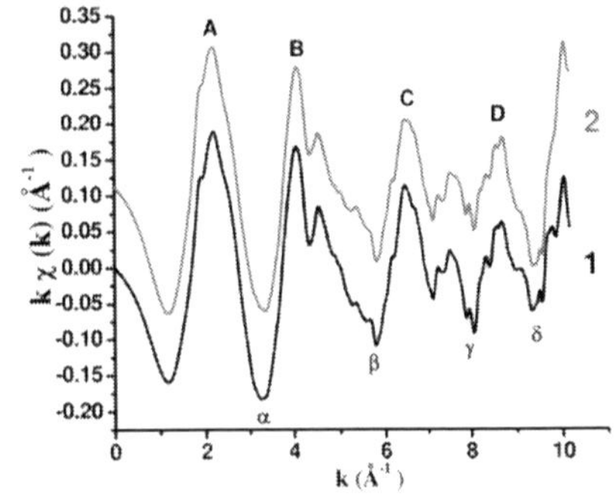

Figure 2. |χ(k)| vs. k spectra of complexes **1** and **2**

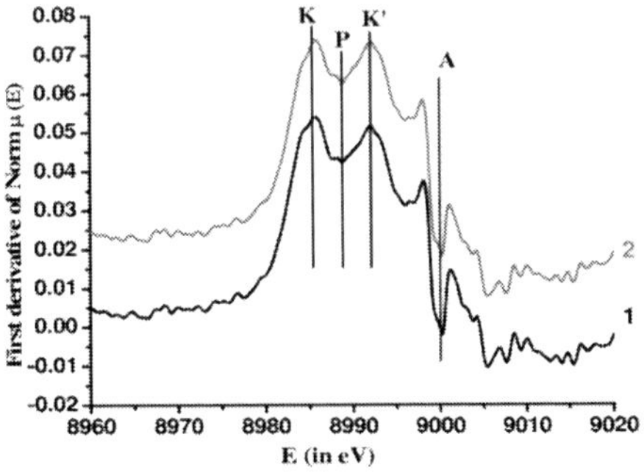

Figure 3. Derivative spectra for the complexes **1** and **2.**

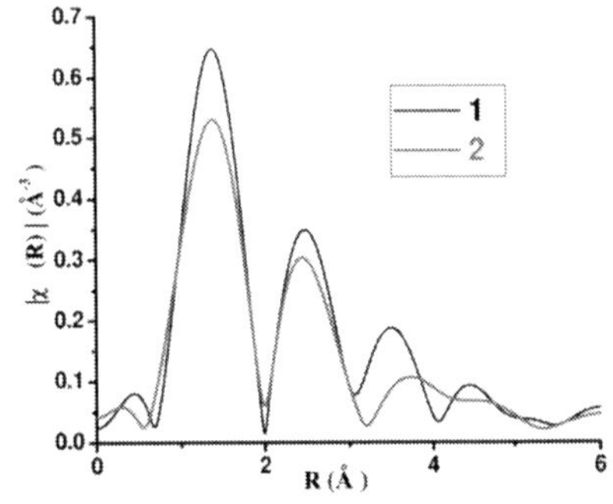

Figure 4. Fourier transformed data of figure 2.

Table 1. Data for the K-edge of copper in the complexes

Complex	E_K (eV)	$E_{K'}$ (eV)	E_A(eV)	Chemical shift (E_K-8980.5)	ENC	Shift of the principal absorption maximum	Edge width (E_A - E_K)
Cu foil	8980.5	8992.3	8997.0	-	-	-	16.5
1	8985.7	8992.2	9000.4	5.2	0.57	3.4	14.7
2	8985.7	8992.2	9000.4	5.2	0.57	3.4	14.7

Table 2. Energy E (eV) and wave vector k ($Å^{-1}$) for EXAFS maxima and minima at the K-edge of copper in the complexes and their corresponding values of n and energy level Q

Structure	n	Q	Complex 1		Complex 2	
			E(eV)	k(Å)	E(eV)	k(Å)
A	0	2.04	18.4	2.2	18.4	2.2
α	1	-	41.4	3.3	41.4	3.3
B	2	6.04	63.9	4.1	63.9	4.1
β	3	-	127.9	5.8	127.9	5.8
C	4	12.00	155.7	6.4	160.6	6.5
γ	5	-	249.5	8.1	249.5	8.1
D	6	20.00	287.8	8.7	287.8	8.7
δ	7	-	328.9	9.3	336.0	9.4

Table 3. Values of first shell bond lengths (in Å) calculated from Levy's [11], Lytle's [12], LSS[13] and Fourier transform methods for the complexes.

	Levy's method	Lytle's method	LSS method	FT method
	R_1	R_s	R_1-α_1	R
Complex 1	1.97	1.68	1.36	1.37
Complex 2	1.97	1.68	1.36	1.38

References

[1] Samrah P, Singha S, Chakrabarty R, Bora S J, Das B K 2007 *Ind J Chem.* **46A** 1929.
[2] Nishiyama H, Furuta A 2007 *Chem. Commun.* 760.
[3] Sano T, Ohashi K, Oriyama T 1999 *Synthesis* 1141.
[4] Das N C, Jha S N, Bhattacharyya D, Poswal A K, Sinha A K and Mishra V K 2004 *Sadhana* **29** 545.
[5] Gaur A, Johari A, Shrivastava B D, Gaur D C, Jha S N, Bhattacharyya D, Poswal A, Deb S K 2011 *Sadhana* **36** 339.
[6] Ravel B and Newville M 2005 *J. Synchrotron Rad.* **12** 537.
[7] Kau L S, Spira-Solomon D J, Penner-Hahn J E, Hodgson K O, Solomon E I, 1987 *J. Am. Chem. Soc.* **109** 6433.
[8] Kumar A, Nigam A N and Shrivastava B D 1981 *X-ray Spectrom.* **10** 25.
[9] Nigam A K and Gupta M K 1974 *J. Phys. F: Metal Physics* **4** 1084
[10] Parsai N, Mishra A, Shrivastava B D 2014 *X-ray Spectrom.* **43** 157
[11] Levy R M 1965 *J. Chem. Phys.* **43** 1846.
[12] Lytle F W 1966 *Adv. X-Ray Anal.* **9** 398.
[13] Lytle F W, Sayers D E, Stern E A 1975 *Phys. Rev. B* **11** 4825.
[14] Stern E A, Sayers D E, Lytle F W 1975 *Phys. Rev. B* **11** 4836

X-ray K-absorption spectroscopic studies of copper complexes

Ashutosh Mishra[1], Vishakha Vibhute[1], P Sharma[2], N Parsai[1] and S Ninama[1]
[1]School of Physics, Devi Ahilya University, Indore - 452001, India
[2]School of Chemical Sciences, Devi Ahilya University, Indore - 452001, India

Email: amishra1960@yahoo.in, vishakhavibhute28@ gmail.com

Abstract. The X-ray absorption spectra at the K-edge of copper have been recorded in five mixed ligand copper complexes having orthophenylenediamine as one of the ligand. Various X-ray absorption near edge structure (XANES) parameters, such as chemical shift, i.e., shift of K-edge of copper in complexes with respect to K-edge of copper metal, energy position of principal absorption maximum and edge-width have been determined. Copper has been found to be in oxidation state +2 in all the complexes. From these XANES data the effective nuclear charge on copper and percentage covalency in these complexes has been estimated.

1. Introduction

The present paper describes the X-ray absorption spectroscopic studies of the following mixed ligand copper complexes having orthophenylenediamine as one of the ligand. The X-ray absorption measurements have been performed at the recently developed BL-8 dispersive EXAFS beam line at 2.5 GeV Indus-**2** synchrotron sources at RRCAT Indore, India. The data obtained has been processed using EXAFS data analysis program Athena. From the experimental measurements, the energies of the K absorption edge, chemicals shifts, edge widths and shifts of principal absorption maximum in the complexes have been estimated. The data obtained from these studies have been utilized to estimate Effective nuclear charge and percentage covalence. Following complexes have been studied:

Complex 1: Cu(3-[4-chlorophenyl]-2 methyl-3H-1,5, benzodiazipine).
Complex 2: Cu(3-[4-phenyldiazenyl]-3H-1,5, benzodiazipine).

2. Experimental details

In the present investigations, the X-Ray absorption spectra have been recorded using synchrotron radiation. The X-Ray spectroscopic setup is available at RRCAT and is called beam line. This beam line has been recently commissioned at the 2.5 GeV Indus-2 synchrotron radiation sources. The beamline used in the present investigation has been designed in the dispersive mode and is called dispersive EXAFS (DEXAFS) BL-8 beamline. The computer software *Athena* version 0.8.056 has been used for analysis of the digital spectral data. All the complexes were prepared according to the standard methods and their purity was checked. For preparing the samples, the samples are mixed uniformly with a binder material like cellulose. Now the pellets of well mixed samples are made using the press (pelletizer).

3. Results and discussion

The energies of the K-absorption edge and the principal absorption maximum, the chemical shifts and the edge width for the complexes are given in Table 1. None of the complexes under investigation shows any splitting of the K-absorption edge, indicating octahedral geometry. Fig.1and Fig 2 shows the XANES curve between for complex1 and complex2 respectively.

3.1 *Chemical shift*

A perusal of Table 1 shows that the K-absorption edge of copper is shifted towards the high energy side with respect to the metal edge in all the complexes studied here and the value of the chemical shift ranges from 6.5 eV to 7.5 eV. The magnitude of the chemical shift is found to depend mostly on valance state, stereochemistry and covalent bonding [1]. The complexes of the present study are all copper complexes and hence valence criterion cannot be applied. Moreover all these complexes are octahedral and hence the stereochemistry is not changing, among the various complexes in the above-mentioned series. Therefore, covalent bonding is the important factor for observed values of chemical shifts in the above complexes.

3.2 *Shift of principal absorption maxima*

The shift of the principal absorption maximum depends upon the type of overlap between metal and the ligand orbital. As seen from table 1, the values of shift of principal absorption maximum in copper complexes are ranging from 10.5 to 11.0 eV. It is worth noting that the order of the shift of the principal absorption maximum is the same as the order of the chemical shift. This is an expected result [1].

3.3 *Edge-width*

The edge-width of the K-absorption edges increase with the increase in covalent character of the bonds provided other factors like molecular geometry etc remain the same [2, 3]. It can be seen from table 1 that the edge-width is observed in the range of 12.0 to13.0 ev in the copper complexes studied.

3.4 *Effective nuclear charge*

In the present work, the effective nuclear charge Z_{eff} has been obtained from the measured chemical shift by using the semi-experimental method outlined by Nigam and Gupta [2] and given in table1. The data in the table 1 indicates that Z_{eff} varies between 0.6 – 0.8 electrons/atom in the copper complexes studied.

3.5 *Percentage covalency*

It is well known [3] that, the chemical shift is also treated as a measure of degree of covalency. Using Clementi's [4] results for 1s orbital energy in different oxidation states of copper atom, shifts in 1s orbital energy in different oxidation number have been calculated [5] and a theoretical graph is plotted between shift in binding energy of 1s and percentage covalency. The percentage covalency of M-L bond in the complexes has been estimated by using this plot and values are given in table1.

Name of the complex	Edge position E_K(eV)	Chemical shift ΔE_K(eV)	Shift of principal absorption maximum (eV)	Edge-width (eV)	Effective nuclear charge electron/ atom	% covalency
Complex 1	8986.9	6.4	10.8	12.9	0.64	71
Complex 2	8987.6	7.1	10.9	13.1	0.78	55

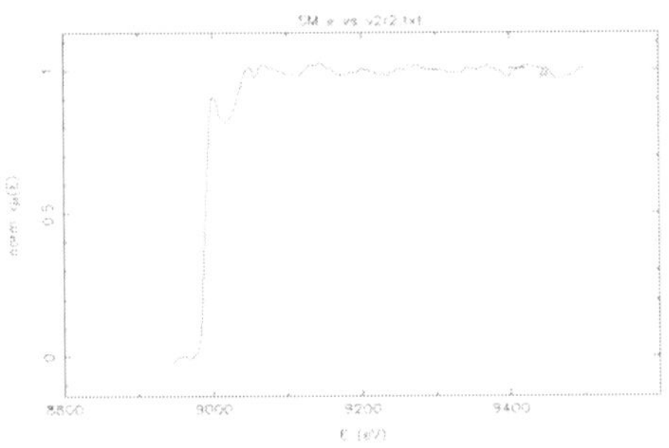

Figure 1: XAFS of complex 1

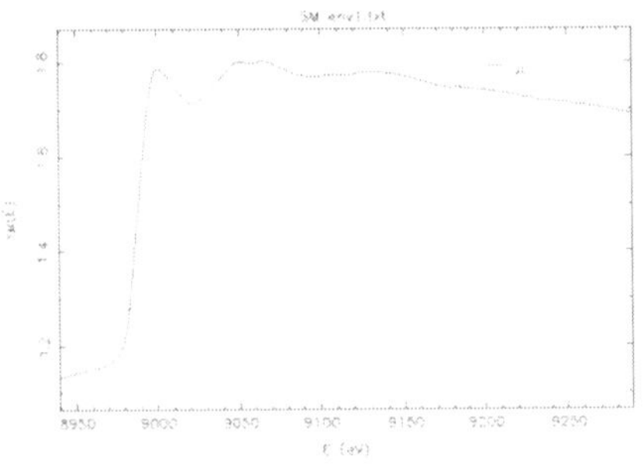

Figure 2: XAFS of complex 2

References

1. Srivastav U. C. and Nigam H. L. 1973 *Coord. Chem. Rev.* **9** 275
2. Nigam A. K and Gupta M. K. 1974 *J Phys* F **4** 1084.
3. Sapre V. B. and Mande C. 1972 *J. Phys. C* **5** 793.
4. Clementi E. 1965 *IBM J. Res. Dev. Suppl.* **9** 2.
5. Khan A. H. and Leydeka A. J. 1964 *Phys Rev. A* **135** 1321.

High temperature Raman study of $BaTi_{0.96}Fe_{0.04}O_3$ and $BaTi_{0.96}Co_{0.04}O_3$ synthesized via solid state route

Ashutosh Mishra, Niyati Mishra, Supriya Bisen and Kanaka M. Jarabana
School of Physics, Devi Ahilya University, Khandwa Road, Indore-452001, India

E-mail: nmishra.sop@gmail.com

Abstract. A finest possible sample of 4% Iron and Cobalt doped ceramics of $BaTiO_3$ (BTO) with possible tetragonal structure via a solid-state route was prepared. Prepared sample was characterized by X-ray diffraction (XRD) using Bruker D8 Advance XRD instrument, the value of 2θ is in between 20^0 to 80^0. Rietveld refinement of both samples shows the information about direct cell parameters, direct cell volume and phase. High temperature Raman spectroscopy was performed using JOBIN-YOVN HORIBA LABRAM HR800 single monochromator and the region of wave number is from 200 cm^{-1} to 800 cm^{-1} from the temperature range of 27 to 207 ^{0}C.

1. Introduction

Barium titanate ($BaTiO_3$) is a ferroelectric oxide that undergoes a transition from a ferroelectric tetragonal phase to a paraelectric cubic phase upon heating above 120 $^{\circ C}$. In cubic perovskite $BaTiO_3$, titanium atoms are octahedraly coordinated by six oxygen atoms. Ferroelectricity in tetragonal $BaTiO_3$ is due to an average relative displacement along the *c*-axis of titanium from its centrosymmetric position in the unit cell and consequently the creation of a permanent electric dipole. The elongation of the unit cell along the *c*-axis and consequently the deviation of the *c/a* ratio from unity are used as an indication of the presence of the ferroelectric phase [1–3]. Ferroelectric properties and a high dielectric constant make $BaTiO_3$ useful in an array of applications such as multilayer ceramic capacitors [4, 5] gate dielectrics [6] waveguide modulators, IR detectors and holographic memory.

2. Synthesis

Both samples of Fe and Co doped barium titanate were synthesized via solid state route. We started from highly pure fine powdered samples of $BaCO_3$, TiO_2, Fe_2O_3 and Co_3O_4 for bulk sample of doped $BaTiO_3$. All these were mixed in the calculated percentage ratio. Samples were grinded for around 16-hours and calcinations were done at 900 ^{0}C and 1100 $_0$C than sintered at 1200^{0}C for 24 hours.

3. Experimental techniques

3.1. X-ray Diffraction

D8 Advance XRD: The XRD measurements were carried out using Bruker D8 Advance X-ray diffractometer. The x-rays were produced using a sealed tube and the wavelength of x-ray was 0.154 nm (Cu K-alpha). The x-rays were detected using a fast counting detector based on Silicon trip technology (Bruker Lynx Eye detector).

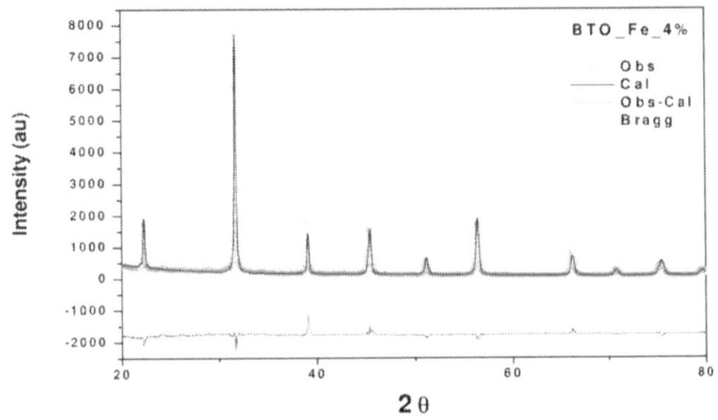

Figure 1. Rietveld refinement fit of room temperature XRD pattern for 4% Fe modified BTO.

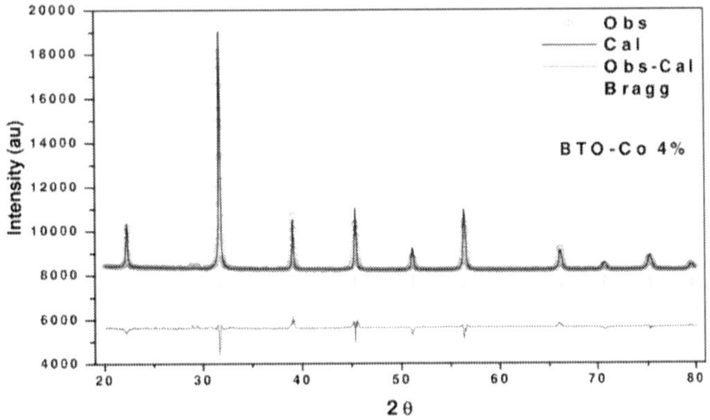

Figure 2. Rietveld refinement fit of room temperature XRD pattern for 4% Co modified BTO.

X-ray diffraction (XRD) is a versatile, non-destructive technique that reveals the detailed information about the chemical composition and crystallographic structure of natural and manufactured materials. In Figure1 (A) and Figure 1(B) XRD patterns reports the single phase tetragonal crystal system of the space group P4mm (99) and pattern matched with the standard pattern JCPDS no. 79-2265.

Table 1. XRD Detailed parameters of Co modified BTO.

Sample	Direct cell parameters			Direct Cell Volume	Phase
	a	b	C		
$BaTi_{0.96}Fe_{0.04}O_3$	3.9882	3.9882	4.0043	63.6510	1
$BaTi_{0.96}Co_{0.04}O_3$	3.9923	3.9923	4.0013	63.7744	1

3.2 Raman Spectroscopy

The Raman spectrum of Fe and Co doped $BaTiO_3$ has been done over a range of temperatures from room temperature up to around 210 ^{0}C.

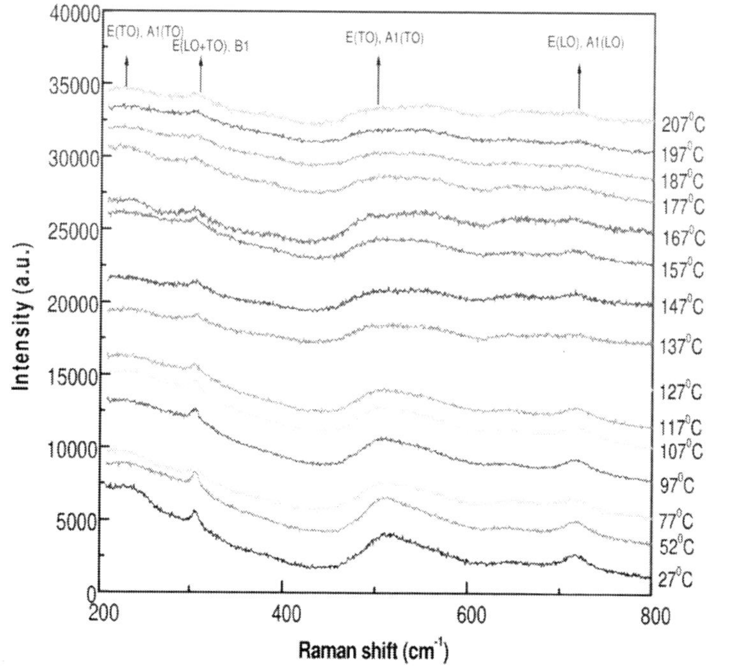

Figure 3. Unpolarised Raman spectra of 4% Fe doped BTO recorded at selected temperature range.

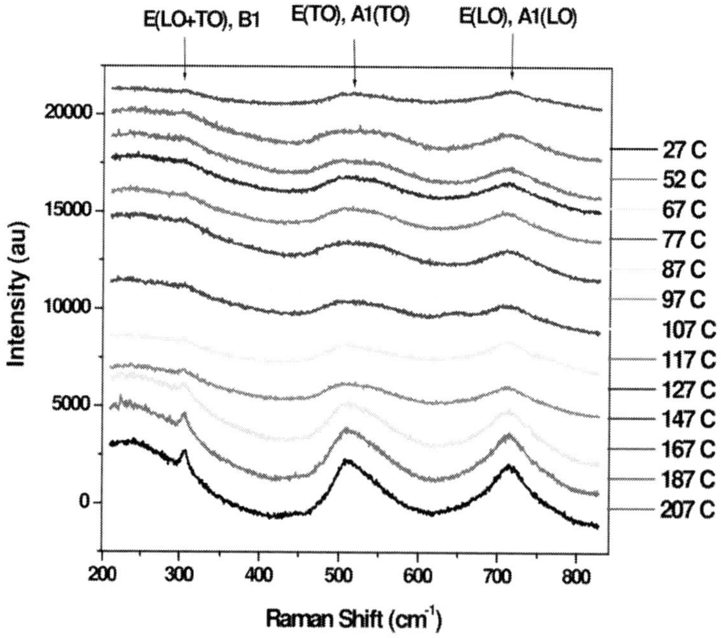

Figure 4. Unpolarized Raman spectra of 4% Co doped BTO recorded at selected temperature range.

The $BaTiO_3$ Raman The study of Raman spectroscopy was carried out by means Jobin-Yvon Horiba Labram HR800 single monochromator. The spectra have been recorded in unpolarized mode in the full range starting from 200 to 800cm^{-1}. Spectra have the broad features characteristics of titanium disorder in the unit cell at all temperatures.

In the bulk 4% Fe doped $BaTiO_3$ spectra in Figure 2(A), the intensities of the [E (LO+TO), B1] peaks at ~307 cm^{-1} and [E (LO), A1 (LO)] peaks at ~ 717 cm^{-1}decrease rapidly as the temperature, increase through the bulk T_c.

In the bulk 4% Co doped $BaTiO_3$ spectra in Figure 2(B) the intensities of the [E (LO+TO), B1] peaks at ~308 cm^{-1} and [E (LO), A1 (LO)] peaks at ~ 713 cm^{-1}decrease rapidly as the temperature, increase through the bulk T_c.

The first is an increase in peak width at the phase transition temperature and the second is the loss of peak intensity with increasing temperature. For $BaTiO_3$, the line width for the [E (LO+TO), B1] peak increases with increasing temperature. It is to be noted that bulk $BaTiO_3$ near the cubic-to-tetragonal phase transition displays a Raman line width.

4. Discussion

The $BaTi_{0.96}Fe_{0.04}O_3$ and $BaTi_{0.96}Co_{0.04}O_3$ ceramics was prepared via solid state route. The XRD data has concluded that all major peaks of Fe and Co doped BTO are matching with reported XRD data of pure BTO, and sample is in single phase, which shows that doping is perfect. The measurement of Raman spectroscopy shows that in the Fe doped BTO ceramic, the intensities of the [E (LO+TO), B1] peaks at ~307 cm^{-1} and [E (LO), A1 (LO)] peaks at ~ 717 cm^{-1} decrease rapidly as the temperature, increase through the bulk's T_c. We interpret the disappearance of the 307 cm^{-1} peak as an indicator of the tetragonal phase and use two characteristics as an indication of the phase transition. And, in the 4% Co doped $BaTiO_3$ ceramic, the intensities of the [E (LO+TO), B1] peaks at ~308 cm^{-1} and [E (LO), A1 (LO)] peaks at ~ 713 cm^{-1}decrease rapidly as the temperature, increase through the bulk's T_c. We interpret the disappearance of the 308 cm^{-1} peak as an indicator of the tetragonal phase and use two characteristics as an indication of the phase transition. The first is an increase in peak width at the phase transition temperature and the second is the loss of peak intensity with increasing temperature.

Acknowledgements

The authors are thankful to Dr. Ajay Gupta, Dr. Mukul Gupta (XRD Lab) and Dr. Vasant Sathe (Raman Lab), UGC-DAE CSR, Indore for providing the experimental facilities and for their precious guidance. One of the author (Niyati Mishra) is thankful to UGC-DAE CSR Indore for providing the CRS project fellowship.

References

[1] Jaffe B, Cook W R and Jaffe H 1971 *Piezoelectric Ceramics*, Vol. **3** (New York: Academic Press)

[2] Lines M E and Glass A M 1977 *Principles and Applications of Ferroelectrics and Related Materials*, (Oxford: Clarendon Press)

[3] Strukov B A and Levanyuk A P 1998 *Ferroelectric Phenomena in Crystals*, (Berlin: Springer-Verlag)

[4] Wang S F and Dayton G O 1999 *J. Am. Ceram. Soc.* **82** 2677

[5] Hennings D, Klee M and Waser R 1991 *Adv. Mater.* **3** 334

[6] Yildirim F A, Ucurum C, Schliewe R R, Bauhofer W, Meixner R M, Goebel H and Krautschneider W 2007 *Appl. Phys. Lett.* **90** 083501

Frequency and temperature dependent dielectric studies of $BaTi_{0.96}Fe_{0.04}O_3$

Ashutosh Mishra, Niyati Mishra, Supriya Bisen, Kanaka M. Jarabana
School of Physics, Devi Ahilya University, Indore-452001, India

Email : supriyabisen0@gmail.com

Abstract. A Finest possible sample of 4% Iron doped $BaTiO_3$ (BTO) with possible tetragonal structure via a solid state route was prepared. Prepared sample was characterized by X-ray diffraction (XRD) using Bruker D8 Advance XRD instrument, the value of 2θ is in between 20^0 to 80^0.Detailed analysis of dielectric constant, dielectric loss, ac conductivity and electrical modulus at various range of frequency and temperature have been done of 4% Fe doped BTO was recorded on hp-Hewlett Packard 4192 A, LF impedance, 5Hz-13Hz analyser.

1. Introduction

Barium titanate is a ferroelectric ceramic material with ABO_3 type perovskite structure [1]. It exhibits ferroelectric properties at and above room temperature, and finally because it can be easily prepared and used in the form of ceramic polycrystalline samples. Due to its high dielectric constant and low loss characteristics, barium titanate has been used in applications, such as capacitors and multilayer capacitors (MLCs) [2]. Doped barium titanate has found wide application in semiconductors, PTC thermistors and piezoelectric devices, and has become one of the most important ferroelectric ceramics.

2. Synthesis of iron doped barium titanate

We started from highly pure fine powdered samples of $BaCo_3$, TiO_2 and Fe_3O_4 for bulk sample of doped $BaTiO_3$. All these were mixed in the calculated percentage ratio. Samples were grinded for 8-hours and heat treatment was given at 900^0C for 24-hours.

3. Experimental technique

3.1. X-ray Diffraction

The X-ray was produced using a sealed tube and the wavelength of X-ray was 0.15nm (Cu-Kα). XRD pattern of the Fe-doped $BaTiO_3$ at 900^0C. The XRD data has concluded that all major peaks of Fe-BTO are matching with reported XRD data of pure BTO, and sample is in single phase. Which shows doping is perfect. Fig 1 shows that all the peaks in the pattern are matching and it's showing purely tetragonal single phase crystal of the $BaTiO_3$ ceramics. The single phase crystal of the $BaTiO_3$ ceramics are tetragonal symmetry in the space group P4mm and pattern matched with the standard pattern PCPDF, no. 81-2203.

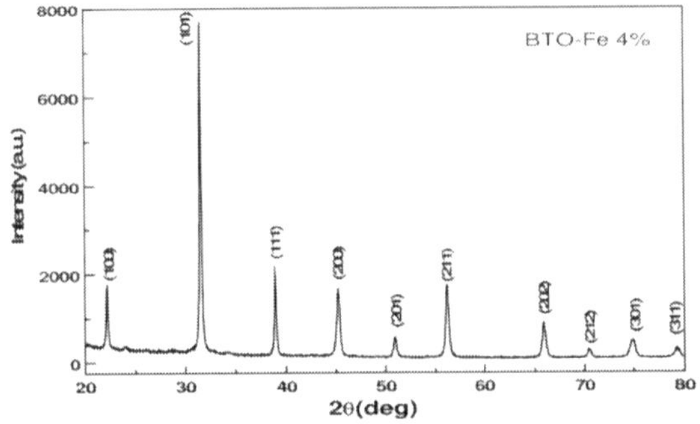

Figure 3.1. XRD pattern of $BaTi_{0.96}Fe_{0.04}O_3$ doped BTO at 900^0C

3.2. Dielectric constant measurement

3.2.1. Dielectric constant

First a pallet of 12mm diameter of bulk sample was prepared and then coated both sides of the sample with low temperature silver paste. Put the sample in between two electrodes in furnace and obtain the value of dielectric constant with variation of temperature manually. In fig 3.2.1(a) after doping Fe in pure BTO, on increasing the temperature on higher side, the transition temperature was shifted from 120^0C to 118^0C. Figure shows the temperature dependence of dielectric constant in Fe ion substituted at the Ti site of BTO. The dielectric studies indicated a shift in the transition temperature (T_c) towards the lower temperature side.

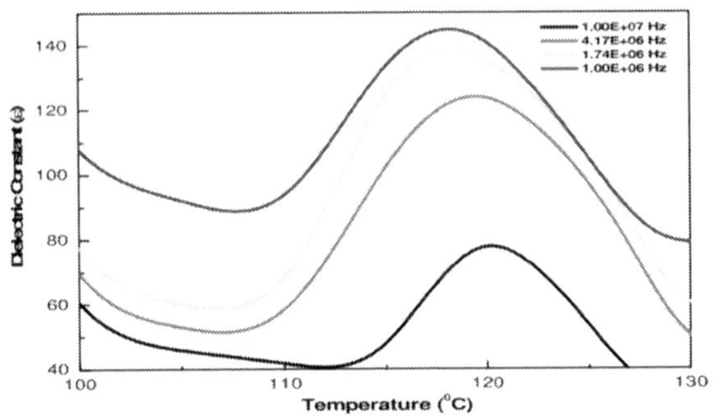

Figure 3.2.1(a). Temperature dependent Dielectric Constant.

Figure 3.2.1(b) & (c) shows a strong increase of ε'' (also ε) at low frequency regime is attributed to interfacial polarization. The magnitude of ε' slowly decreases with increase of frequency before leaving off a constant value at higher frequency. This is a commonly observed feature in ceramic materials.

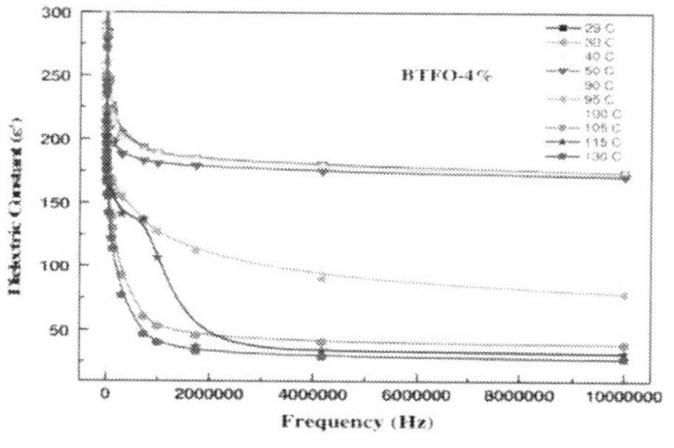

Figure 3.2.1(b). Frequency dependent Dielectric Constant real (Ɛ') at different temperature for BTFO bulk-sample.

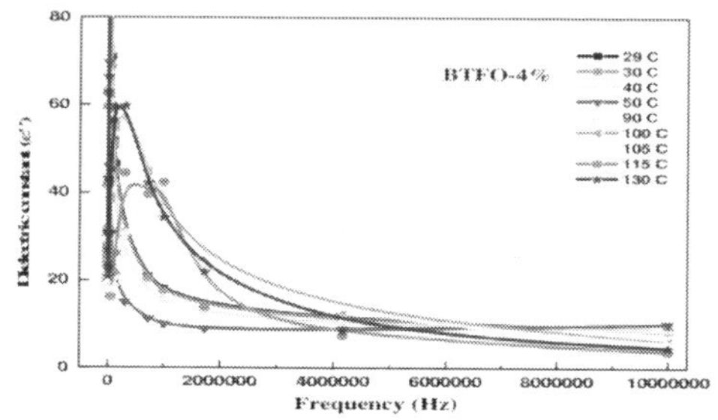

Figure 3.2.1(c). Frequency dependent Dielectric Constant imaginary (Ɛ") at different temperature for BTFO bulk-sample.

3.2.2. Dielectric loss

The dielectric loss (tan δ) is an important parameter that determines the suitability of a dielectric material in microelectronics. The Fe doped $BaTiO_3$ sample have shown thermal activated sharp loss peak. The peak position of sample $BaTiO_3$ shift towards higher frequency with the increase in temperature. The high dielectric loss at lower frequency in Fe doped $BaTiO_3$ sample is the effect of interfacial loss but dielectric loss is below 0.1 at higher frequency.

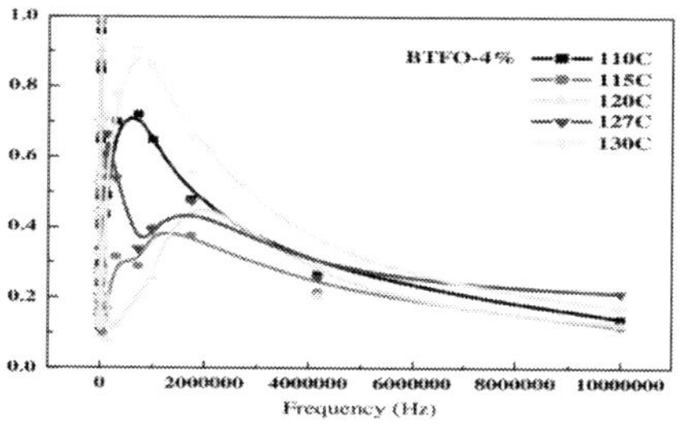

Figure 3.2.2. Variation of Dielectric loss (tan δ) with frequency at selected temperature.

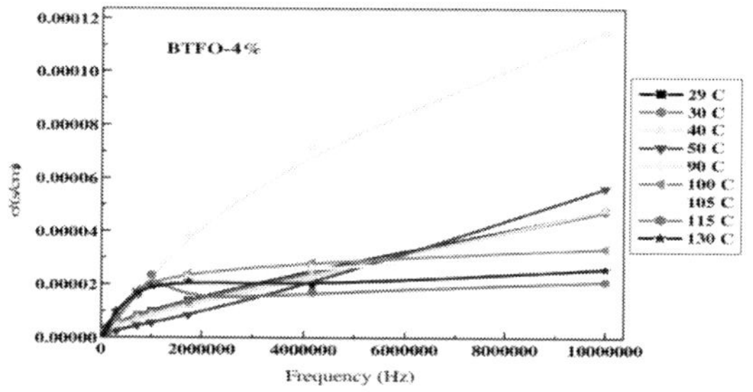

Figure 3.3.3. Variation of σ' (f) at selected temperature

3.3.3. AC conductivity

Fig 3.3.3 shows the frequency dependence of the real part (σ') of ac conductivity at selected temperatures in the range 29C to 130C for BTFO-bulk sample. BTFO sample showed nearly a frequency independent conductivity at lower frequencies. BTFO showed a conductivity minimum in between the frequency independent and frequency dependent conductivity regimes. This minimum becomes prominent at higher measurement temperature.

3.3.4. Electrical modulus

The main advantage of applying modulus formalism is to suppress the electrode effect on electric conductivity contribution BTFO-bulk sample. Fig 3.3.4(a) and (b) shows sharp increases of M' below 100Hz followed by a slow increase with frequencies and finally saturated above 100KHzThe imaginary part of modulus (M") shows a sharp peak at lower frequency and the peak position shift to higher frequencies with the increase of measurement temperature.

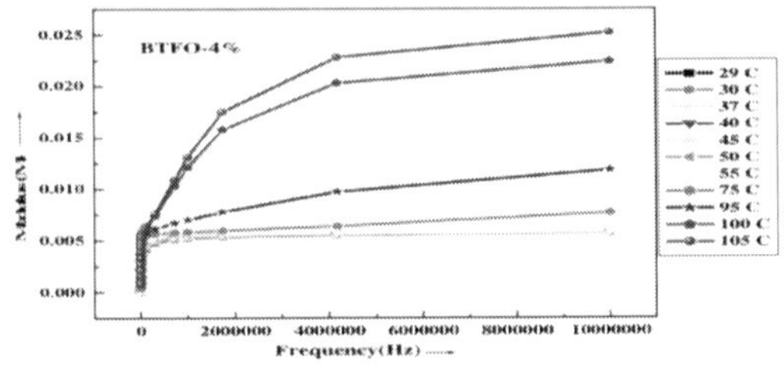

Figure 3.3.4 (a). Frequency dependent modulus (M') at different temperature for BTFO-bulk sample.

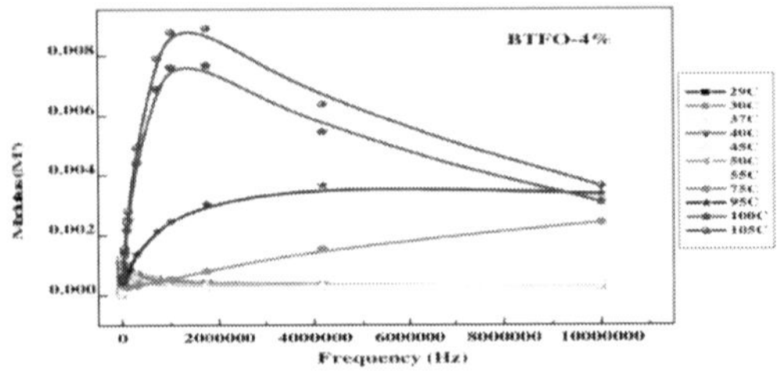

Figure 3.3.4 (b). Frequency dependent modulus (M") at different temperature for BTFO-bulk sample.

4. Discussion

The XRD pattern showed that this material crystallizes in a perovskite-type single phase tetragonal structure with space group p4mm. In dielectric constant measurement, the transition temperature of Fe-BTO is occurring at 118^0C. The reported value of transition temperature of pure BTO is 120^0C. The high dielectric loss at lower frequency in Fe doped $BaTiO_3$ sample is the effect of interfacial loss but dielectric loss is below 0.1 at higher frequencies. The conductivity of BTFO sample showed nearly a frequency independent conductivity at lower frequencies. The imaginary part of modulus (M") shows a sharp peak at lower frequency (below 100Hz) and the peak position shifts to higher frequency with the increase of measurement temperature. In conductive relaxation process the real part of modulus (M') generally exhibits a sharp monotonic increase with frequency.

Acknowledgement

The authors are thankful to prof. Ajay Gupta, Dr. Mukul Gupta, UGC-DAE CSR for XRD measurement, Dr. A.M. Awasthi & Mr. Suresh Bhardwaj, UGC-DAE CSR for Dielectric measurement for providing the experimental facilities and for the precious guidance. One of the authors (Niyati Mishra) is thankful to UGC-DAE CSR Indore for providing the research fellowship.

References

[1] Jona F and Shirane G 1993 *Ferroelectric Crystals* (New York: Dower publications)

Journal of Physics: Conference Series **534** (2014) 012011 doi:10.1088/1742-6596/534/1/012011

[2] Maga D, Igor P and Sergei M 2000 *J. Mater. Chem.* **10** 941
[3] Jona A, Kundu T K, Pradhan S K and Chakrvorty D 2005 *J. Appl. Phys.* **97** 44311
[4] Ismailzade I H and Ismailov R M 1980 *Phys. Stat. Sol. (a)* **59** K 191
[5] Smolenski G A and Isupov V A 1954 *Dokl. Akad. Nauk.* SSSR **9** 653

Effect of addition of non ionic surfactant on the gas sensing properties of CuO films

Iqbal Singh and Taminder Singh [1]
Post Graduate Department of Physics, Khalsa College, Amritsar

E-mail: iqbalsgh@yahoo.com

Abstract. Non ionic surfactant (Poly ethylene glycol (PEG-400)) has been used as dispersing and stabilizing agent along with citric acid in the sol-gel reaction to synthesize nanocrystalline CuO structures The decomposed gel sample calcined at temperature of 400℃ has been assembled on glass substrate using an organic binder to form thick film sensor. The X-ray diffraction study of samples shows the formation of CuO as dominant phase in as decomposed gel and calcined CuO powder samples. Surfactant addition in precursor results in the decrease of crystallite size and leads to uniform distribution of CuO aggregates as noticed in the SEM micrographs. The activation energy of charge carrier in thick film samples has been investigated by two probe technique and is found to be of the order of 0.24 eV. Gas sening measurements on the samples performed at room temperature for analyte (ammonia gas) show an increase in response (4 times) as compare to sample formed by same procedure without surfactant.

1. Introduction

The chemical route using surfactants to synthesize metal oxide nanoparticles has been attracted considerable attention in last decade [1-5]. Surfactant molecules are widely used for controlling size, shape of grains and acts as polymer which plays an important role in synthesizing self assembled nano as well as micro structures. The surfactant molecules absorb on specific crystal planes and initiate an anisotropic grain growth. In this paper, the PEG (non ionic surfactant) assisted CuO powder has been synthesized by sol gel auto combustion. The samples are systematically characterized and obtained results have been investigated to study the effect of non-ionic surfactant on CuO based thick films.

2. Experimental

All the chemicals of analytical reagent grade are used as precursors. To prepare 100 mL of precursor solution with metal nitrate to citric acid (MN:CA) molar ratio of 1:1, at pH 7 required amount of cupric nitrate and citric acid dissolved in water. 10 mL of 0.5 M PEG-400 solution is added drop wise to previously formed citrate complex. The solution is thermally dehydrated in an oven at temperature of 80 ± 5 °C and the detailed combustion mechanism has been reported earlier [6]. The fine powder is converted into thixotropic paste by mixing it with an organic solvent diethnolamine. A uniform layer of the thixotropic paste is then deposited on ultrasonically cleaned glass slides. The drying of films is carried out at 400 °C in the muffle furnace for 1 hour to burn out the organic binder. The samples have been characterized for phase analysis by X-ray diffraction (XRD), for morphological investigations by FESEM and for activation energy conduction by temperature-conductivity measurements respectively.The samples have been assigned names as mentioned in Table 1.

3. Results and discussion

XRD diffractogram of P1 and P2 samples shown in Figure 1 are found to be polycrystalline in nature and all prominent reflections can be indexed to monoclinic CuO phase. It has been observed that addition of PEG lowers intensity peaks corresponding to the atomic planes (110), (020) and (220) of CuO phase. Using Scherrer's formula, (equation 1) average crystallite size of samples has been calculated (Table 1) and found to be 51.2 nm in P2 sample. The lattice parameters of samples have been calculated using the formula given in equation 2, 3 and found to be in good agreement with ICDD data card 41-25. It has been noticed that lattice parameter values found to be lower in magnitude as compared to those reported in literature suggesting more number of defects which enhanced with PEG addition in precursors.

$$D = \frac{0.9\lambda}{\beta \cos\theta} \quad (1)$$

$$\frac{1}{d^2} = \frac{1}{\sin^2\beta}\left(\frac{h^2}{a^2} + \frac{k^2\sin^2\beta}{b^2} + \frac{l^2}{c^2} - \frac{2hl\cos\beta}{ac}\right) \quad (2)$$

$$V = abc\sin\beta \quad (3)$$

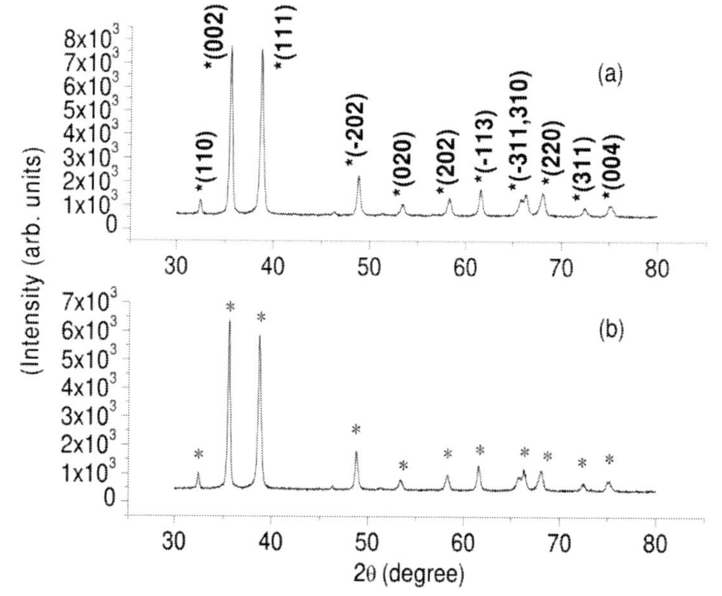

Figure 1. XRD pattern, of CuO powder sample calcined at 400 °C for (a) P1 and (b) P2.

FESEM images of the P1 and P2 samples are shown in Figure 2. The images reveal that particle agglomerates get smaller and uniform in size for PEG assisted samples. The particle size and the agglomeration in the material depend mainly on the way the combustion propagates and the amount of disintegration occurs. The transformation of facets like morphology of particles to spherical might be due to templating nature of the non-ionic, polymer type PEG surfactant used in the synthesis of several porous materials [7].

Figure 2. FESEM images of CuO powder sample calcined at 400 °C for (a) P1 and (b) P2.

The electrical conductivity of film samples has been found to increases with temperature indicating, their semiconducting behaviour and suggesting a thermally activated conduction mechanism. The activation energy of conduction in different samples has been calculated from temperature conductivity measurements and corresponding values are recorded in Table 1. The increase of activation energy in case of P2 sample can be understood from the increase in large number of grain boundaries. It is expected that increase of defects as suggested by the XRD analysis of the PEG doped samples consequently increases the scattering of charge carrier [7,8]. The higher value of activation energy of conduction in case of P2 sample can be understood from the increase in large number of grain boundaries. The gas sensing performance of films the resistance time curves of the samples are shown in the Figure 3. The response have been calculated using equation 4 and found to be 3.69 for P1 film whereas it is 25.6 for P2 film. The response time for the P1 and P2 sensors has been found to be 250 and 35 seconds. The room temperature appreciable response in case of surfactant assisted film sensor may be attributed to the large specific surface area as depicted by FESEM, images. The response of PEG assisted films has been found to increase almost 4 times with respect to P1 films

$$S = \frac{R_a}{R_g} \quad (4)$$

where R and R are the CuO film resistances measured in ammonia and air respectively.

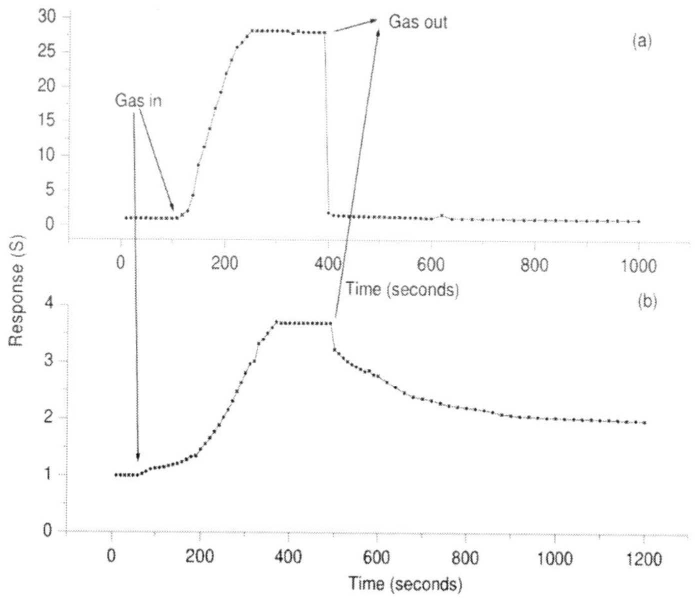

Figure 3. Room temperature gas sensing performance of the CuO films for (a) P1 and (b) P2.

3. Conclusions

Nanocrystalline PEG-400 doped CuO powder based thick films have large number of defects which increases the activation energy of conduction and provide the more active favourable sites for the gas adsorption hence higher response.

Table 1. Values of the lattice constants, crystallite size (D) from XRD, activation energy (E_g)

Property/Sample	a (Å)	b (Å)	c (Å)	β (Degree)	V (Å^3)	D (nm)	E_g (eV)
P1	4.682	3.424	5.114	99.111	80.949	63.40	0.222
P2	4.684	3.427	5.117	99.306	81.087	51.28	0.242

References

[1] El-Sheikh S M, Harraz F A and Abdel-Halim K S 2009 *J. Alloys Compds.* **487** 716.
[2] Abdel Rafea M and Roushdy N, 2009 *J. Appl. Phys. D* **42** 015413
[3] Kaung Q, Lao C S, Li Z, Liu Y Z, Xie Z X, Zheng L S and Wong Z L, 2008 *J. Phys. Chem. C* **112** 11539.
[4] Zhang J, Liu J, Peng Q, Wang X and Li Y 2006 *Chem. Mater.* **18** 867.
[5] Gou X, Wang G, Yang J, Park J and Wexler D 2008 *J. Mater. Chem.* **18** 965.
[6] Bedi R K and Singh I 2009 *Curr. Nanosci.* **5** 273.
[7] Bilgin V, Kose S, Atay F and Akyuz I 2005 *J. Mater. Sci.* **40** 1909.
[8] Bedi R K and Singh I 2010 *Appl. Mater. Interfaces* **2** 1361.

Composition dependence of the magnetic properties of strontium hexaferrite doped with rare earth ions

Taminder Singh[1,4], M S Batra[2], Iqbal Singh[1] and Arun Katoch[3]
[1]Department of Physics, Khalsa College, Amritsar, Punjab, India
[2]Department of Chemistry, Khalsa College, Amritsar, Punjab, India
[3]BKKSJ College of Engineering, Amritsar, Punjab, India

E-mail: dr.taminder@gmail.com

Abstract. Rare earth substituted ferrite $Sr_{1-x}RE_xFe_{12}O_{19}$ (where RE = La, Gd and Dy, x = 0.0, 0.10, 0.20 and 0.30), have been prepared by employing the ceramic technique and subsequent heat treatment. The magnetic properties of the calcined samples were characterized with a Vibrating Sample Magnetometer (VSM). The samples were sintered at 1150°C for 8 hours. The samples were characterized for magnetic properties such as specific saturation magnetization M_S, specific remanence magnetization M_r, and coercivity H_c as well as microstructure. It has been observed that all these parameters depend on the composition and heat treatment of the prepared samples. The coercively H_c exhibits an increase as the RE content increases in $Sr_{1-x}RE_xFe_{12}O_{19}$ ferrite. With increasing RE content the M_S and M_r decrease due to dissolution of RE ion into hexaferrite lattice.

1. Introduction

M-type hexagonal (Ba M and Sr M) ferrites are perspective materials for developing of high density data storage devices due to their high uniaxial magneto crystalline anisotropy and chemical stability [1-3]. In comparison to alloy films, the hexagonal ferrites have lower saturation magnetization but higher anisotropy field. Their magnetic properties depends on the preparation condition , such as sintering temperature, sintering atmosphere and the soaking time as well as the type of the substituted ions. The hexagonal Strontium ferrite (SrM) is considered to be an excellent candidate for magnetic recording media and characterized with high magneto crystalline anisotropy, moderate hard magnetic properties and high chemical stability, compared with other magnetic materials. The common processing methods of hexagonal ferrites are conventional ceramic process of solid-state reaction[4], co-precipitation method [5], sol-gel process[6] and molten salt method [7,8] etc. The conventional ceramic process includes the mixing the raw materials, calculation, milling, pressing and sintering at 1150- 1200°C [9]. In a fine particle form, Strontium ferrite is suitable for high -density recording media. In the present paper an attempt has been made to investigate the temperature and composition dependence of magnetic properties of SrM ferrites containing three rare earth elements: La, Gd and Dy.

2. Experimental

All the samples of Sr-hexaferrites ($SrFe_{12}O_{19}$) of nominal composition (x = 0.0, 0.10, 0.20 and 0.30) were synthesized starting from ball-milling mixtures of $SrCO_3$, Fe_2O_3 and Dy–rare earth ion for 12h. After drying at 60°C for 6h, the powder mixture was heated at temperature of 800-1150 °C for 4 hrs in a lid –covered alumina crucible with a heating rate of 5° C/min in air. In order to make the sintered magnet, the strontium ferrite powder were wet mixed in acetone medium with addition of 4% polyvinyl alcohol (PVA) binder solution by using a ball mill. The mixture was reground again for 5 hrs and the final fine powder was pressed in disk-shaped pellets with thickness ranging from 2 to 4mm and with diameter from 7 to 9mm and subsequently pellets were sintered in a resistance heated furnace for 3hrs at beach specified level of sintering temperature from 950 to 1250°C. The line intercept method [10] was used to measure the grain size. The magnetic measurements (saturation magnetization M_S, remanence magnetization, M_r and Coercivity, H_c were performed by a vibrating sample magnetometer operating up to a maximum field of 10kOe, in air, at room temperature.

3. Results and discussion

The microstructures of prepared samples is shown in Figure 1. It indicates that the M-type ferrite grains are homogeneous hexagonal shaped crystals [11].

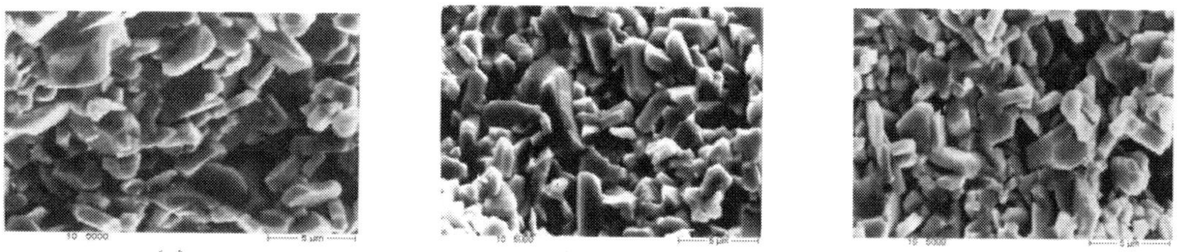

(a) (b) (c)

Fig. 1 SEM images for the $Sr_{1-x}RE_xFe_{12}O_{19}$ where RE = (a) La, (b) Gd and (c) Dy

The average grain size estimated from SEM has been found to be approximately 1μm and is almost dependent on the composition x (x=0.0, 0.10, 0.20 and 0.30). The complex permeability spectra of polycrystalline ferrite depends not only on the chemical composition of the ferrite but also on the post-sintering density and the microstructures such as grain size and porosity.The variation of saturation magnetization, M_S and remanence, M_r with composition for all the series has been shown in Figure 2.

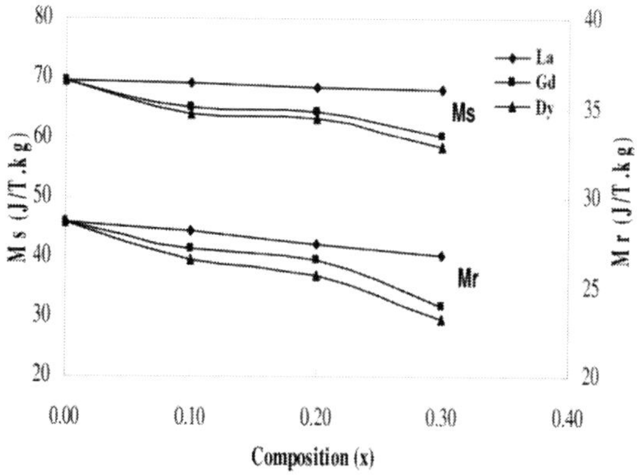

Fig. 2 Variation of saturation magnetization (M_S) and remanence (M_r) of $Sr_{1-x}RE_xFe_{12}O_{19}$ with composition X (X=0.0,0.10,0.20 and 0.30)

It is clear that the values of M_S and M_r decrease with increasing substitution. For the same substitution content, the values of M_S and M_r of La doped strontium hexaferrite are higher than those of Gd-doped and Dy-doped samples. The decrease in M_S and M_r values of all the series may be due to magnetic dilution with changing of Fe^{3+} (high spin) valence state to Fe^{2-}(low spin) state on 2a site by substitution of the Sr^{2+} site with RE^{3+} ions and existence of spin canting which may be responsible for promoting reduction of superexchange fields [12]. The Fe^{3+} - O^{2-} - Fe^{3+} exchange interaction is disrupted and weakened by Fe^{2+} ion canted spins, which could be produced by substitution of the rare earth ions into the hexaferrite. These result in lower magnetization [12, 13]. Moreover the decrease in M_s and M_r are found to be maximum at a composition of x = 0.30 in case of Gd and Dy- substituted strontium hexaferrite. This decrease could be due to the existence of nonmagnetic phases observed for higher rare earth concentration in case of Gd and Dy substitutions. It seems that the main role of these nonmagnetic phases are to isolate Sr-ferrite particles from each other thus reducing exchange interaction between them, and are known to have a detrimental effect on M_s and M_r.The variation of H_c with composition for all the series is shown in Figure 3. It is observed that the value of H_c increases with increasing concentration of rare earth substitution.

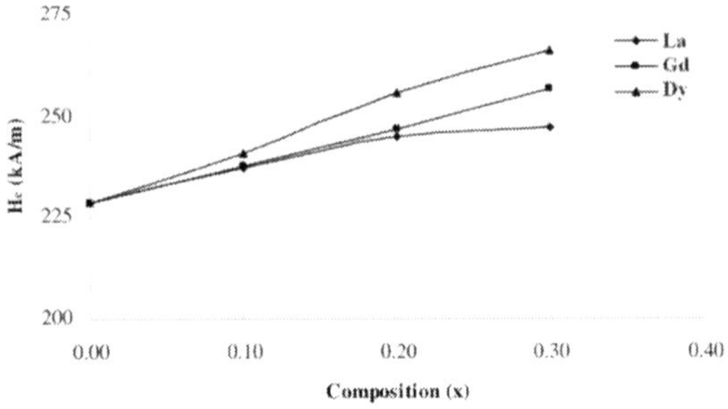

Fig. 3 Variation of coercive field, H_c of $Sr_{1-x}RE_xFe_{12}O_{19}$ with composition x.

The increase in H_c for all the series can be attributed to enhancement of magneto crystalline anisotropy [14] with anisotropic Fe^{2+} ions locating on a 2a site and the grain size reduction which is usually found on rare earth substitutions and is evident in SEM micrographs [15,16]. However for the same substitution content, the H_c value of La-doped strontium hexaferrite is correspondingly smaller than those of Gd-doped ferrites and Dy-substituted ferrites due to the smaller magneto crystalline anisotropies and larger grain size. This suggests that the effect of Fe^{2+} anisotropy on 2a site in the Gd and Dy doped samples could be stronger than those in La doped samples [15, 16]. A similar result has been also reported by Ounnunkad [17] in Ba (La or Pr) ferrites. The increase in coercivity, H_c with increase in density depends upon the oxidation of the grain and their size. Figure 4 shows the variation of the Curie temperature T_c (K) with composition for all the series. It is observed that the value of Curie temperature T_c(K) decreases with increasing RE content in each case. This behaviour is based on the number of magnetic ions present in the two sub-lattices and their mutual contraction due to the substitution of ions. It is known for SrM hexaferrite that the magnetic moments of Fe^{3+} ions have collinear arrangement due to the existence of super exchange interaction. Thus the super exchange interaction determines the orientation of magnetic moments of Fe^{3+} ions. Substitution of Sr^{2+} ions by RE^{3+} ions associated with a valence change of one Fe^{3+} ion per formula unit to Fe^{2+} ion can reduce the strength of this interaction due to Fe^{2+} ions. Since the Curie temperature T_c(K) is determined by the overall strength of the exchange interactions, the weakening of exchange interactions result in decrease of Curie temperature as concentration of RE ions is increased. However for the same substitution content, the T_c(K) value of La-doped strontium hexaferrite is correspondingly higher than those of Gd-doped ferrites and Dy-doped ferrites. It may be due to the small separation between the spin inside the domain and hence the exchange interaction will be large for La-doped sample as compare to the Gd and Dy doped samples.

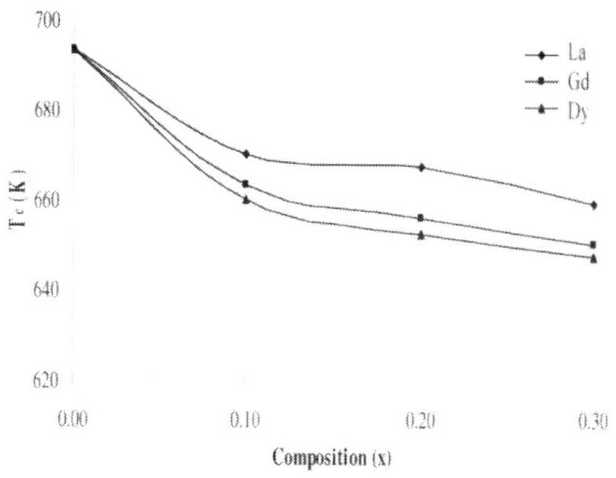

Figure 4 Variation of Curie temperature (T_c) with composition (x).

References

[1] Zaitsev D D, Kasin P E, Tretyakov Yu D and Jansen M 2004 *Inorg. Mat.* **8** 881
[2] Hilton T L, Parker M, Coffrey K R, Umphress R and Howard J K 1994 *Appl. Phys.* **75** 5960
[3] Saui X and Kryder M H 1993 *Appl. Phys. Lett.* **63** 1582
[4] Harberey F and Kockel A 1976 *IEEE Trans. Magn.* **12** 983
[5] Lee J H and Won C W 1995 *J. Kor. Inst. Met. Mater.* **33** 21
[6] Zhang H, Li L, Zhou J, Ma Y Z and Gui Z 2001 *J. Eur. Ceram. Soc.* **21** 149
[7] Hayashi Y, Kanazawa T and Yamaguchi T 1986 *J. Mater. Sci.* **21** 2876
[8] Kimura T, Takahashi T and Yamaguchi T 1980 *J. Mater. Sci.* **15** 1491
[9] Dong Kim Si and Jung Sik Kim 2006. *Mater.* **307** 295
[10] Rana M U and Abbas T 2002 *J. Magn. Magn. Mater.* **246** 110
[11] Gonzalez-Carreno T, Morales M P and Serna C J 2000 *Mater. Lett.* **43** 97
[12] Liu X, Zhong W, Yang S, Yu Z, Gu B and Du Y, 2002 *Phys. Status Solidi (a)* **193** 314
[13] Liu, X, Zhong W, Yang S, Yu Z, Gu B and Du Y, 2002 *J. Magn. Magn. Mater* **238** 207
[14] Tellez Blanco, Grössinger J C, Müller R, Kupferling M, Wiesinger M, Wang G, Ponton J F and Harris I R 2002 *Proc. of International Workshop on Rare Earth Magnets and their Applications* (Newark, Delaware, USA) p 192
[15] Sauer C, Kobler U, Zinn W and Stablein H 1978 *J. Phys. Chem. Solids*, **39** 1197
[16] Dho J, Lee E K, Park Y Y and Hur N H 2005 *J. Magn. Mater.* **285** 164
[17] Ounnunkad S 2006 *Solid State Commun.* **138** 472

X-Ray diffraction and Fourier transformation infrared spectroscopy studies of copper (II) thiourea chloro and sulphate complexes

Ashutosh Mishra[1], Jagrati Dwivedi[1], Kritika Shukla[1], Pramod Malviya[2]
[1]School of Physics, Devi Ahilya University, Indore-452001, India
[2]Physics Department, Govt. P. G. College, Nagda, India

Email: amishra1960@yahoo.co.in, hemu.dwi@gmail.com

Abstract. Copper (II) thiourea complexes were synthesized by chemical route method. To analyze the samples, X-Ray diffraction and Fourier transformation infrared spectroscopy method have been used. XRD analysis shows that sample is crystalline in nature and having particle size in the range of nanometres. FTIR spectroscopy shows the organic and inorganic compounds present in the sample. The X-ray diffraction pattern (XRD) using Bruker D-8 advance instrument and Infrared spectra of the complexes were obtained by KBr disc technique by using VERTEX 70 (Bruker).

1. Introduction

The aim of this paper is to determine the crystal structure and lattice parameter by X-ray diffraction technique. Complexes prepared for this experiment are $Cu(NH_2CSNH_2)_2Cl_2$ and Cu $(NH_2CSNH_2)_2(SO_4)_2$. Thiourea, having a considerably wide range of applications, is a functional organic compound similar to urea, except that the oxygen atom is replaced by a sulfur atom. The properties of urea and thiourea differ significantly because of the difference in electronegativity between sulfur and oxygen. Thiourea ligands and their metal complexes exhibit a wide range of biological activity including anticancer, antimicrobial, antibacterial, antifungal, antimalarial and antituberculosis.[1-2]

X-ray diffraction (XRD) is a versatile, non-destructive technique that reveals detailed information about the chemical composition and crystallographic structure of natural and manufactured materials or sample. For diffraction, only short wavelength X-rays in the range of few angstroms to 0.1 angstrom are used because the wavelength of X-rays is comparable to the size of atoms. The energetic X-rays can penetrate the materials and provide information about the crystal structure. X-ray diffraction is based on constructive interference of monochromatic X-rays and a crystalline sample. [3] In infrared spectroscopy, IR radiation is passed through a sample. Some of the infrared radiation is absorbed by the sample and some of it is passed through the sample (transmitted). The resulting spectrum represents the molecular absorption and transmission, creating a molecular fingerprint of the sample. It can identify unknown materials, it can determine the quality or consistency of a sample, and it can determine the amount of components in a mixture.[4]

2. Experimental

2.1. Powder method for XRD

X-ray powder diffraction (XRD) is a rapid analytical technique primarily used for phase identification of a crystalline material and can provide information on unit cell dimensions. The analyzed material is finely ground, homogenized, and average bulk composition is determined. X-ray diffraction is based on constructive interference of monochromatic X-rays and a crystalline sample. The interaction of the incident rays with the sample produces constructive interference (and a diffracted ray) when conditions satisfy Bragg's Law. These diffracted X-rays are then detected, processed and counted. By scanning the sample through a range of 2θ angles, all possible diffraction directions of the lattice should be attained due to the random orientation of the powdered material. Conversion of the diffraction peaks to d-spacings allows identification of the mineral because each mineral has a set of unique d-spacings. Typically, this is achieved by comparison of d-spacings with standard reference patterns. [5-6]

2.2. Synthesis of the ligands

Stirred solution of Urea /Thiourea and $CuCl_2$.and $CuSO_4$ in methanol (10 mL) was added drop wise to Malonic acid in methanol. After the addition was completed, the stirring was continued for 12 hour. The precipitate was filtered and washed with methanol, then dried in air. [7]

$$CuCl_2 + 2(NH_2CSNH_2) \rightarrow Cu\,(NH_2CSNH_2)_2Cl_2$$
$$Cu\,(SO_4)_2 + 2(NH_2CSNH_2) \rightarrow Cu\,(NH_2CSNH_2)_2(SO_4)_2$$

2.3. Physical measurement

The X-ray diffraction pattern (XRD) using Bruker D-8 advance (IUC, Indore), X-ray powder diffractometer was obtained with Ni-filtered Cu Kα radiation (λ=1.54A°) having voltage 40kV and current 100mA. The X-ray was detected using a fast counting detector based on silicon strip technology (Bruker Lynx Eye detector.1).

Infrared spectra of the complexes were obtained by KBr disc technique by using VERTEX 70 (Bruker) (IUC, Indore) in the range of 4000-400 cm-1. Its innovative design results in the highest flexibility and highest performance. This advanced DigiTect technology prevents external signal disturbance and guarantees the highest signal-to-noise ratio.

3. Results and discussion

The diffraction pattern of complexes recorded between 2θ ranging from 10° to 80°. The crystalline size of the samples is estimated using the Scherrer's formula. According to Scherrer's equation [2], the particle size is given by $t = 0.9\ \lambda/B\ \cos\theta$, where t is the crystal thickness (same units asλ), B is half width (in radians) of the diffraction the line, θ is the Bragg angle and λ is the wavelength. Lattice parameter for simple cubic crystal structure is determined by $a^2= \lambda^2(h^2+k^2+l^2)\ /4\sin\theta$. The determined particle size and lattice parameter are shown in table 1 and XRD pattern shown in figure 1.

Table 1. Particle size and Lattice parameter of copper (II) thiourea complexes

S. No.	**Complexes**	**Particle size t (nm)**	**Lattice parameter a (Å)**
1.	$[Cu(II)(NH_2CSNH_2)]Cl_2$	20.83	8.02
2.	$[Cu(II)(NH_2CSNH_2)](SO_4)_2$	07.43	14.35

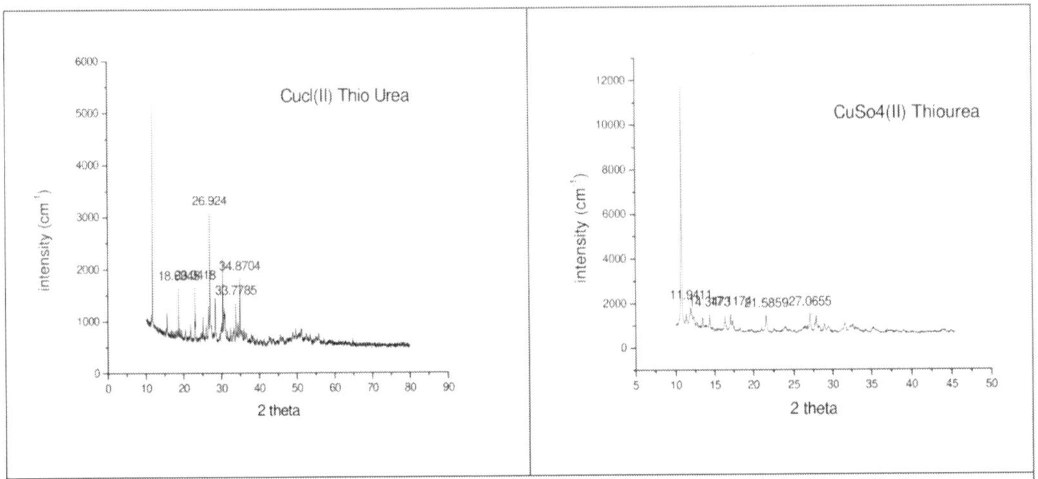

Figure 1. X-Ray diffraction pattern of $CuCl_2$ (II) thiourea and $CuSO_4$ (II) thiourea.

Figure 2 represents selected Infra red absorption bands of thiourea and its Cu(II) complexes, The infra red spectra of free legend, i.e., thiourea and its complexes are complicated as a number of vibrations are observed. The assignments of the important vibrational band have been made by analogy to other reported complexes. In table 2, the data of copper complexes have been reported.

Table 2. Selected Infra red absorption bands of thiourea and its complexes with Cu

Thiourea		**$CuCl_2$+ Thiourea**		**$CuSO_4$+ Thiourea**	
Band intensity assignment		**Band intensity assignment**		**Band intensity assignment**	
3364	(N-H)	3282	(N-H)	3303	(N-H)
3163	(C-H)	2767	(C-H)	2781	(C-H)
1605	(C=N)	1541	(C=N)	1564	(C=N)
729	(C-O)	1631	(C-O)	1633	(C-O)
593	(C-N)	1622	(Cu-O)	1627	(Cu-O)
		1174	(C=S)	1085	(C=S)

4. Conclusion

On the basis of the above discussion, table 3.1 Figure 3.1 shows the suggested structure for copper (II) complexes are cubic and crystalline in nature. Range of particle size is 7.43 to 20.83nm and range of lattice parameter is 8.02 to 14.35Å. FTIR clearly showing in thiourea (N-H), (C-H), (C=N) and (C-O) groups are present .The wavelength lie between 3364 to 593cm^{-1}. In $CuCl_2$+thiourea (N-H), (C-H), (C=N), (C-O), (Cu-O) and (C=S) groups are present. The wavelengths lie between 3382 to 1174 cm^{-1}. Also FTIR showing in $CuSO_4$+thiourea (N-H), (C-H), (C=N), (C-O), (Cu-O) and (C=S) groups are present. The wavelengths lie between 3303 to 1085 cm^{-1}.

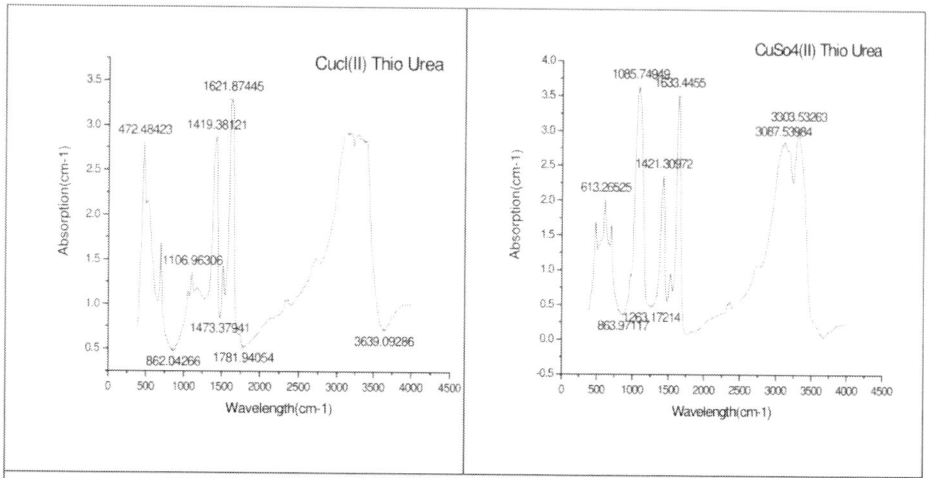

Figure 2. FTIR representation of $CuCl_2$ (II) Thiourea and $CuSO_4$ (II) Thiourea.

Acknowledgments

The author are greatly thankful to Dr. M. Gupta and Dr. T. Shripathi, UGC-DAE Indore for allowing me to use the facility, to prepare the samples to be studied and also thankful for providing XRD and FTIR facility.

References

[1] Halim Nur Illane Mohamad, Kassim K, Fadzil A H and Yamin Bohari M 2012 *The Malaysian Journal of Analytical Sciences* **16** 56
[2] Alkan C, Tek Y and Kahraman D 2011 *Turkish Journal of Chemistry* **35** 769
[3] Cullity B D 1978 *Elements of X-ray Diffraction* (Addison Wesley).
[4] Connes J and Connes P 1966 *J Opt Soc Am* **56** 896.
[5] David W I F, Shankland K, McCusker L B and Baerlocher Ch 2002 *Structure determination form powder diffraction data IUCr Monographs on crystallography* (Oxford)
[6] Bearden J A 1967 *Rev Mod Phys* **39** 78
[7] Ohlan R, et al. 2008 *Org. Commun.* **1**, 24

X-ray diffraction and extended X-ray absorption fine structure study using synchrotron radiation of cobalt (II) complexes

Ashotosh Mishra, Kritika Shukla, Jagrati Dwivedi, Samrath Ninama
School of Physics, Devi Ahilya University, Indore-452001 India

E-mail: amishra@yahoo.co.in, kratika.shukla14@gmail.com

Abstract. XRD and EXAFS investigation of cobalt (II) macro cyclic complexes were carried out. These complexes were synthesised by chemical rout method. On the analysis of the complexes with X-ray diffraction (XRD) and extended X-ray absorption fine structure (EXAFS), XRD analysis shows that complexes are crystalline in nature and having particle size and lattice parameter in the rang of few micro meter and EXAFS technique extract the local structure of complexes. The nearest neighbouring atom distance commonly known as 'bond length' were calculated using Fourier transform method. The bond lengths determined from these methods were also compared with the bond length obtained from several other known technique.

1. Introduction:

The nature of a complex sample is best revealed by the application of several different experimental techniques, with each individual measurement providing both unique and complementary information. The most common technique for characterizing abundant soil minerals is X-ray diffraction (XRD), which relies on long range ordering of atomic planes to probe crystalline structure at a length scale of approximately 50 Å or more [1]. EXAFS have been used extensively in the investigation of local atomic structures such as the number and type of neighboring atoms, inter-atomic distances and disorder. Since the application of EXAFS does not require the materials to have long-range order. It is well suited for determining the local structures of both non-crystalline materials. The well marked EXAFS feature on the high energy side of the K-absorption edge up to several hundred eV, have been observed in the cobalt complexes. From the knowledge of EXAFS, we have calculated the bond length for the following cobalt complexes with the help of Levy's, LSS, Lytle and Fourier transform methods.

1. Co (II) (2-[(2[1-(2-hydroxypheny)ethyl] amino phenyl) ethanimidoyl]chloride (Co_2M_{21}2HEAPEtCl) - **complex 1**.
2. Co (II) (2-[(2[1-(3-hydroxypheny)ethyl] amino phenyl) ethanimidoyl] chloride. (Co_2M_{21}3HEAPEtCl)-**complex 2**.

2. Experimental Technique:

2.1. Preparation of Schiff base

The ligand of 2-[(2-[1-(hydroxyphenyl)ethy1] aminophenyl) ethanimidoyl] phenol was synthesized by mixing an ethanolic solution of 0.01 mole of O-Phenylenediamine (1.08g) with same solvent of 0.01 mole of 2-hydcoxyacetophonone (1.36g). The reaction mixture was refluxed for two hours, then the formed precipitate was filtered, washed several time with ethanol and dried in a desiccators over calcium hydroxide. The yield of the reaction was 73%. The following structure represents the formation of the Schiff base.

2.1. *Preparation of complex*

Co (II), with 2-[(2[1-(2-hydroxypheny)ethyl] amino phenyl) ethanimidoyl] phenol were synthesized by mixing-ethanolic solution. (50ml) of the 0.01mol of $CoCl_2.6H_2O$ respectively and few drops of ammonia solution were added to adjust the pH until the complexes isolated. The reaction mixture was refluxed for three hours. And left cool and filtered by suction. The precipitates were washed several times with ethanol, then ether. The complexes were dried in desiccators over anhydrous calcium chloride. The yields were 75% and 78% respectively [5-6]

$$MX2 + 2L \rightarrow M (L) 2 + 2X$$

In this equation, M is metal Co (II) and X is Cl and L Ligand: 2-[(2[1-(2-hydroxypheny) ethyl]aminophenyl)ethanimidoyl]phenol.

(XRD) measurements were carried out with Cu Kα radiation using a BRUKER D8 Advanced with a rotating anode scanning (0.01 step in 2q) over the angular range 10° - 80° at room temperature generating X-ray by 40 kV and 100 mA power settings. Monochromatic X-rays of $\lambda = 1.5406$ A $K\alpha_1$ line from a Co target were made to fall on the prepared samples. The X-ray absorption spectra have been recorded using synchrotron radiation .The X- ray spectroscopy setup is available at Raja Ramanna Centre for Advanced Technology (RRCAT) and is called beam line .This beam line BL-8 has been recently commissioned at the 2.5 GeV Indus-2 synchrotron radiation sources.

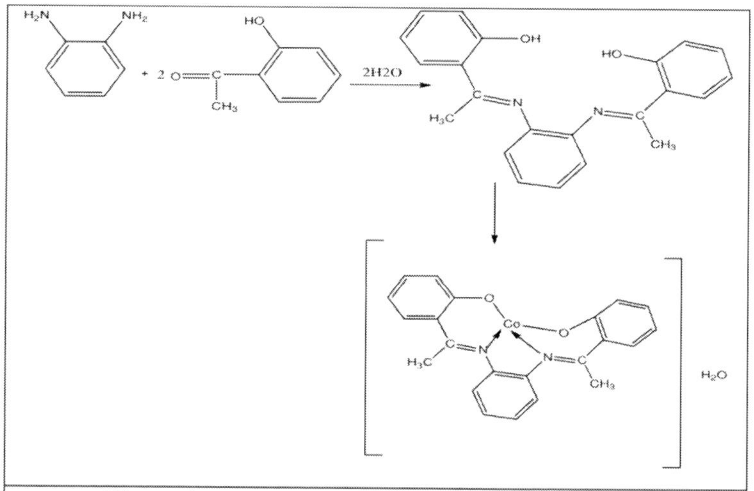

Figure 1. Chemical scheme for Co (II) complexes.

3. Result and discussion:

The pattern has been indexed using JCPDF software and lattice parameters have been using the Braggs law. The Braggs condition is:

$$2d\sin\theta = n\lambda$$

Here d is lattice spacing and given by

$$a^2 = \lambda^2(h^2+k^2+l^2) / 4\sin^2\theta$$

In cubic structure a=b=c , α=β=γ=90. .The particle size of the determined by Debye sherrar formula

$$t = 0.9\lambda / B\cos\theta$$

where, λ=1.54 Å, B=Width. Calculated Lattice parameter and particle size are shown in table 1

Table 1. Lattice parameter and particle size

Complex	Particle size (nm)	Lattice parameter(Å)
Complex 1	13.99	5.6
Complex 2	10.07	16.79

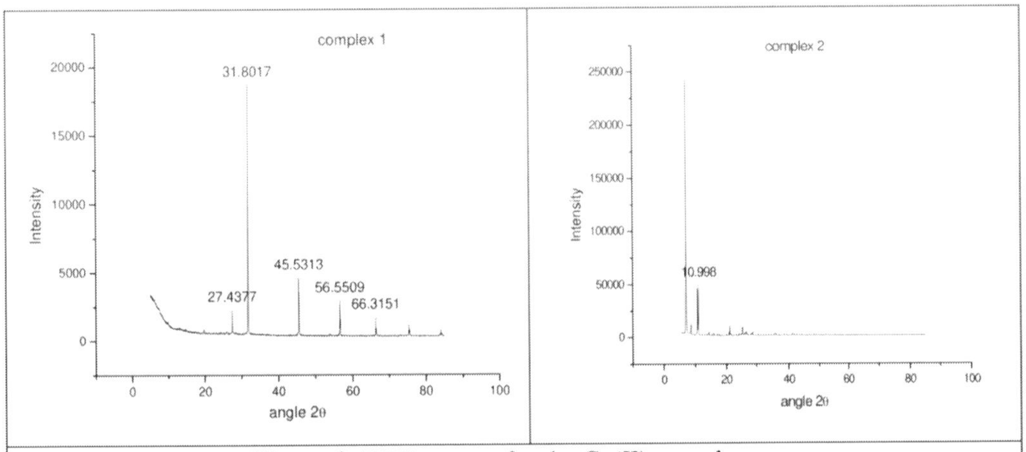

Figure 2. XRD pattern for the Co(II) complexes.

(i) Levy's method
In Levy's method, the bond lengths are given by $R_1 = [151/\Delta E]^{1/2}$ Å,
Where, ΔE is the difference in eV of the energies of the EXAFS maximum B and minimum β and R_1 is the radius of the first coordination sphere [2].

(ii) Lytle's method
The energy values (E) of the EXAFS maxima, according to Lytle for p symmetry, i.e., Q = 2.04, 6.0, 12.0, and 20.0. To evaluate the radius R_s of equivalent polyhedron through the relation [3]
$R_s = [37.60 / M]^{1/2}$

(iii) Lytle, Sayers and Stern's (LSS) method
In the LSS method for determination of the nearest neighbor distances, n versus k graph is plotted. The slope of n versus k plot, gives the value of $2(R_1 - \alpha_1) / \pi$ where R_1 is the bond length [3-4].

(iv) Fourier transform method
Bond length has also been determined by the Fourier transformation method for the cobalt complexes studied however, determined only the phase uncorrected bond lengths by this method. No attempt has been made to employ the fitting procedures by which phase corrected bond length can be determined, because the required crystallographic data is not available for any of the complexes studied.

Energies of the maxima and minima position are tabulated in table 2. The bond lengths are calculated using Levy's, LSS and Lytle methods. The values of bond length R is tabulated in table 2 The bond lengths obtained in Co(II) Complexes by LSS, Levy's, and Lytle method with F.T. method are comparable each other.

Table 2: Average values of the bond length in (Å) for the Co (II) Complexes.

Complexes	R_{LSS}	R_{Levy}	R_{Lytle}	$R_{F.T.}$
Complex 1	1.41	2.72	1.97	1.52
Complex 2	1.45	3.06	1.92	2.0

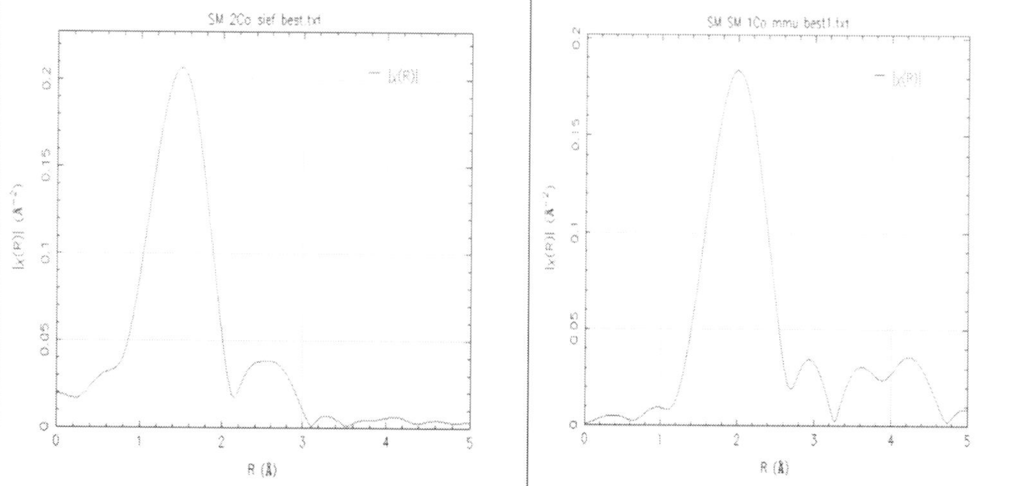

(a) complex 1 (b) complex 2

Figure 2: F.T.magnitude for Co(II) complexes

4. Conclusion

The XRD studies done on the two -macro cyclic complexes of cobalt showed that they are crystalline in nature and lattice parameter in the range of angstrom.

As pointed out above, the LSS method, Levy's, Lytle and the Fourier transformation method give the value of bond length R, are comparable to each other. This distance is called the phase uncorrected bond length. It can be seen from table.2 that the phase uncorrected bond lengths obtained from these two methods agree with each other within the limits experimental error.

Acknowledgments

The author are grateful to Dr. Mukul Gupta UGC DAE CSR Indore, for their support in carrying out measurement and also thankful to Dr. S N Jha for providing facility of synchrotron.

References

[1] Cullity B D 1978 *Elements of X-Ray Diffraction* (Addison-Wesley)
[2] Levy R M, 1965 *J. Chem. Phys.* **43** 1846
[3] Lytle F W, Sayers D E and Stern E A, 1975 *Phys. Rev. B* **11** 4825.
[4] Lee J D 1991*Concise of Inorganic Chemistry 4th ed.* (Chapman & Hall, London).
[5] Sandler S R and Karo W, 1971 *Organic Functionally Group Preparations II*, (New York: Academic press).
[6] Geissman T A 1968 *Principles of Organic Chemistry* (Freeman and Company, New York).

Influence of high temperature sintering on the structural and magnetic Properties of $Mn_{1-x}Zn_xFe_2O_4$

M Satalkar[a], **N Ghodke**[b] **S N Kane**[a,1]

[a]School of Physics, D. A. University, Khandwa road Campus, Indore – 452001, India
[b]UGC-DAE CSR, University Campus, Khandwa Road, Indore – 452 001, India

[1]Corresponding author, E-mail: kane_sn@yahoo.com (S. N. Kane)

Abstract. Nano crystalline $Mn_{1-x}Zn_xFe_2O_4$ (x = 0.0 – 1.0) spinel ferrite have been synthesized by sol-gel auto-combustion method. Prepared samples were annealed at 1000 °C for 3 hrs to study the effect of high temperature sintering on their structural and magnetic properties. The x-ray diffraction (XRD) analysis reveals the presence of cubic spinel phase and, secondary phase of β-Fe_2O_3. Variation in the structural parameters obtained from XRD analysis are interpreted in terms of changes in the Zn concentration. Obtained magnetic properties can be ascribed to the solubility of cations in the ferrite lattice and distribution of cations on tetrahedral and octahedral sites.

1. Introduction

Mn–Zn spinel ferrites are very important soft magnetic materials because of their high initial magnetic permeability, saturation magnetization, electrical resistivity and low power losses [1,2]. These materials are extensively used as inductors, transformers, antenna rods, loading coils, deflection yokes, choke coils, recording heads, magnetic amplifiers, electromagnetic interference devices (EMI), power transformers and splitters [3,4]. Mn–Zn ferrites are also important in biomedicine as magnetic carriers for bio-separation, enzymes and proteins immobilization [5–8]. Recently, with the development of high frequency, low power miniaturized electronic devices; special focus has been placed on preparation of high performance Mn–Zn ferrite powders. Variation of composition, a specific `synthesis process (e. g.- sol-gel auto combustion method, co-precipitation, solid state reaction etc.) and, post preparation thermal treatments can be used to control structural parameters (e. g. - average grain diameter, lattice parameter etc.), as well as the soft magnetic properties of the ferrites [9]. Recently more attention has been paid to the sol-gel auto combustion [10,11], which allows producing ultra-fine powders with chemically homogeneous composition, uniform size and good reactivity [12,13], allowing one to tune magnetic properties by selectively controlling the structural properties [9]. In the present work, we report the synthesis, structural and, magnetic characterization of thermally annealed $Mn_{1-x}Zn_xFe_2O_4$ (with x = 0.0, 0.15, 0.20, 0.35, 0.50, 0.85 and 1.0) at 1000 °C for 3hrs.

2. Experimental details

Manganese – Zinc based ferrite of nominal composition $Mn_{1-x}Zn_xFe_2O_4$ (with x = 0.0, 0.15, 0.20, 0.35, 0.50, 0.85 and 1.0) in powders form were prepared using sol-gel auto combustion method. For synthesizing Manganese – Zinc based ferrite, citrate-nitrate precursors [Zinc Nitrate – $Zn(NO_3)_2.6H_2O$,

Manganese Nitrate – $Mn(NO_3)_2.6H_2O$, Ferric Nitrate – $Fe(NO_3)_3.9H_2O$ and Citric acid - $C_6H_8O_7$] were mixed in stoichmetric ratio with citric acid as fuel. Metal salt to fuel ratio was taken as 1:1. Citric acid primarily acts as a chelating agent for metal ions of varying ionic sizes, which helps in preventing their selective precipitation to maintain compositional homogeneity among the constituents. Afterwards it also serves as a fuel in the combustion reaction [14]. De-ionized water was used to dissolve all the precursors and pH was maintained at 7, by adding ammonia solution. Then the solution was heated at 120 °C in air till the loose powder was formed, called as '*dry gel or as-burnt powder*'. In order to investigate high temperature sintering effect on structural and, magnetic properties of $Mn_{1-x}Zn_xFe_2O_4$ spinel ferrite powders the dry gel powders were annealed at 1000 °C for 3hrs. X-ray diffraction (XRD) was used to determine the formed phases and crystalline size. XRD measurements were done at room temperature in θ-2θ configuration by Bruker D8 advance diffractometer utilizing CuK_α (λ= 1.54056 A°) radiation. The x-rays were detected by using a fast counting detector based on Silicon strip technology (Bruker LynxEye detector).

The lattice parameter (***a***) corresponding to [311] reflection was obtained by: $a = d\sqrt{h^2+k^2+l^2}$
where d – inter-planer spacing and, (h, k, l) - miller indices.

X-ray density (ρ_{XRD}) of the studied samples was obtained by: $\rho_{XRD} = 8M_w/N(a)^3$
where M_w – Molecular weight, N – Avagrado's Number, a - Lattice parameter.

Scherrer's grain diameter is obtained by the line width of [311] reflection by: $D = 0.9\lambda / \beta \cos\theta$
where, λ - wavelength of the x-ray used, β - line width, θ - peak position (in 2θ scale).

Hoping length for site A and B were calculated by: $L_a = \frac{a\sqrt{3}}{4}$ and, $L_b = \frac{a\sqrt{2}}{4}$

where L_a and L_b are respectively hoping length for site A and B , a – lattice parameter.

Room temperature hysteresis loops of the synthesized samples were measured at 50 Hz by conventional induction technique by applying maximum field – $H_{max} \approx \pm$ 130 kA/m (1600 Oe). Loops were analyzed by a Matlab® based program [15] to obtain coercivity (H_c).

Satutation magnetization (M_s) values were obtained by plotting magnetization against inverse of magnetic field (1/H) and, extrapolating the data to 1/H = 0. Anisotropy constant (K_1) was calculated by: $K_1 = [H_c \times M_s] / 0.96$

where H_c – coercivity in Oe. and M_s – saturation magnetization in emu/cc.
Magnetic moment (n_B) for a particular composition in Bohr magneton (μ_B) can be calculated by using the relation: $n_B = [M_w \times M_s] / 5585$

where M_w is the molecular weight of a particular ferrite composition and M_s is saturation magnetization (emu/g) and, 5585 is the magnetic factor.

3. Results and discussion

Figure 1 depicts the XRD patterns of the annealed samples. The X-ray diffractogram recorded for the ferrite powder obtained on calcinations at 1000 ° C for 3 hrs corresponds to that of spinel structure (JCPDS 74-2401). Presence of secondary phase- β Fe_2O_3 (JCPDS 85-0599 and 76-1821) was also confirmed with the help of XRD patterns. Intermediate phase can be formed by the excessive (relative to zinc) Fe content in the samples due to the zinc loss during auto combustion [16]. At higher temperature (1150 and 3000 ° C) the phases again build into the spinel structure [17]. Table 1 depicts the calculated values of lattice parameter (***a***), average Scherrer's grain diameter (***D***), x-ray density (ρ_{XRD}), cell volume (V), hopping length for A-site (L_A) and, B-site (L_B) for the annealed $Mn_{1-x}Zn_xFe_2O_4$ samples. Perusal of table 1 shows that ***a*** ranges between 0.8335 nm – 0.8428 nm. The variation of ***a*** can be ascribed to the replacement of an ion with higher ionic radius (Mn^{2+} = 0.093 nm) by an ion with lower ionic radius (Zn^{2+} = 0.074 nm). Scherrer's grain diameter ranges between 28.94 – 84.67 nm. Crystallite size of the studied ferrite decrease as Zn substitution increases upto x = 0.50 and

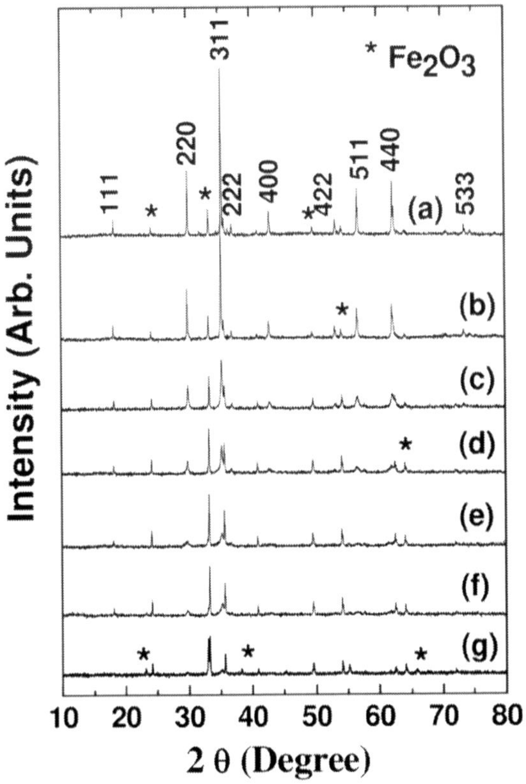

Figure 1. X-ray diffraction patterns of $Mn_{1-x}Zn_xFe_2O_4$ (annealed at 1000 °C for 3hrs), where (a) x = 1.00, (b) x = 0.85, (c) x = 0.50, (d) x = 0.35, (e) x = 0.20, (f) x = 0.15, (g) x = 0.0

then increases for x = 0.85. X-ray density increases from 5273.39 to 5374.26 kg/m^3 for x = 0.00 – 0.35 and 5314.05 to 5390.88 kg/m^3 for x = 0.85 – 1.00 which is ascribable to the fact that decrease in the volume of the unit cell overtakes the decrease in mass of the unit cell. Behaviour of hopping length (*distance between magnetic ions*) for A-site (L_A) and B-site (L_B) with Zn-content is similar to that of the variation of ***a*** with Zn-content. This variation may be attributed to the fact that the larger Mn^{2+} ions (ionic radius = 0.093 nm) are being replaced by smaller Zn^{2+} ions (ionic radius = 0.074 nm).

Table 1: Zn-content dependence of lattice parameter (***a***), grain diameter (***D***), x-ray density (ρ_{XRD}), cell volume (***V***), hoping length for site A (L_a) and hoping length for site B (L_b) for $Mn_{1-x}Zn_xFe_2O_4$.

x	*a* (nm)	*D* (nm)	ρ_{XRD} (Kg/m^3)	Cell Volume (*V*) (nm^3)	L_a (nm)	L_b (nm)
0.00	0.8344	84.67	5273.39	0.5809	0.3613	0.2950
0.15	0.8339	67.65	5318.74	0.5799	0.3611	0.2948
0.20	0.8344	64.78	5321.11	0.5809	0.3613	0.2950
0.35	0.8335	43.41	5374.26	0.5791	0.3609	0.2947
0.50	0.8424	28.94	5240.49	0.5977	0.3648	0.2978
0.85	0.8428	57.75	5314.05	0.5987	0.3649	0.2980
1.00	0.8406	67.80	5390.88	0.5940	0.3640	0.2972

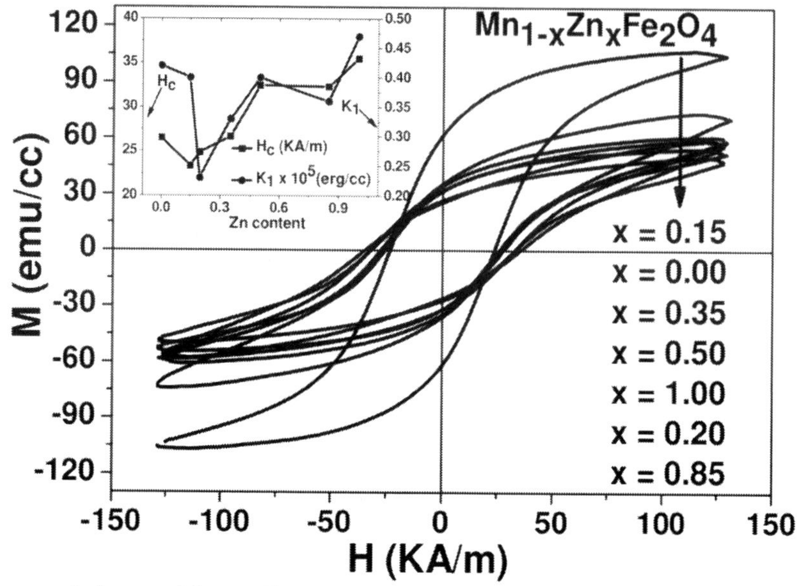

Figure 2. Hysteresis loops of the studied samples with varying Zn content (x). Inset: Variation of H_c and K_1 with Zn content.

Fig. 2 depicts the hysteresis loops of the annealed $Mn_{1-x}Zn_xFe_2O_4$ samples where the changes in the magnetic properties with varying Zn content can be observed whereas the inset shows the variation of H_c and K_1 as a function of Zn content. The variation in the magnetic properties of Mn-Zn ferrites depends on the solubility of cations (Mn^{2+} or Zn^{2+}) in the ferrite lattice and the occupying positions of tetrahedral or octahedral sites [4]. Fig. 3 (left panel) shows the variation of saturation magnetisation (M_s) and magnetic moment (n_B) for a particular composition in Bohr magneton (μ_B) with Zn doping. M_s values are obtained by plotting M versus 1/H and extrapolating the data to 1/H =0 (Fig. 3 right panel). The variation in M_s and n_B can be ascribed to the migration of Mn^{2+} (replaced by Zn^{2+}) ions to tetrahedral and octahedral sites and Fe^{3+} ions from octahedral site to tetrahedral site. When the nonmagnetic zinc ion is substituted into the manganese ferrite lattice, it has a strong preference for the tetrahedral site than that of ferric ion and thus reduces the amount of Fe^{3+} on the A site. Due to the anti-ferromagnetic coupling, the net result in an increase in magnetic moment on the B lattice and an increase in M_s (for x = 0.15). Decrease in M_s and n_B can be ascribed due to the spin canting effect on one sub-lattice when a non- magnetic ion occupies the other sub-lattice.

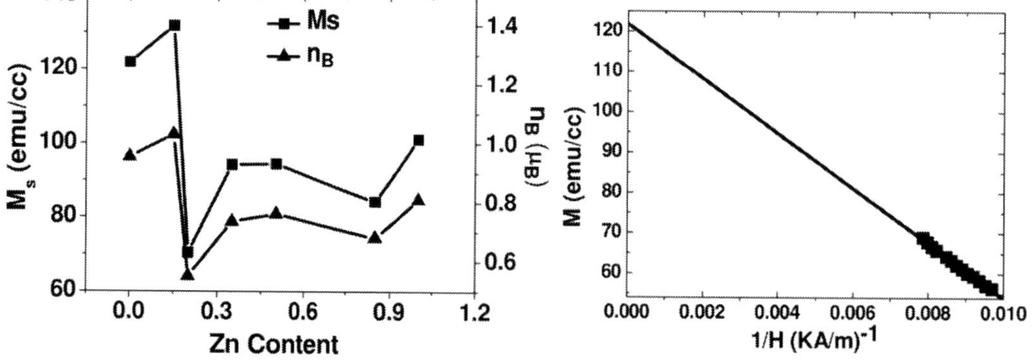

Figure 3. Variation of M_s and, n_B with Zn content (left panel). M versus 1/H curve showing the extrapolation of the data to 1/H =0 (Right panel) for x = 0.0

To summarize, sol-gel auto-combustion method was used to synthesize nano-crystalline $Mn_{1-x}Zn_xFe_2O_4$ spinel ferrite. XRD of the studied specimens reveals the formation of cubic spinel structure with a secondary phase of β-Fe_2O_3. Variation in the structural parameters are observed as the doping content (Zn) increases. Lattice parameter and Schereer's grain diameter respectively ranges between 0.8335 – 0.8428 nm and. 28.94 – 84.67 nm. Coercivity ($\boldsymbol{H_c}$), anisotropy constant ($\boldsymbol{K_1}$), saturation magnetization ($\boldsymbol{M_s}$) and, magnetic moment ($\boldsymbol{n_B}$) changes with Zn^{2+} concentration. Observed increase in $\boldsymbol{M_s}$ and $\boldsymbol{n_B}$ is due to the increase in Fe^{3+} ions on octahedral site and decrease of Fe^{3+} ions on tetrahedral site and decrease in $\boldsymbol{M_s}$ and $\boldsymbol{n_B}$ in the system may be because of the non-collinear spin canting effect.

Acknowledgments
This work is supported by MPCST project No. 1783/CST/R & D/Phy & Engg Sc/2013. Research fellowship to MS form UGC-DAE CSR project No. CSR-I/CSR51/2011-12 is gratefully acknowledged. Authors are grateful to Dr. M. Gupta for doing XRD measurements.

References

[1] Ott G, Wrba J, Lucke R, 2003 *J. Magn. Magn. Mater.* **535** 254–255.
[2] Ghazanfar U, Siddiqi S A, Abbas G, 2005 *Mater. Sci. Eng. B* **118** 84.
[3] Keluskar S H, Tangsali R B, Naik G K, Budkuley J S, 2006 *J. Magn. Magn. Mater.* **305** 296–303.
[4] Hessien M M, Rashad M M, El-Barawy K, Ibrahim I A, 2008 *J. Magn. Magn. Mater.***320** 1615–1621.
[5] Arulmurugan R, Jeyadevan B, Vaidyanathan G, Sendhilnathan S, 2005 *J. Magn. Magn. Mater.* **288** 470.
[6] Papazoglou P, Eleftheriou F, Zaspalis V T, 2006 *J. Magn. Magn. Mater.* **296** 25.
[7] Upadhyay C, Verma H C, Rath C, Sahu K K, Anand S, Das R P, Mishra N C, 2001 *J. Alloys Compd.* **326** 94.
[8] Costa A C F M, Tortella E, Morelli M R, Kiminami R H G A, 2003 *J. Magn. Magn. Mater.* **256** 174.
[9] Ghosh A, Satalkar M, Rathod S, Nag S P, Vyas P, Kane S N, Ghodke N, Prasad R, and Dwivedi R 2013 *Int. J. Mod. Phys.: Conf. Ser.* **22** 28.
[10] Satalkar M, Kane S N, Ghosh A, Ghodke N, Barrera G, Celegato F, Coisson M, Tiberto P, Vinai F, 2014 *J. Alloys Comp.* http://dx.doi.org/10.1016/j.jallcom.2014.01.248.
[11] Batra A S, Satalkar M, Kane S N, Ghosh A and Ghodke N, 2014 *AIP Conf. Proc.* **1591** 537.
[12] Bueno A R, Gregori M L, Nobrega M C S, 2007 *Mater. Chem. Phys.* **105** 229–233.
[13] Verma A, Chatterjee R, 2006 *J. Magn. Magn. Mater.* **306** 313–320.
[14] Kapse D, Ghosh S A, Raghuwanshi F C, Kapse S D, 2009 *Mater. Chem. Phys.* **113** 638.
[15] Kane S N, Gercsi Z, Ghodke N, Khinchi S S and Gupta A, "A digital hysteresis loop tracer for studying soft magnetic materials", in Proc. of the DAE Solid State Physics Symposium vol. 50, edited by Aswal V K, Bhushan K G and Yakhmi J V, Mumbai, Prime Time Education, Mumbai, India, 2005, pp 287-288.
[16] Irena Szczygiel Katarzyna Winiarska, 2011 *J. Therm Anal Calorim* **104** 577-583.
[17] Hu P, Yang H, Pan D, Wang H, Tian J, Zhanga S, Wang X, Volinsky A A, 2010 *J. Magn. Magn. Mater.* **322** 173-177.

Effect of Ni doping on optical, electrical and magnetic properties of Nd orthoferrite

Sajad Ahmad Mir[1], M Ikram[1] and K Asokan[2]
[1]Solid State Laboratory, Department of Physics, National Institute of Technology, Srinagar, J&K-190006, India.
[2] Materials Science Division, Inter University Accelerator Centre, New Delhi-110067, India

E-mail: sajadphysics@gmail.com

Abstract. $NdFe_{1-x}Ni_xO_3$ (NFNO) (x=0.0 and 0.3) perovskite structures were synthesized by ceramic method. The variation of dielectric behavior with frequency at room temperature reveals an order of enhancement in both dielectric constant and dielectric loss with 30 percent Ni substitution. UV-Visible spectroscopy data exhibit that optical energy band gap decreases with Ni substitution from 4.3eV to 4.0eV for the sample with 30% Ni doping. Magnetic disorder over some range between 1.5K and 100K is observed in SFNO samples and is explained in milieu of interaction among Nd and Fe/Ni sub-lattice moments.

1. Introduction:

Perovskite type rare-earth transition metal oxides (RTO) comprise a special place in solid state physics as they provide hefty scope for optical, electronic and magnetic tuning depending on the atoms involved, inter-atomic distances and the bonding strength. Some of these oxides are high dielectric ceramics, some are antiferromagnetic insulators and others are pauliparamagnetic metals etc [1-2]. The varied electric and magnetic properties of RTO render these materials important in advanced technologies such as gas sensors[3], catalysts[4] and solid oxide fuel cells[5] etc. A profound study from last few years has shown that transition metal ion plays an important role in rare-earth transition metal oxides through its electronic ground state and associated magnetic properties. Thus, they have been under thorough investigation over couple of decades. As NdFeO3 and NdNiO3 belongs to RTO family and much emphasis has been given to study their optical, electric and magnetic properties in past few years[6-8].Both these oxides conform to orthorhombically distorted perovskite structure with space group D_{2h}^{16}-*Pbnm* possessing 4 formula units per elementary cell[9-10].

With *Pbnm* space group, $NdFeO_3$ (NFO) is a room temperature insulator and involves three different interactions which decide the outcome of its several physical properties such as optical, electric and magnetic etc. The strong interaction between Fe-Fe results in long range antiferromagnetic (AF) ordering with a Neel temperature of 690K [11] but weak ferromagnetism (F) is observed due to spin canting. The relatively less strong Nd-Fe interaction causes spin-reorientation transition between 100K and 200K [12] and polarization of Nd^{3+} ion magnetic moments below 25K[9].The weakest of the three interactions between Nd-Nd results in long range AF ordering of Nd^{3+} magnetic moments at 1.5K[13].On the other hand with *Pbnm* space group, NdNiO3 (NNO) is a room temperature metal but near metal-insulator boundary and exhibits a metal-insulator transition at the same Neel temperature of 200K[14,15] and a metal-semiconductor temperature at 130K[16]. Ni magnetic moments order in such a way so that quadrupled magnetic unit cell contains same number of ferromagnetically and anti ferromagnetically coupled nearest neighbors [17].

Thus replacement of Fe in NFO by Ni will develop a new system which lies between two end members of NFNO series and lend some interesting physical properties like optical, electrical and magnetic etc. This motivates us to take studies of pure and 30 percent Ni doped NFO and observe how these properties will evolve and mutate.

2. Experimental Details

NFNO (x=0 and 0.3) samples were prepared by using ceramic method, the details of which our group have been published elsewhere [18]. X-ray diffraction (XRD) measurements are carried on samples by Bruker D8 Advanced X-ray diffractometer using Cu K_α radiation for phase identification and analyzed in the similar way as discussed in detail by our group somewhere else [19].Optical measurements are recorded by Perkin Elmer Lamda 950 spectrometer in the wave length range of 200-800 nm. Dielectric measurements are performed with Agilent 4285A precision LCR meter in the frequency range of 0.02 kHz to 1MHz at room temperature. Temperature dependent zero field cooled (ZFC) and Field cooled (FC) dc magnetizations are measured using Quantum Design Vibrating sample magnetometer (VSM).

3. Results and Discussion

From X-ray diffraction (XRD) results, the samples are in single phase having distorted orthorhombic structure and space group D_{2h}^{16}-*Pbnm*. There is an increase in strain and a decrease in unit cell volume with Ni doping similar to that calculated here [20].

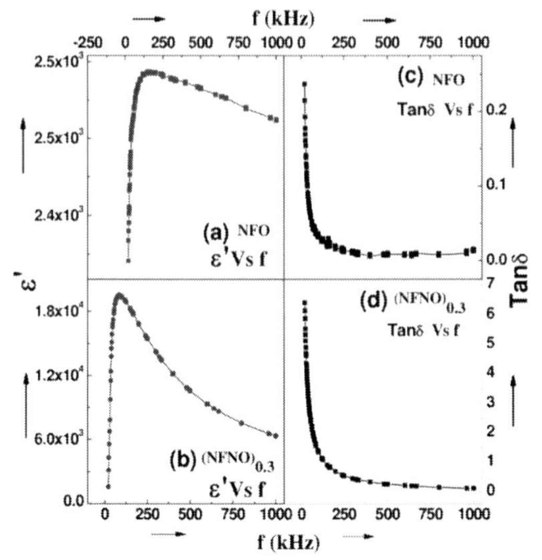

Figure 1. Variation of ε′ and tanδ as a function of frequency at room temperature for $NdFeO_3$ and $NdFe_{0.7}Ni_{0.3}O_3$.

Figure1 (a,b) represent the frequency dependent dielectric constant (ε′) and figure1(c,d) the dielectric loss (tanδ) of NFNO samples over the frequency range from 0.02 kHz-1 MHz at room temperature. High ε′ in these samples of the order of 103 -104is observed. The high ε′ in Ni-Zn ferrites of the order of 10^3 prepared by ceramic method has been already reported in literature [21, 22]. The high ε′ in NFNO structure may possibly be attributed to the local charge accumulation which takes place at the interface between grain and grain boundary under the effect of external field. This leads to polarization which results in a decrease in dielectric constant with increase in frequency and is in accordance with Maxwell-Wagner type of polarization [23-24]. However, ε′ decreases till it attains some constant value. It is because local charge is made available through electronic charge exchange between Fe^{2+} and Fe^{3+} and after certain frequency limit the charge exchange between Fe^{2+} and Fe^{3+} cannot follow the alternating nature of field. It is evident from Fig. 1(b,d) that 30 percent Ni doping can change the ε′ and tanδ of NFO significantly. For example, the ε′ of $NdFe_{0.7}Ni_{0.3}O_3$ $[(NFNO)_{0.3}]$

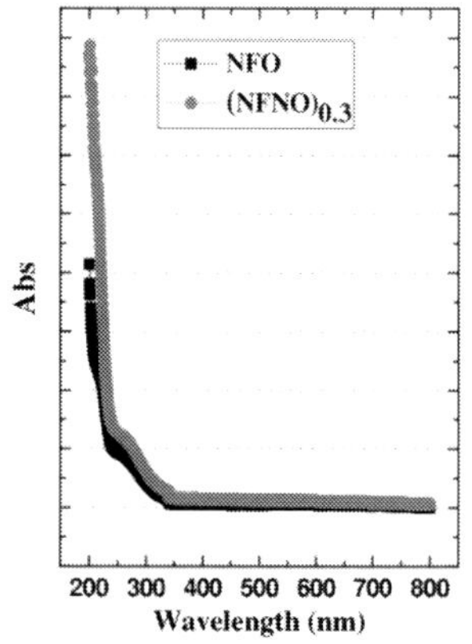

Figure 2. UV-Visible spectograph of Wavelength Vs Absorption of $NdFeO_3$ and $NdFe_{0.7}Ni_{0.3}O_3$.

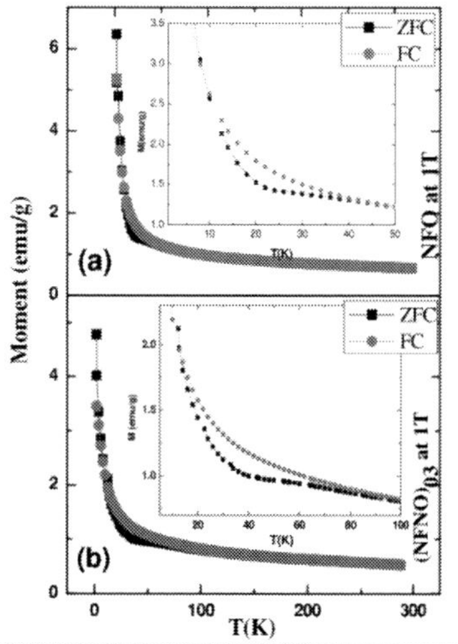

Figure 3. Temperature dependence of the magnetization after ZFC and FC measured at 1T. **a** $NdFeO_3$. **b** $NdFe_{0.7}Ni_{0.3}O_3$.The insets envisage the disordered state of both **a** and **b.**

measured at 100 Hz is about ten times larger than that of NFO. The other possible reason of the large ε' of $(NFNO)_{0.3}$ corresponds directly to the small resistivity at x = 0.3, suggesting that the increase in the ε' may be ascribed to the leakage current caused by Ni doping. The tanδ data presented in figure 1(b-d) is also in accordance to this notion. The tanδ decreases with increasing frequency without indication of loss peak. Similar demeanor was reported in Bi-Cr ferrites by F. Chang et al [25].

Figure 2 shows the absorption spectrum of NFNO over the wavelength range from 200-800nm at room temperature. The value of absorption is high between 200-234nm in case of NFO and 202-237 nm in case of $(NFNO)_{0.3}$. The absorption coefficient for these materials is calculated by Tauc relation [26]:

$$\mu h\upsilon = A(h\upsilon - E_g)^n$$

Where, μ is absorption coefficient, A is constant different for different transitions, $h\upsilon$ is energy of incident photon, h is the planks constant, n is the index which assumes the different values 1/2, 3/2, 2 and 3 corresponding to direct allowed, direct forbidden, indirect allowed and indirect forbidden electronic transitions respectively responsible for reflection and E_g the optical band gap. Using Tauc plot (not shown here), $(\mu h\upsilon)^{1/2}$ versus $h\upsilon$ by extrapolating linear portion of the curve to $(\mu h\upsilon)^{1/2}=0$ axis yields E_g values for NFO and $(NFNO)_{0.3}$ as 4.3 eV and 4.0eV respectively. One of the possible reasons for this red shift observation in Eg with Ni doping could be orthorhombic distortion [27].

In any RTO family there are two magnetic sublattices namely T and R. In NFO family the Fe^{3+} sublattice is G_z type AF ordered below 100K with a small F_x type F component due to spin canting structure of Fe^{3+} sublattice [28-30] and Nd^{3+} sublattice changes from disordered state at higher temperature to AF ordered state at lower temperature of 1.5K [14,31]. It is in-between these two temperatures (1.5K and 100K) under 1T magnetic field an anomalous mode of magnetization is observed. ZFC and FC modes of magnetization as a function of temperature in 1T field for NFO and $(NFNO)_{0.3}$ are shown in figure 3. It is evident from the figure 3 that there is hardly any bifurcation seen between ZFC and FC magnetizations at higher and lower temperatures. However, as shown in inset of figure 3 in case of NFO the bifurcation is evident from 10-40K and in $(NFNO)_{0.3}$ from 14-72K which suggests a disorder state arises due to intricate interaction between Fe^{3+}/Ni^{3+} and Nd^{3+} moments. Another possible reason may be the heterogeneous mixture of F and AF order instead of distinct F or AF order. Similar explication was reported in Pr ferrites by A. Bashir et al [19]. Also, as rare-earth ion subsystem acquires magnetization due to an interaction with iron subsystem, so the shift of disordered state in the temperature range with the Ni doping at Fe site may likely be due to weaker AF exchange between rare earth and transition metal sub-lattices [32].

4. Conclusions

In conclusion, the present study confirms colossal room temperature dielectric constant and dielectric loss of ceramically synthesized NFNO with an order of increase in these quantities at 30% Ni doping. A decrease in optical energy band gap is observed with Ni doping that we propose is related to an orthorhombic distortion. At 1T it is clear ZFC and FC magnetization curves of NFNO within the temperature range of 10-40K [in case of NFO] and 14-72K [in case of $(NFNO)_{0.3}$] have not overlapped indicating the existence of disordered or inhomogeneous mixture of F and AF ordered state over these temperature ranges.

Acknowledgement

Authors wish to thank the IUAC, New Delhi and the UGC-DAE Consortium for Scientific Research, Indore for providing experimental facilities. One of the authors (SAM) thanks Director, National Institute of Technology, Srinagar for encouragement and Dr R J Choudhury, UGC-CSR, Indore for stimulating discussion.

References

[1] Torrance J B, Lacorre P, Asavaroengchai C and Matzger R 1991 *J. solid. Chem.* **90** 168
[2] Hearne G R, Pasternak M P, Taylor R D and Lacorre P 1995 *Phys. Rev. B* **51** 11495
[3] Niu XS, Du WM, Du WP 2004 *Sens. Actuators* B **99** 399
[4] Haralambous K J, Loizos Z and Spyrellis N 1991 *Mater. Lett.* **11** 133
[5] Minh N Q 1993 *J. Am. Ceram. Soc.* **76** 563
[6] Torrance J B, Lacorre P, Nazzal A I, Ansaldo E J and Niedermayer Ch 1992 *Phys. Rev. B* **45** 8209
[7] Blasco J and Garcia J 1994 *J. Phys. Chem. Solids* **55** 843
[8] Xinshu Niu, Honghua Li and Guoguang Liu 2005 *J. Mol. Catalys A: Chem.* **232** 89
[9] Bartolome J, Palacios E, Kuzmin M, Bartolome F, Sosnowska I, Sonntag R, Lukina M 1997 *Phys. Rev. B* **55** 11432
[10] Zhou J S and Goodenough J B 2004 *Phys. Rev. B* **69**, 153105
[11] Treves D 1965 *J. Appl. Phys.* **36** 1033
[12] Belov K P, Kadomtseva A M, Ovchnnikowa T L, Timofeeva V A, Uksov V V 1971 *Sov. Phys.-Solid State* **13** 518
[13] Prezenioslo R, Sosnowska I, Fischer P, Marti W, Bartolome F, Bartolome J, Palacois E and Sonntag R 1996 *J. Magn. and Magn. Mater* **160** 370
[14] Bartolome F, Kuzmin M D, Bartolome J, Blasco J, Garcia J and Sapina J 1994 *Solid State Commun.* **91** 177
[15] Barman S R, Chainani A and Sarma D D 1994 *Phys. Rev. B* **49** 8475
[16] Vassiliou J K, Hornbostel M, Ziebarth R P and DiSalvo F J 1989 *J. Solid State Chem.* **81** 208
[17] Garcia-Munoz J L, Rodriguez-Carvajal J and Lacorre P 1992 *Europhys. Lett.* 20, 241
[18] Kumar R, Choudhary R J, Ikram M, Shukla D K, Mollah S, Thakur P, Chae K H, Angadi B and Choi W K 2007 *J. Appl. Phys.* **102** 073707
[19] Bashir A, Ikram M, Kumar R, Lisboa Filho P N and Thakur P 2010 *Matrl Sc. and Eng. B* **172** 242
[20] Bashir A, Ikram M and Kumar R 2009 *J. Phys. Condens. Matter* **21** 325501
[21] Abdeen A M 1999 *J. Magn. Magn. Mater.* **192** 121
[22] Paik H, Hwang H, No K, Kwon S and Cann D P 2007 *Appl. Phys. Lett.* **90** 042908
[23] J.C. Maxwell 1873 *Electricity and Magnetism* **1** 328
[24] Wagner K W 1973 *Ann. Phys,* **40** 817
[25] Chang F, Zhang N, Yang F, Wang S and Song G 2007 *J. Phys. D: Appl. Phys.* **40** 7799
[26] Davis E A and Mott N F 1970 *Philos. Mag.* **22** 903
[27] Mir F A, Banday J A, Chong C, Dahoo P and Najar F A (to appear) Eur. Phys. J. Appl. Phys
[28] Sosnowska I, Steichele E and Hewat A 1986 *Physica B* **136**, 394
[29] Przenioslo R, Sosnowska I and Fischer P 1995 *J. Magn. Magn. Mater.* **140-144** 2153
[30] Bedekar V, Jayakumar O D, Manjanna J and Tyagi A K 2008 *Mater. Lett.* **62** 3793
[31] Sosnowska I and Fischer P 1987 *Phase Trans.* **8** 319
[32] Tsymbal L T, Bazaliy Y B, Derkachenko V N, Kamenev V I, Kakazei G N, Palomares F J and Wigen P E 2007 *J. Appl. Phys.* **101** 123919

Origin of the satellite observed on the high energy side of $L\gamma_{2,3}$ diagram lines

Rajeev Trivedi[1], Uma Shrivastava[2] and B. D. Shrivastava[2]
[1]Alpine Institute of Technology, Ujjain - 456010. India
[2]School of Studies in Physics, Vikram University, Ujjain - 456010.India

Email: rajeevtrivedi80@gmail.com

Abstract. On the high energy side of the unresolved doublet $L\gamma_{2,3}$in the L spectra of some elements, a satellite $L\gamma_{2,3}'$ has been reported in literature, which has been ascribed to the transition L_1M_x - $M_xN_{2,3}$, on the basis of energy considerations alone, by earlier workers. In the present work, apart from energy, calculations have also been done for the intensity of the probable transitions which may give rise to the satellite in the elements from $_{40}Zr$ to $_{47}Ag$ and from $_{57}La$ to $_{63}Eu$. Firstly, the probability of creation of a single hole K state has been calculated. Double hole state L_1M_x is created by Coster-Kronig transitions (K - L_1M_x) as well as by shake off process. The probabilities of these two processes have been calculated. The total probability of creation of the double hole state thus obtained has been statistically distributed among the 24 allowed transitions from the set of L_1M_x levels. By taking each of the transitions as a Gaussian line, a composite spectrum has been computed. This theoretical spectrum has been compared with the available satellite energy data. The transitions which give rise to this satellite have thus been identified in the elements studied.

1. Introduction

The wavelength tables of Cauchois and Senemaud [1] show that $L\gamma_{2,3}'$ satellite has been observed on the high energy side of the unresolved doublet $L\gamma_{2,3}$ and has been reported in $_{40}Zr$, $_{41}Nb$, $_{42}Mo$, $_{45}Rh$ and $_{47}Ag$. In 1986 this satellite has been reported by Shrivastava et al. [2] on the high-energy side of $L\gamma_2$ line in the elements $_{57}La$, $_{59}Pr$, $_{60}Nd$, $_{62}Sm$ and $_{63}Eu$. Soni [3] has assigned the transition schemes L_1M_X-M_XN_{23} to the satellite $L\gamma_{2,3}'$. Shrivastava et al. [2] have also tried to explain the origin of this satellite. Both Shrivastava et al. and Soni's assignments are based on energy considerations alone.

In the present work, we have calculated both energy and intensity of the various transitions belonging to the transition scheme $2s_{1/2}^{-1}3x^{-1}$-$3x^{-1}4p^{-1}$, i.e., L_1M_x - $M_xN_{2,3}$ and have assigned transitions to the satellites. There are 24 transitions which are allowed by selection rules. All these transitions have been taken into consideration in the present study.

2. Method of calculation

2.1. Creation of two hole states

Any two hole state, e.g. one useful as initial state for the emission of $L\gamma_{2,3}'$ satellite can be created by two processes: (a) K-L_1M_x Coster - Kronig transition, i.e., conversion of one hole state L_1 to a two hole state $L_1M_x(x = 1 - 5)$ through the Auger transition K - L_1M_x and (b) by shake-off process, i.e., an electron from M-shell of the atom may escape out simultaneous to the formation of L_1 vacancy.

2.2. Calculation of energies

We have calculated the energies of the transitions using the equation

$$E(L_1M_x - M_xN_{2,3}) = E(K\alpha_1) + E(L_3 - M_xN_{2,3}) - E(K - L_1M_x) \quad ..... (1)$$

The energy of $K\alpha_1$ line has been taken from the tables of Cauchois and Senemaud [1] and energies of Auger transitions L_3 - $M_xN_{2,3}$ and K - L_1M_x have been taken from the tables of Larkins [4].

2.3. Calculation of relative intensities of transitions

The relative intensities of the transitions have been calculated by first calculating the probability of initial single hole creation in the K shell by following the method of Moores et al. [5]. For calculating the probability of convertion of single hole K state to double hole state L_1M_x through Coster - Kronig transition (K - L_1M_x) the tables of Chen et al. [6] have been used. The probability of creation of double hole state L_1M_x through shake-off process has been calculated from the work of Carlson and Nestor [7]. The total probability of creation of L_1M_x double hole state has been found out by taking the sum of Coster-Kronig probability and shake- off probability. The cross section for a set of L_1M_x levels, with x denoting any one sub-shell of M shell, so calculated, has been assumed as the total probability of all the transitions from the set. This has been statistically distributed among all the allowed transitions from this set of levels, considering first of all the multiplets of super multiplets from various (2S+1) L levels of the set and then using the tables of White and Eliason as given by Condon and Shortley [8] for relative probabilities of the transition of each multiplet.

2.4. Computation of the theoretical spectrum

A composite spectrum formed by spectral lines emitted by the 24 transitions has been computed by taking each as a Gaussian line. Theoretical spectrum is thus obtained and compared with the available satellite energy data. The theoretical spectra for elements $_{40}Zr$ and $_{60}Nd$ are reproduced in figs.1 and 2 as representative spectra.

3. Results and discussions

3.1. $L\gamma_{2,3}'$ satellite in elements $_{40}Zr$ to $_{47}Ag$.

Neither Soni's [3] nor Shrivastava et al. [2] transition assignments for $L\gamma_{2,3}'$ satellite seem to be correct. According to the present work $^3D_3$-$^3F_4$(A), $^3D_3$-$^3D_3$(C), $^3D_2$-$^3F_3$(D),$^3D_1$-$^3F_2$(E), $^3D_3$-$^3P_2$(F), $^3D_2$-$^3P_1$(H), $^3P_2$-$^3D_3$(I) and $^3P_2$-$^3P_2$(K) are mainly and collectively responsible for giving rise to the $L\gamma_{2,3}'$ satellite in the elements $_{40}Zr$ to $_{47}Ag$ (the abbreviations A, B, C, D..... have been given to the transitions so that they may be marked on the figs. 1 and 2). There are some minor contributor and some negligible contributor transitions, which also add up to give rise to the spectral line of $L\gamma_{2,3}'$ satellite.

3.2. $L\gamma_{2,3}'$ satellite in the elements $_{55}Cs$ to $_{63}Eu$

Shrivastava et al.'s [2] transition assignments for $L\gamma_{2,3}'$ satellite seem to be incorrect. The present assignments based on both energy and intensity calculations appear to be more plausible. Thus, satellite $L\gamma_{2,3}'$should arise mainly because of $^3P_2$-$^3D_3$(I), $^3P_2$-$^3P_2$(K), $^3P_0$-$^3D_1$(S) and $^3P_1$-$^3P_2$(U) transitions, while $^3D_3$-$^3F_4$(A), $^1D_2$-$^1F_3$(B), $^3D_3$-$^3D_3$(C) and $^3P_1$-$^3P_2$(U) contribute to it as minor contributors. Also negligible contributor transitions $^3D_2$-$^3D_3$(O), $^3D_2$-$^3D_1$(Q) and $^3D_1$-$^3D_2$(R) make very small contribution to the intensity of $L\gamma_{2,3}'$ satellite. Thus, these 11 transitions are responsible for the occurrence of $L\gamma_{2,3}'$ satellite in $_{55}$Cs to $_{63}$Eu .

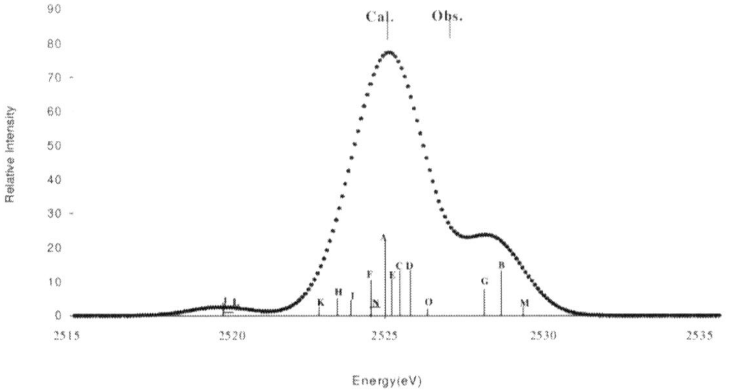

Figure 1. The computed $2s_{1/2}^{-1}3x^{-1}$ - $3x^{-1}4p^{-1}$ (L_1M_x - $M_xN_{2,3}$) spectrum of $_{40}$Zr. The experimentally observed $L\gamma_{2,3}'$ satellite positions and theoretically calculated positions are shown at the tops.

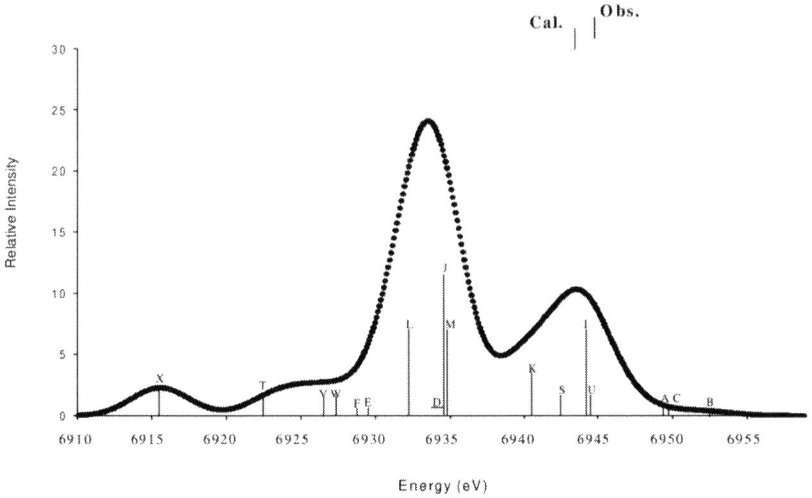

Figure 2. The computed $2s_{1/2}^{-1}3x^{-1}$ - $3x^{-1}4p^{-1}$ (L_1M_x - $M_xN_{2,3}$) spectrum of $_{60}$Nd.. The experimentally observed $L\gamma_{2,3}'$ satellite positions and theoretically calculated positions are shown at the tops.

Table 1. Theoretical and experimental energies (in eV) for the $L\gamma_{2,3}'$ satellite in the elements $_{40}Zr$ to $_{47}Ag$

S.No.	Elements	Present calculated value	Additional higher energy peak	Experimental value[1]	Soni's calculated value[3]	Shrivastava et al. calculated value[2]
1	$_{40}Zr$	2525.2	2527.9	2527.21	2527.0	2525.2
2	$_{41}Nb$	2684.4	2690.7	2690.6	2690.8	2687.5
3	$_{42}Mo$	2855.7	2859.0*	2860.8	2861.3	2855.3
4	$_{43}Mo$	3034.8	3038.3	-	-	-
5	$_{44}Mo$	3214.0	3217.5	-	-	-
6	$_{45}Rh$	3393.0	3396.7	3392.2	3392.4	3392.1
7	$_{46}Rh$	3584.0	3587.1	-	-	-
8	$_{47}Ag$	3781.2	3784.5	3783.4	3783.5	3777.5

Table 2. Theoretical and experimental energies (in eV) for the $L\gamma_{2,3}'$ satellite in the elements $_{55}Cs$ to $_{63}Eu$.

S. No.	Elements	Calculated value*		Experimental value[1-2]	Shrivastava et al's calculated value[2]
		Strongest peak in the middle	Higher energy side peak		
		Present value	Present		
1	$_{55}Cs$	5577.2	5594.7	-	-
2	$_{56}Ba$	5838.2	5847.5	5854.4	5844.7
3	$_{57}La$	6105.0	6114.1	6118.7	6110.3
4	$_{58}Ce$	6371.4	6287.3	-	-
5	$_{59}Pr$	6649.0	6659.2	6664.0	6652.9
6	$_{60}Nd$	6933.8	6944.0	6945.5	6937.7
7	$_{61}Pm$	7223.8	7239.5	-	-
8	$_{62}Sm$	7523.6	7534.2	7535.7	7528.3
9	$_{63}Eu$	7829.4	7841.2	7851.1	7831.3

References

[1] Cauchois Y and Senemaud C 1978 *Wavelengths of X-ray emission lines and absorption edges* (Pergamon, Oxford)
[2] Shrivastava B D, Jain R K, Mishra A and Singh D 1986 *J.Phys. B* **19** 3839
[3] Soni S N 1980 *J. Phys. B* **13** 2859
[4] Larkins F P 1977 *Atom. Data Nucl. Data Tables* **20** 311
[5] Moores D L, Golden L B and Sampson D H 1980 *J. Phys. B* **13** 385
[6] Chen M H, Crasmann B and Mark H 1979 *At. Data. Nucl. Data Tables* **24** 13
[7] Carlson T A and Nestor Jr C W 1973 *Phys. Rev. A* **8** 2887.
[8] Condon E U and Shortley G H 1967 *The theory of atomic spectra* (Cambridge Univ. Press) pp 241

Investigation of the thermal properties on $Dy_5Ge_2Si_2$

S. Shanmukharao Samatham[1], D. Venkateshwarlu[1], Mohan Gangrade[1], Swati Pandya[2] and V. Ganesan[1]

[1]Low Temperature Laboratory, UGC-DAE Consortium for Scientific Research, University Campus, Khandwa Road, Indore-452001, M.P., India
[2] Currently at Science and Humanities Department, BITS Edu. Campus, Varnama, Vadodara, Gujarat, India

E-mail: vganesan@csr.res.in

Abstract. The calorimetric and thermopower measurements on $Dy_5Ge_2Si_2$ are presented. The antiferromagnetic ordering resulted in a sharp peak in specific heat and paramagnetic temperature as a shoulder. Magnetic entropy (S_{mag}) is estimated from the magnetic heat capacity, C_{mag} and is in good agreement with the theoretically calculated value. A small note on correct assessment of S_{mag} is presented. Thermoelectric power is analysed in line with the specific heat and a temperature region of magnetic correlations is identified. Charge carrier density n~2.5×10^{23} cm^{-3} is calculated using Sommerfeld parameter γ and diffusion thermopower constant σ_D.

1. Introduction

The study of rare-earth based ternary alloys, especially R_5T_4 (R-rare-earth, T-transition metal), compounds has been paid attention for a decade. Initially, these were studied mainly in the context of magnetic refrigeration usage, for example, $Gd_5Ge_2Si_2$ [1] for its giant magnetocaloric effect at its magnetic transition temperature 276 K. Later, in the fundamental physics point of view, these are reputed for the first order based kinetics of magnetic glasses and frozen magneto-structural order (Gd_5Ge_4) [2]. Of the latest years, Dy based alloys are being revisited in the context of existence of Griffiths-like phase. $Dy_5Ge_2Si_2$ compound exists in Orthorhombic crystal structure with space group Pnma [3]. Though extensive studies have been carried out on these alloys, the physical properties like heat capacity and thermoelectric power are still intriguing and need to be addressed with deeper insight.

This papers presents detailed analysis of calorimetric and thermopower characterizations of the title compound. A small note on the assessment of proper background of specific heat to calculate the magnetic heat capacity/entropy is added. The sensitivity of heat capacity to any kind of magnetic transitions in the title compound by the application of magnetic fields is also highlighted.

2. Experimental Methods

Polycrystalline $Dy_5Ge_2Si_2$ was prepared by arc-melting the constituent elements in Ar atmosphere using Ti ball as a getter. The elements Dy, Ge and Si are taken in the stoichiometric molar ratio of

5:2:2. The melted ball was flipped many times to ensure the homogeneity. Specific heat, C_p(T) measurements were carried out by relaxation time calorimetric method using 14 T/2.0 K Physical Property Measurement System (QD, USA) down to 2 K and in magnetic fields up to 1 T. Thermoelectric power measurements were done using home-made setup coupled to 4 K Closed Cycle Refrigerator down to 4 K by differential sandwich method [4].

3. Results and discussion

From the magnetic behaviour, *M* (T), it was reported that $Dy_5Ge_2Si_2$ behaves as an antiferromagnet (AFM) below 57 K [3]. The employed Curie-Weiss fit to the inverse susceptibility in Ref. [5] resulted in the positive paramagnetic temperature (θ_p=38 K) which is unusual for AFM. Generally, as per Curie-Weiss law, a negative paramagnetic temperature is expected for AFM and a positive value for ferromagnet (FM). Hence, it was speculated that the positive θ_p in $Dy_5Ge_2Si_2$ indicates coexistence of the short-range FM and long-range AFM interactions [5].

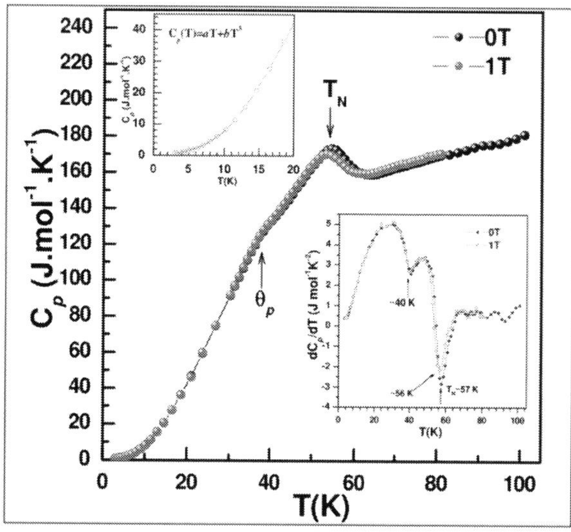

Figure 1. Specific heat C_p(T) of $Dy_5Ge_2Si_2$ from 2-100 K at 0 T and 1 T. The magnetic transitions at T_N and T_p are around 57 K and 40 K, respectively. Upper inset depicts the fit of specific heat equation.1. Down inset shows the temperature derivative of C_p, dC_p/dT with clearly identifying transition temperatures, T_N & θ_p.

In this paper, efforts have been put forward to understand the competing interactions through specific heat and thermoelectric power and also a note on the correct estimation of magnetic entropy of $Dy_5Ge_2Si_2$. Fig.1 presents the specific heat data on $Dy_5Ge_2Si_2$, collected in the absence of magnetic fields, from 2-100 K. The peak at T_N=57 K is due to AFM ordering and the specific heat value at peak reduces and smears out in field as evident from the 1 T data. The upper inset to the figure is the fitting of the date to the equation (1) for the low temperature data below 10 K.

$$C_p(T) = \gamma T + (\beta + \delta)T^3 \qquad (1)$$

Here, γT is the electronic heat capacity, βT^3 is the lattice specific heat and δT^3 is for the AFM magnetic contribution. The free fitting parameters are γ and δ only as β is directly taken from $Y_5Ge_2Si_2$ (will be discussed later). The obtained Sommerfeld parameter, γ = 0.15598 J. mol^{-1}.K^{-2} which is unusually high when compared to normal metals [6] and indicates the strong correlation of electrons because of the complex magnetic interactions in the compound. The estimation of magnetic contribution in AFM compounds is not straight forward as it is for other magnetic materials. The complexity arises because of its spin wave spectrum dispersion relation $\omega_q = \alpha_a (2J's a^2/\hbar) q$ which is similar to that of phonons [7]. Here, J' is the exchange, a is lattice parameter and $2\pi/q$ is the wave length, α_a dependent on the crystal structure details and s is spin quantum number. Using the above relation, C_{AFM} ~T^3. Hence, we have adopted Debye temperature of $Y_5Ge_2Si_2$, an isostructural nonmagnetic analogue of $Dy_5Ge_2Si_2$, which is 290 K [8]. β is calculated using $\theta_D(K) = \sqrt[3]{1944 p/\beta}$ where p is the number of atoms in a formula unit

(p=9 for $Dy_5Ge_2Si_2$). The calculated β~0.0007173 $J.mol^{-1}.K^{-4}$ is a fixed parameter in the equation (1). Now, the estimated magnetic contribution is δ~0.0056427 $J.mol^{-1}.K^{-4}$ that is 8 times greater than β indicating the dominant magnetic contribution in $Dy_5Ge_2Si_2$.
Magnetic entropy has been estimated using the formula $S_{mag} = \int (C_{mag}/T)dT$. C_{mag} is the magnetic heat capacity assessed by subtracting the sum of C_{el} and C_{ph} from the total specific heat C_p, i.e., $C_{mag}=C_p-(C_{el}+C_{ph})$, figure 2*a*. For estimating the C_{ph}, instead of simple T^3 dependence, the actual Block Gruniesen function is being used with Debye temperature 290 K of $Y_5Ge_2Si_2$. C_{el} is γT of $Dy_5Ge_2Si_2$. Theoretically calculated magnetic entropy for Dy^{+3} is $S_{mag} = R\ln(2J+1)$ =23.05 $J.mol^{-1}.K^{-1}$ with J=15/2 [L (=4) +S (7/2)]. It is evident from the figure 1 that S_{mag} is 21.85 $J.mol^{-1}.K^{-1}$ at 100.85 K which essentially meets the theoretical value at 300 K, figure 2*b*. Hence, the experimentally assessed value is in good agreement with the theoretical value without any reduction whereas the reduction in S_{mag} was shown in Ref. [5]. It is because of the over subtraction of C_{el} by assuming the C_{ph} and C_{el} of $Dy_5Ge_2Si_2$ are the same as those for $Y_5Ge_2Si_2$. However, γ (1.6758 $J.mol^{-1}.K^{-2}$) [8] of $Y_5Ge_2Si_2$ is one order higher than that for $Dy_5Ge_2Si_2$ (0.15598 $J.mol^{-1}.K^{-2}$). Generally, C_{mag} of magnetic rare-earth based compounds is estimated by subtracting the heat capacity ($C_{ph}+C_{el}$) of their non-magnetic counterparts of the same structure (either Y or La based alloys). In this process, at times, C_{el} of non-magnetic structural analogues is not the same as that of magnetic ones. Hence, the problem of over/under estimation of C_{mag} arises because of under/over consideration of C_{el} of nonmagnetic analogues.

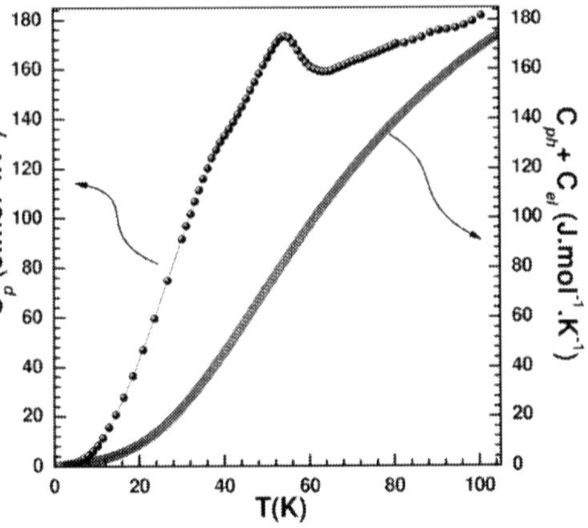

Figure 2*a*. Left side: $C_p(T)$, Right side: the estimated $C_{el}=\gamma T$ and C_{ph} using θ_D of $Y_5Ge_2Si_2$ and with Bloch Grueniesen expression.

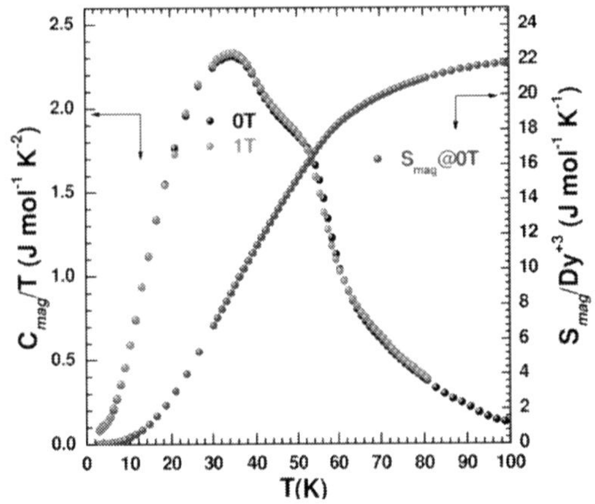

Figure 2*b*. Left side: Camg/T as a function of T at 0 T and 1 T. Right side: The calculated S_{mag}/Dy^{+3}, 21.85 $J.mol^{-1}.K^{-1}$ at 100.85 K

Figure 3 shows the thermoelectric power, σ(T) of $Dy_5Ge_2Si_2$ down to 8 K from 100 K. Data from 40 K to 80 K is fitted to the linear equation to estimate the diffusion coefficient of thermopower, σ_D. Sign of σ_D is negative, hence indicating the electrons as the charge carries in the sample. The estimated absolute σ_D = 0.02926 $\mu V.K^{-2}$. The value of Fermi energy from the diffusion coefficient of thermopower [9] can be estimated as $E_F = \pi^2 k_B^2 / 2e\sigma_D$ and E_F is found to be at around 1.26 eV. The density of states (DOS) at E_F is calculated using $\gamma = (\pi^2 k_B^2/3) N(E_F)$ where γ is Sommerfeld coefficient. $N(E_F)$~3.065x10^{23} $(eV)^{-1}.cm^{-3}$. The carrier density of charge carries is being estimated using the following formula $n = (2/3) E_F N(E_F)$ and n~2.5x10^{23} cm^{-3} which indicates a good metallic behaviour.

The manifestation of FM short-range interactions in the AFM material resulted in the small hump in specific heat in the proximity of θ_p. It is more evident in the thermoelectric power as shown in figure 3*b*. It changes its slope from the linear behaviour at around 42.5 K and low temperature slope change is at 31.87 K. This region can, perhaps, be attributed as the region of pronounced FM correlations in $Dy_5Ge_2Si_2$. The low temperature fall is at around 10 K.

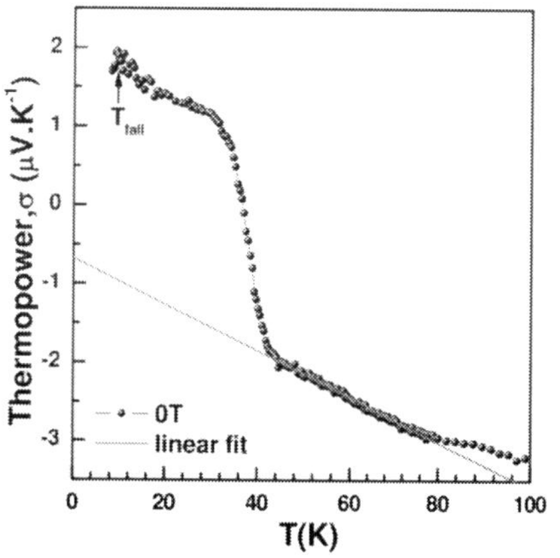

Figure 3*a*. Thermopower of $Dy_5Ge_2Si_2$ down to 8 K. The linear fit is diffusion thermopower and diffusion constant, σ_D=0.02926 μV.K^{-1}.

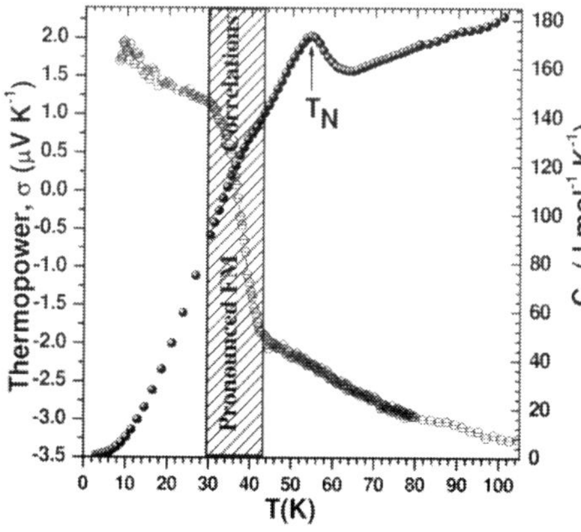

Figure 3*b*. The correlation between the thermopower and specific heat. Pronounced FM correlations region is in the proximity of θ_p.

In conclusion, calorimetric and thermopower measurements on $Dy_5Ge_2Si_2$ were carried out. The magnetic transitions like antiferromagnetic (T_N) and paramagnetic temperature (θ_p) are seen in specific heat. The correctly estimated magnetic entropy by considering C_{el} of $Dy_5Ge_2Si_2$ and lattice heat capacity of its nonmagnetic isostructural analogue $Y_5Ge_2Si_2$ is presented. The estimated S_{mag} is in good agreement with the theoretically obtained value. The magnetic correlations in the proximity of paramagnetic temperature are seen in the thermopower and are corroborated with specific heat.

Acknowledgments

Authors thank Dr. A. K. Sinha, Dr. P. Chaddah and Prof. Ajay Gupta for their encouragement. Cryogenics staff, especially Er. P. Saravanan is thanked for the supply of cryogens. Mr. Sankararao Y. and Mr. Durgesh Singh are thanked for their help in laboratory. S. S. Samatham thanks CSIR-India for financial support (File no. 9/838(0049)/2013-EMR-I). DST-India is thanked for LTHM facility.

References

[1] Pecharsky V. K. and Gschneidner K. A. Jr., 1997 *Phys. Rev. Lett.* **78** 4494
[2] Sharath Chandra L. S., Pandya Swati, Vishwakarma P. N., Jain Deepti and Ganesan V. 2009 *Phys. Rev.* B **79** 052402
[3] Ivtchenko V. V., Pecharsky V. K. and Gschneidner K. A. Jr., 2000 *Adv. Cryog.* Eng. **46A** 405
[4] Sharath Chandra L. S., Lakhani A., Jain D., Pandya S., Vishwakarma P. N., Gangrade M. and Ganesan V., 2008 *Rev. Sci. Instrum.* **79** 103907
[5] Nirmala R., Morozkin A. V. and Malik S. K., 2007 *Phys. Rev.* B **75** 094419
[6] Kittel Charles, 2009 *Introduction to Solid State Physics*, (New Delhi: Wiley) p. 157
[7] Gopal E. S. R. 1966 *Specific heats at low temperatures* (New York: Plenum Express)
[8] Pacharsky A. O., Pecharsky V. K., Gschneidner K. A. Jr. 2004 *J. Alloys and Compd.* **379** 127
[9] Barnard R. D, 1972 *Thermoelectricity in Metals and Alloys* (London: Taylor & Francis)

Investigation of electronic phase segregation in $La_{0.75}Ca_{0.15}Sr_{0.10}MnO_3$ manganite

R.Venkatesh[1], D. Venkateshwarlu[1] and Nini Pryds[2]
[1]Low Temperature Laboratory, UGC-DAE Consortium for Scientific Research, University Campus, Khandwa Road, Indore-452001, M. P., India.
[2]DTU Energy conversion,Department of Energy Conversion and Storage,Technical university of Denmark, Frederiksborgvej 399, Building 779, room S25, 4000 Roskilde.

E-mail: rvenkateshphysics@gmail.com

Abstract. The effect of electronic phase segregation in a broad metal-Insulator transition (MIT) observed in $La_{0.75}Ca_{0.25-x}Sr_xMnO_3$ (x=0.1) composition is investigated using heat capacity, magnetization, electrical resistivity and magnetoresistance measurements. The negative magnetoresistance of 65% in an applied magnetic field of 12T and 15% in 1T with a broad working range of 18K around 300K which is beneficial for room temperature colossal magnetoresistance (CMR) applications. The broad transition in temperature dependent zero field resistivity measurement is analyzed in the light of percolation model indicates the abundance of insulating/metallic clusters in metallic/insulating region. A significant difference between the metallic fraction around the MIT and the ferromagnetic phases observed around the Curie temperature demonstrates the interplay between volume of itinerant and polaronic electronic phases.

1. Introduction

Manganese perovskites are extensively investigated for the past few decades for their colossal magnetoresistance (CMR) property exhibited in a sprawling range of metal to insulator transition temperatures (T_P) with a concomitant magnetic phase transition indicated by Curie temperature (T_C) The colossal magnetoresistance (CMR) observed in manganites around the Curie temperature T_C are attributed to the interplay between spin, lattice, charge and orbital degrees of freedom.[1].CMR property shows a strong dependence on the nature and order of magnetic/electronic phase transitions and it is reported to be controlled by the electronic phase segregation of the system and its relative volume change at T_P [2]. The manganite $La_{0.75}Ca_{0.25}MnO_3$ has a significant colossal magnetoresistance around its curie temperature, and the order of transition in $La_{0.75}Ca_{0.25}MnO_3$ changes from first to second order upon the application of a magnetic field.[3]. In this study an unusually broad phase transition was observed in $La_{0.75}Ca_{0.25-x}Sr_xMnO_3$ (x= 0.10) composition. This composition is investigated using Magnetic, magnetoresistance, heat capacity measurements and analyzed in the light of percolation model is reported in this work.

2. Experimental details

Polycrystalline powder samples of $La_{0.75}Ca_{0.25-x}Sr_xMnO_3$ (x=0.10) were prepared by the conventional solid-state reaction method. Stoichiometric amounts of La_2O_3, $CaCO_3$, $SrCO_3$ and MnO_2 were mixed well and calcined initially at 1200° C for 24hrs and then cooled to room temperature. The second calcinations took place at 1250^OC for 24 hrs and cooled back to room temperature. After this second calcination two batches of samples were extracted as LCSM1250 and LCSM1400 whose processing details are reported elsewhere [4].The phase purity and crystal structure of the powder samples were checked by X-ray powder diffraction (XRD) using Cu K_α radiation in a Bruker D8 Advance powder diffractometer, and reported in our earlier work [4]. In brief, both LCSM1250 and LCSM1400 are orthorhombic structured (*pnma* space group), $La_{0.75}Ca_{0.15}Sr_{0.10}MnO_3$ composition, but processed in different sintering temperatures 1250^OC (12hrs) and 1400^OC (24hrs) respectively. The magnetic measurements were taken with fields up to 1T using a Lake Shore 7407 vibrating sample magnetometer. Our heat capacity measurements were made with a custom made differential scanning calorimeter (DSC) [5] and the magnetoresistance measurements were made using Quantum design PPMS upto a magnetic field of 12T and down to a temperature of 2K

3. Results and Analysis

The temperature dependent magnetization in Zerofield cooled (ZFC) and Field cooled (FC) conditions for the sample LCSM1250 and LCSM1400 in an applied magnetic field of 0.01T is shown in figure 1. The maximum change in slope observed in the magnetization *vs* temperature plot at 0.01T is used to determine the Curie temperature (Tc). The Tc value of LCSM1400 is 276 K indicated by sharp dM/dT curve in figure 1 agree very well with the earlier studies of Guo *et al* [6]. A prominent splitting between the ZFC and FC behavior is observed in LCSM1400 upto a Curie temperature of 276K after which it merges into a single curve. In the case of LCSM1250 the splitting is visible upto 295K which

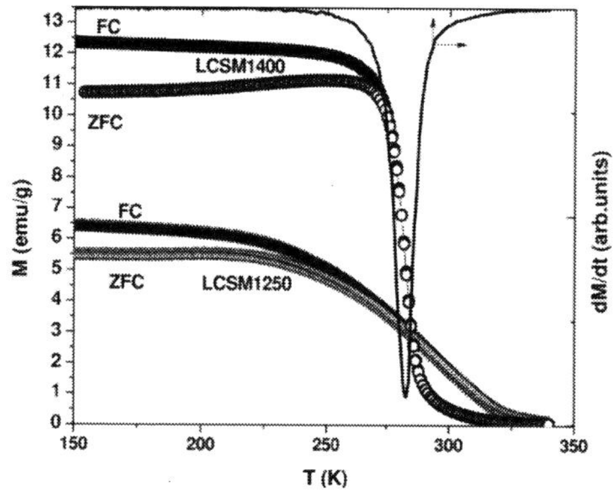

Figure 1. M vs T measurements in LCSM 1400 and LCSM1250 with right axis showing T_C for LCSM1400

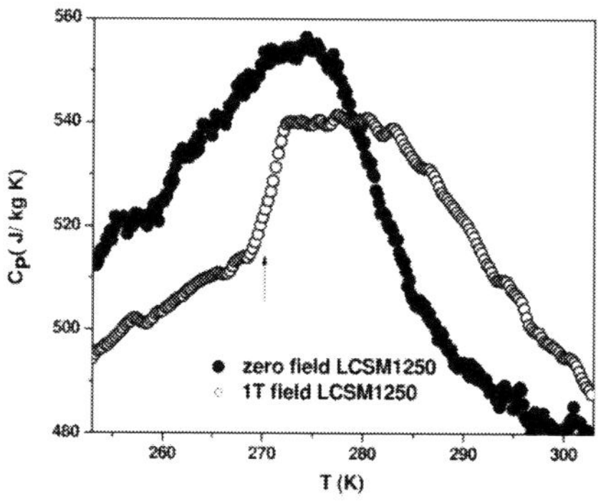

Figure 2. C_P vs T measurements in zero and 1T with arrow pointing the slope change below the curie temperature in LCSM1250 (measured using DSC)

is above the Curie temperature of 287 K. The broad magnetic transition observed in LCSM1250 is very much interesting for room temperature CMR applications and hence the LCSM1250 is further investigated using heat capacity and magnetoresistance measurements. The heat capacity of sample LCSM1250 in zero magnetic fields and in an applied field of 1T is shown in figure 2. The zero field heat capacity of LCSM1250 shows a peak at 275K which indicates the position of the Curie temperature (T_C). There is a significant difference between the T_C observed in magnetization measurements and zero field heat capacity measurement. The spread of transition temperatures in different measurements occurring in LCSM type perovskite manganites has already been well documented by Bahl et al [7]. On application of magnetic field of 1T the transition gets broadened with an additional slope change below the magnetic transition temperature as indicated by an arrow in figure 2.

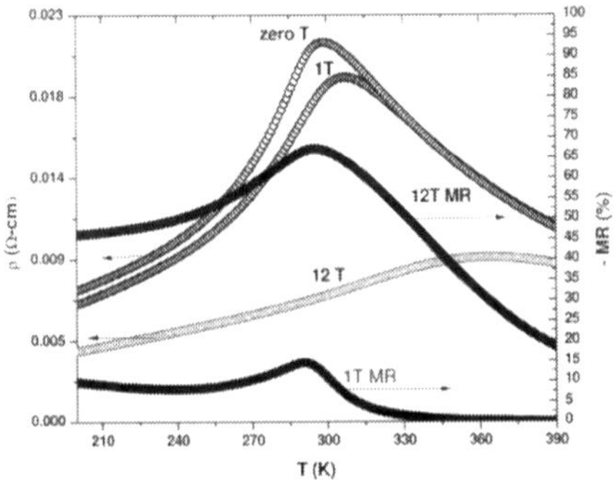

Figure 3. Temperature dependent electrical resistivity with right axis giving the percentage of magnetoresistance in 1T and 12T

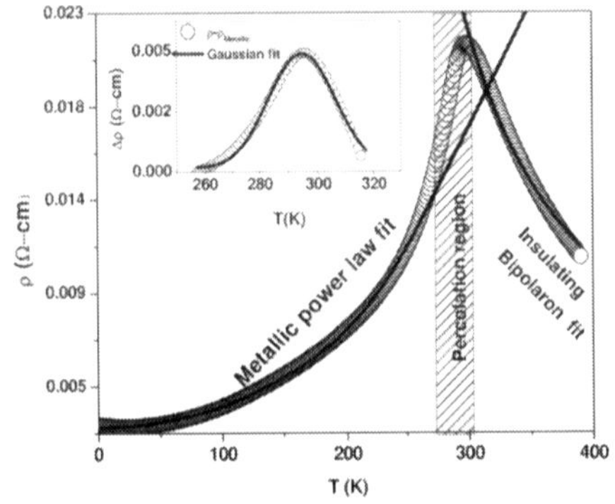

Figure 4. Zero field ρ vs T plot with power law fit for low temperature and bipolaron fit for high temperature, Inset shows MIT with Gaussian fitting

Temperature dependence of electrical resistivity in an applied magnetic field of 0T, 1T and 12T is shown in figure 3. A metal to Insulator transition temperature (T_P) occurring at 299K in the absence of magnetic field and shifts to a higher temperature of 307 K in 1T.The temperature further increases to 350K in an applied magnetic field of 12T. The magnitude of resistance decreases in an applied magnetic field and this percentage of negative magnetoresistance is 15 and 65% in the magnetic field of 1T and 12T respectively. The magnetoresistance is almost constant with a variation of less than 2% for a temperature range of 18K around the MR_{peak} temperature. In the case of LCSM1250 relatively constant magnetoresistance in the range of 286-304K indicates that it is a promising candidate for room temperature CMR applications.

4. Discussions

The nature and electronic properties of this broad MIT transition in LCSM1250 is further investigated by using the temperature dependence of zerofield resistivity as shown in figure 4. The low temperature metallic region below the MIT transition and upto 275 K is fitted using metallic power law with $\rho_m(T)= \rho_0+ \rho_2T^2+ \rho_5T^5$ where ρ_0 is the residual resistivity, ρ_2 and ρ_5 are electron-electron and electron-phonon scattering coefficients respectively. The higher temperature insulating region above the MIT from 305-390K could be fitted using Bipolaron hopping model using $\rho(T)= AT[\exp(\Delta/2k_BT)]$ where A is resistivity constant, Δ is the bipolaronic binding energy and k_B is the Bolzmen constant. The bipolaronic binding energy comes out to be 206 meV which is higher than the Bipolaron binding energy reported in disordered thinfilms with a broad transition.[8].The temperature region of 276-305K around the MIT where itinerant electron model and bipolaron hopping model fails is considered to be the percolation region as indicated by shaded region in figure 4. On extrapolating and subtracting the metallic power law fit,the remaining MIT transition peak is fitted with Gaussian as shown in inset of figure 4 by keeping the width of Gaussian (Γ) as a free parameter. The obtained value of Γ is found to be 27 K.The metallic fractions in the whole temperature range can be estimated from the equation $\gamma(T)=1/2(\mathrm{erfc}([T-T_P]/\Gamma))$ as shown in figure 5. According to the double exchange mechanism proposed for the manganites, MIT as obtained from electrical resistivity measurements usually coincides with the ferromagnetic to paramagnetic transitions obtained from magnetic measurements[8]. A comparison with the metallic fraction obtained in percolation model with that of the ferromagnetic fraction obtained from M vs T measurements (figure 5) shows a mismatch around MIT. $La_{0.75}Ca_{0.25-x}Sr_xMnO_3$ (x=0.1) composition exist in the close proximity of structural transition from orthorhombic to rhombohedral structure as shown in inset of figure5 . Especially LCSM1250 crystallizing in orthorhombic structure with Jhan-Teller distortion posses inhomogeneous strain and enhanced magnetoelastic coupling [4] when compared to LCSM1400 which dictates the volume of itinerant/ polaronic phases and hence the electronic phase segregation leading to the broadening of MIT.

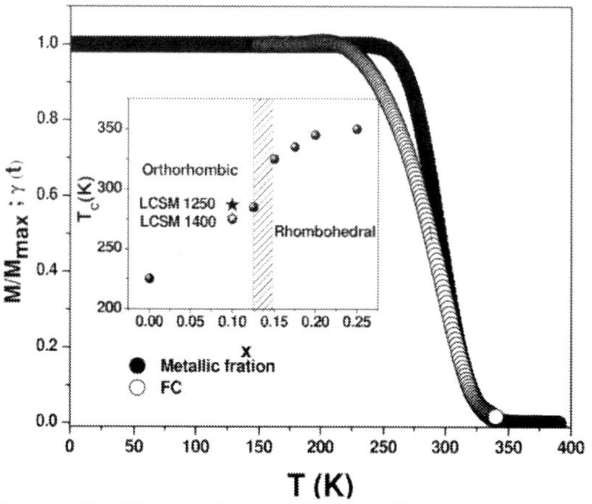

Figure 5. Comparison of metallic fraction with magnetization measurements;(inset) phase diagram of LCSM reproduced from Guo *et* al [6].

5. Conclusion

The broad metal to Insulator transition observed in $La_{0.75}Ca_{0.15}Sr_{0.10}MnO_3$ (LCSM1250) is a promising candidate for room temperature CMR applications. The MIT is analyzed in the light of percolation model. The metallic fraction obtained using electrical resistivity measurements are compared with ferromagnetic phase estimated from magnetic measurements which indicates that the broadening of phase transition can be attributed to the electronic phase segregation (itinerant and polaronic) in and around the MIT.

Acknowledgements

Author R.V thank DST-India for supporting LTHM facilities in UGC-DAE CSR Indore, Director Dr.A.K.Sinha, Centre-Director Dr.V. Ganesan for their encouragement, inmates and staff of low temperature laboratory especially S. Shanmukharao Samatham for discussions and cryogenic staffs especially Er. P.Saravanan for continuous supply of cryogens.

References

[1] Tokura Y and Tomioka Y 1999 *J. Magn.Magn Mater* **200** 1
[2] Rivadulla F, Otero-Leal M, Espinosa A, de Andre´s A, Ramos C., Rivas J, and Goodenough J B 2006 *Phys. Rev. Lett.* **96** 016402 .
[3] Radaelli P G, Iannone G, Marezio M, Hwang H Y, Cheong S W, Jorgensen J D and Argyriou D N 1997 *Phys. Rev. B* **56** 8265.
[4] Venkatesh R, Theil Kuhn L, Pryds N, Bahl C R H and Bohr J 2012 *Mater. Chem. Phys.* **132**192.
[5] Jeppesen S, Linderoth S, Pryds N, Theil Kuhn L and Buch Jensen J 2008 *Rev. Sci. Instrum.* **79** 083901.
[6] Guo Z B, Yang W, Shen Y Tand Du Y W 1998 *Solid State Commun.* **92** 10589.
[7] Bahl C R H, Bjørk R, Smith A, Nielsen K K 2012 *J. Magn.Magn Mater.* **324** 564.
[8] Egilmez M, Chow K H and Jung J 2008 *Apps. Phys. Lett.* **92** 162515.

Effect of magnetic field on the first order Martensite phase transition in $Ni_{55}Fe_{20}Al_{25}$ shape memory alloy

Archana Lakhani
UGC-DAE, Consortium for Scientific Research, University Campus, Khandwa Road, Indore-452001, India

E-mail: archnalakhani@gmail.com

Abstract. The temperature and magnetic field effect on the transport and magnetic properties of a ferromagnetic shape memory alloy $Ni_{55}Fe_{20}Al_{25}$ has been studied. The resistivity and thermo magnetization results provide the evidence of martensite transformation and the presence of coexisting austenite and martensite phases by first order magnetic phase transition (FOMT) at different fields. Martensite transformation temperature (T_m) shifts towards higher temperature with the application of magnetic field which reveals the sensitivity of this alloy. Ni-Fe-Al alloys are potential new candidates for ferromagnetic shape memory alloys.

1. Introduction

Ferromagnetic shape memory alloys (FSMAs) have been the subject of current interest due to the magnetic field induced first order phase transition (FOPT). Similar to various intermetallic alloys and manganites, FSMAs also show field induced structural transition giving rise to functional properties such as giant magnetoresistance, giant magnetocaloric effect, magnetostriction and superelasticity etc. They are technologically potential candidates due to their possible usage in actuators, sensors and in the magnetic refrigeration technology [1,2]. Magnetic field induced transition materials have revitalized the interest in FOPT not only due to their possible applications but also due to the interesting fundamental studies around FOPT. In FSMAs, this field induced martensite transformation (MT) is associated with change in structure and magnetization both. [3]

The temperature at which MT/FOMT occurs in various compounds can be optimized by the careful choice of composition and thermal treatment. Ni-Fe-Al shape memory alloys are originated from Ni-Al binary shape memory alloys. The doping of Fe in Ni-Al system increases the ductility and reduces the MT temperature in Ni-Fe-Al alloys. The site occupancy of Fe in Ni-Fe-Al shape memory alloys plays a decisive role in defining the structural and magnetic properties of these systems. The effect of site disorder on the MT in $Ni_{55}Fe_{20}Al_{25}$ has been studied by Abhyankar et al. by Neutron Diffraction and Magneto transport measurements [4,5]. Ni-Fe-Al alloys show a transition from high temperature cubic austenite to low temperature tetragonal maretensite phase. Magnetically it undergoes a para to ferromagnetic phase transition at T_C [6].

Here we present a resistivity and thermomagnetization study on an annealed and subsequently quenched sample with composition $Ni_{55}Fe_{20}Al_{25}$ in order to distinguish the effect of temperature and field on the MT and on the coexisting austenite and martensite phases associated with para and ferromagnetic phases respectively.

2. Experimental

In Polycrystalline sample with composition $Ni_{55}Fe_{20}Al_{25}$ is prepared by arc melting of the constituent elements with purity better than 99.99% in pure argon atmosphere. The ingot was melted several times to ensure homogeneity. As-prepared ingot was annealed at 1100^0C in an evacuated quartz tube followed by subsequent quenching. Resistivity measurements were performed by standard four probe technique using a commercial cryostat from Oxford Instruments, UK in the temperature range of 5-300K and up to 8T magnetic fields. The magnetization measurements were performed using a 14-T VSM-PPMS from Ms. QD, USA.

3. Results and Discussion

The Figure 1 shows resistivity plots as a function of temperature during cooling and heating cycles without magnetic field and in three different measurement protocols, viz: zero field cooled warming (ZFCW), field cooled cooling (FCC) and field cooled warming (FCW) at 8T field. The arrows in cooling and heating cycles show the start and finish of martensite (M_s and M_f) and austenite (A_s and A_f) phases. The temperatures corresponding to M_s, M_f, A_s and A_f in absence of magnetic field are 155K, 77K, 83K and 163K respectively. The phase coexisting region reclines between ~ 160-80K with a hysteresis of approximately 80K. Thermal hysteresis in FSMAs is considered to be the signature of temperature induced first order martensite phase transition. On increasing the field from zero to 8T, there is not much change in the width of hysteresis whereas martensite transition temperature (T_m) increases from 120K to 127K. Here T_m is defined as $T_m = [(M_s+M_f)/2+(A_s+A_f)/2]/2$.

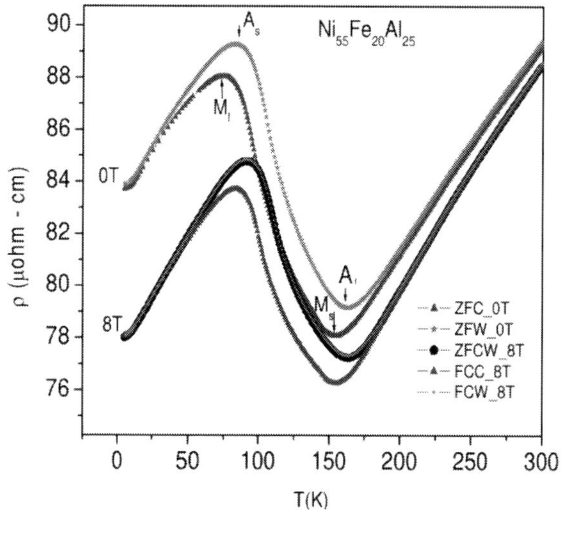

Figure 1. ρ vs T plots for $Ni_{55}Fe_{20}Al_{25}$ sample obtained in ZFC, FCC and FCW manner in zero field and in 8T fields.

Figure 2 shows magnetization as a function of temperature in three different measurement protocols viz: zero field cooled warming (ZFC), field cooled cooling (FCC) and field cooled warming (FCW) at H= 500 Oe, 1T and 8T. For clarity, cooling and heating cycles of Magnetization at 500 Oe are shown in the inset of figure 2. The thermal hysteresis is observed in the cooling and heating cycles of M(T) curves also, although it is more pronounced in ρ(T) curves. The characteristic temperatures defined for the onset and finish of martensite and austenite phases are marked by arrows in the inset shown. The T_m gradually increases with increase in field from 500 Oe to 8T from 100K to 117K. These values are smaller than the T_m values obtained from resistivity measurements but the change in martensite transition temperature ($\Delta T_m = 7K$) in the span of 8T is same. There is almost 1K shift in T_m per Tesla field which reveals the sensitivity of this alloy with magnetic field.

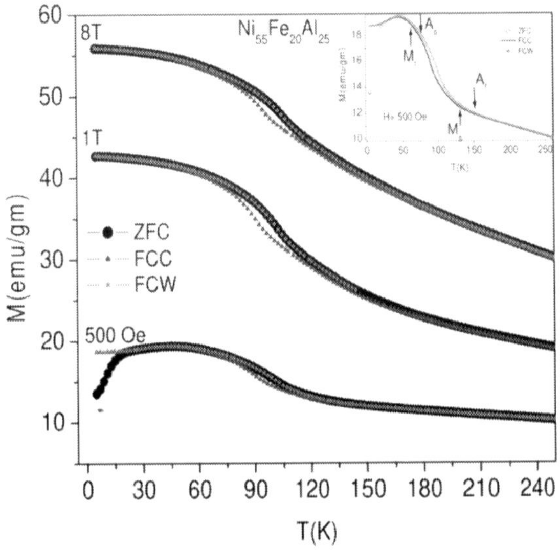

Figure 2. M vs T plots for $Ni_{55}Fe_{20}Al_{25}$ sample obtained in ZFC, FCC and FCW manner at applied fields of 500 Oe,1T and 8T.

At 500 Oe, the magnetic moment increases by ~ 10 emu/g after the martensite phase transition, indicating transition from low magnetic state to higher magnetic state as reported earlier [5]. Isothermal magnetization curve at 5K is shown in figure 3 revealing the ferromagnetic nature of the state in the system. The saturation magnetization value (M_s) at 10T field is ~ 57 emu/g which is almost equal to the saturation magnetization (~59 emu/g) obtained for the melt quenched ribbon sample at 7T discussed by Kaul et al. [5]

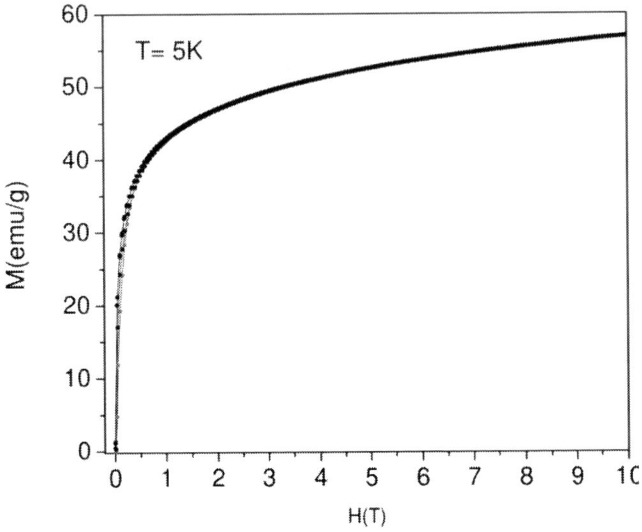

Figure 3. Magnetization as a function of field at T= 5K for $Ni_{55}Fe_{20}Al_{25}$ alloy.

The sensitivity of austenite to martensite phase transformation to applied magnetic field can be understood by considering the effect of magnetic field on the free energies of the respective phases. On lowering the temperature, when martensite state becomes the low energy state, the system undergoes a first-order austenite to martensite phase transition. Since the martensite state exhibits a ferromagnetic behavior, the application of magnetic field lowers its free energy, and therefore the martensite state attains the status of low energy state at a relatively higher temperature. This results in an enhancement of martensite transition temperature (T_m) on increasing the magnetic field.

In conclusion we have studied the effect of field and temperature on an annealed and subsequently quenched shape memory alloy of $Ni_{55}Fe_{20}Al_{25}$ by resistivity and magnetization measurements and demonstrated the field sensitivity of the alloy which reveals its technological importance. The physical significance behind the enhancement of T_m is explained in terms of change in free energy of the two magnetic states.

4. Acknowledgment

I thank Pallavi Kushwaha for help in annealing the sample, Rajeev Rawat for Resistivity measurements and Alok Banerjee for the support. Department of Science and Technology (DST), India is acknowledged for funding the 14T VSM system from Quantum Design USA.

5. References

[1] Kainuma R , Imano Y, Ito W, Sutou Y, Morito H., Okamoto S., Kitakami O., Oikawa K., Fujita A., Kanomata T *et al.* 2006 *Nature* **439** 957.
[2] Lakhani A, Dash S., Banerjee A., Chaddah P., Chen X., and Ramanujan R. V. 2011 *Appl. Phys. Letters* **99** 242503 and reference there in.
[3] Siruguri V, Babu P D, Kaushik S D, Biswas A, Sarkar S K, Madangopal K and Chaddah P 2013 *arXiv:1307.2486v2.*
[4] Abhyankar A.C, Annie Santhoshini B., Kaul S.N. and Nigam A.K., 2008 *Advanced Materials Research* **52** 77.
[5] Kaul S. N., Annie Santhoshini B and Abhyankar A.C, 2006 *Appl. Phys. Letters.* **89** 093119.
[6] Mukhopadhyay P.K. and Kaul S.N. 2008 *Appl. Phys. Letters* **92** 101924.

XRD, LPF and FTIR investigation of Mn-Bi alloy

Ashutosh Mishra, Harsha Patil, Garima Jain
School of Physics, Devi Ahilya University, Khandwa Road, Indore, 452001, India

Email: amishra1960@yahoo.co.in,harshapatil9@gmail.com

Abstract. High purity MnBi low temperature phase has been prepared and analyzed using X-ray diffraction, Lorentz-Polarization Factor and Fourier transforms infrared measurement. After synthesis of samples structural characterization has done on samples by X-ray diffraction, which shows that after making the bulk sample is in no single phase MnBi has been prepared by sintering Mn and Bi powder. The X-ray diffraction measurements were carried out using Bruker D8 Advance X-ray diffractometer. The X-rays were produced using a sealed tube and the wavelength of x-ray was 0.154nm (Cu K-alpha).and x-rays were detected using a fast counting detector based on Silicon strip technology (Bruker LynxEye detector). By Lorentz-Polarization Factor is affecting the relative intensity of diffraction lines on a powder form. The infrared absorption spectra of the alloys and intermetallic compound were measured at room temperature, in the wave number range 4000 to 400 cm^{-1} by a computerized spectrometer type Jasco FTIR-300 (JAPAN) using the KBr pellet technique. And by FTIR which shows absorption peaks of MnBi alloys.

1. Introduction

MnBi is a ferromagnetic inter-metallic compound with hexagonal crystal structure and the binary compound MnBi crystallizes into two phases, the lower temperature phase and the higher temperature phase. But at present the MnBi binary compound is in low temperature phase (LTP) [1]. The structural properties of these MnBi compounds have been studied by using X-ray diffraction, Lorentz-Polarization Factor and Fourier transforms infrared measurement of the low temperature phase (LTP) [2].

2. Experimental

One can start from highly pure fine powdered samples of Mn (Manganese) and Bi (Bismuth). Both of these were mixed in the calculated molar ratio of 45:55. These samples of different percentage of MnBi alloys were prepared by Solid state route method. The X-ray diffraction measurements were carried out using Bruker D8 Advance X-ray diffractometer. The X-rays were produced using a sealed tube and the wavelength of x-ray was 0.154nm (Cu K-alpha).and x-rays were detected using a fast counting detector based on Silicon strip technology (Bruker LynxEye detector).The diffraction pattern was scanned over the range 20 to 50 degree (2θ). The samples were characterized at room temperature using cukα radiation. In this diffractometer, a radiation detector records the position and intensities of the various reflected line as a function of 2θ. Analysis of X-ray diffraction gives information about the different planes present in the specimen, lattice parameters and phase purity.

Lorentz factor is related to the volume of sample irradiated as a function of angle. The polarization factor accounts for increase scattering at low angles. For powder diffraction all the components of the Lorentz correction combine to give, when combined with the polarization correction, there is a pronounced enhancement of the low and high angle reflections relative to those at intermediate angles.

Lorentz-Polarization Factor is combination of two factors the Lorentz factor and the polarization factor that influence the intensity of the diffracted beam. And Lorentz-Polarization Factor is a Constant factor [4]. The Lorentz-polarization factor varies strongly with Bragg angle θ, and the overall effect is that the intensity of reflections at intermediate Bragg angles is decreased compared to those at high or low angles. Now, Lorentz-Polarization factor is given by the relation-

$$LPF = (1+ \cos^2(2\Theta)) / (\sin^2\Theta \cos\Theta)$$

The infrared absorption spectra of the alloys and intermetallic compound were measured at room temperature, in the wave number range 4000 to 400 cm^{-1} by a computerized spectrometer type Jasco FTIR-300 (JAPAN) using the KBr pellet technique. The samples were investigated as fine particles, which were mixed with KBr in the ratio (2:200mg powder to KBr respectively); the weighted mixture was then subjected to a pressure of $5t/cm^2$ to produce clear homogeneous discs.

3. Results and discussion

3.1. X-ray diffraction studies:

XRD pattern of the available highly pure fine powder Mn and Bi and powdered samples such as MnBi in the atomic ratio 45:55, and shows the XRD pattern of MnBi compound before and after sintering (in figure 1). XRD peaks indicate the (hkl) values, and these values are showing the hexagonal crystal structure after indexed.

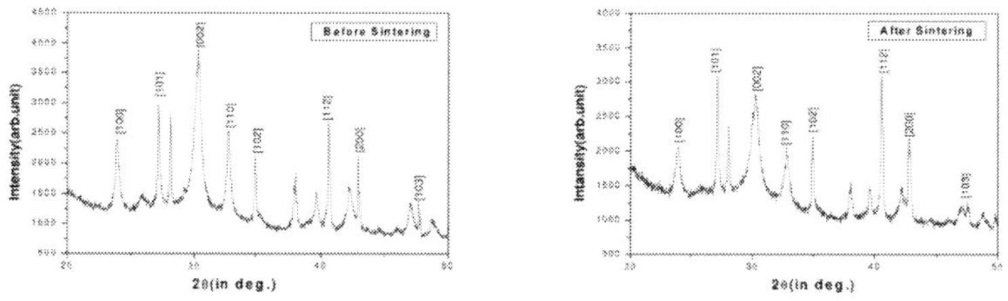

(a) Sample-1 (b) Sample-2

Figure1. XRD pattern of MnBi alloys.

A close observation indicates that there is a systematic change in the FWHM of XRD peaks. The lattice parameter (a=1.42Å) and (c=6.11Å) of the MnBi alloys have been same before and after sintering. The particle size (t) of the MnBi intermetallic compound has been reported in Table1.

Table 1: Particle size

MnBi alloy	Particle size	MnBi alloy	Particle size
Before Sintering	t (Å)	After sintering	t(Å)
Sample 1	301.84	Sample 2	309.56

3.2. Lorentz-polarization factor studies:

The Lorentz-polarization factor is that which control X-ray intensity with respect to diffraction angle. Consider the X-ray intensity recorded on a piece of film as a crystal is rotated in an X-ray beam through the diffraction condition. The reflection is not infinitely narrow for a real crystal and the area under the curve is the integrated intensity. The integrated intensity is affected by the height and width of the curve [5}. These quantities are dependent on the Bragg angle as well as on the structure of the studied sample. When the Lorentz-polarization factor plotted with the 2θ angle, the pattern is reported in Figure 2. In Figure 2 the LPF pattern of MnBi compound before and after sintering have been shown. In this pattern the LPF decreases as the 2θ angle increases. In infrared spectroscopy, IR radiation is passed through a sample. Some of the infrared radiation is absorbed by the sample and some of it is passed through or transmitted. The resulting spectrum represents the molecular absorption and transmission, creating a molecular fingerprint of the sample. Like a fingerprint no two unique molecular structures produce the same infrared spectrum.

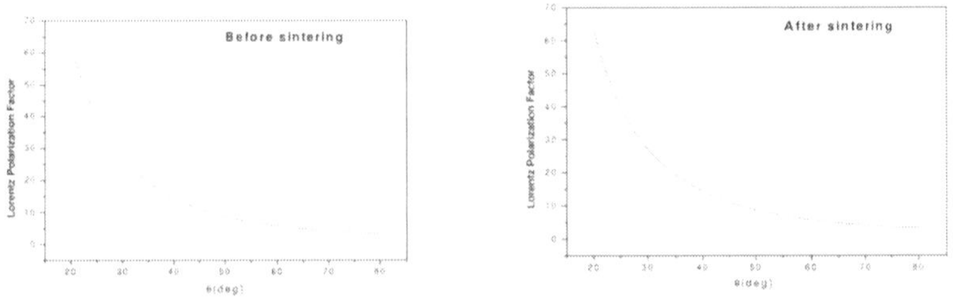

(a) Sample-1 (b) Sample-2

Figure 2. Lorentz-polarization factor pattern of MnBi alloys.

3.3. Fourier transformation infrared spectroscopy studies:

The absorption peaks of MnBi alloys are in the Fourier Transform infrared Spectroscopy data. This is shown in figure 3 for both samples in which before sintering and after sintering.

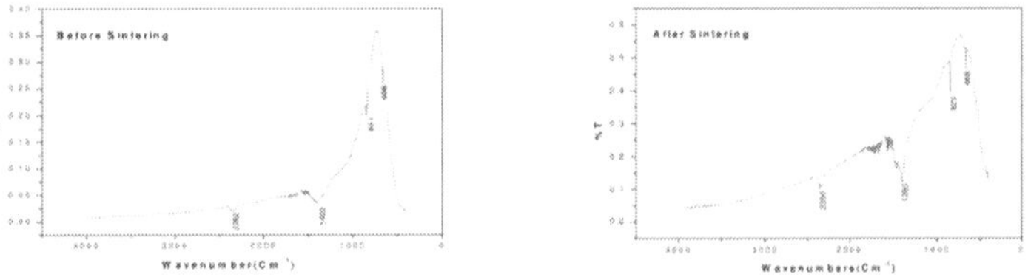

(a) Sample-1 (b) Sample-2

Figure3. Fourier transform IR spectra of MnBi alloys.

4. Conclusion

By the analysis of the X-ray diffraction pattern, one can conclude that both the samples are in crystalline form. Particle size is decreasing as one goes from before sintering to after sintering at the same temperature. Lorentz polarization factor decreases when the angle of diffraction increases. The absorption peaks of MnBi alloys are in the Fourier Transform Infrared Spectroscopy data. The fingerprints have no two unique molecular structures produces in the same infrared spectrum.

Acknowledgement

The authors are thankful to Dr. M. Gupta, UGC CSR, Indore for XRD measurements and Dr. T. Shripathi, UGC CSR, Indore for FTIR measurements.

References

[1] Wohlfarth E P 1980 *Ferromagnetic Materials* (North-Holland).
[2] "Heusler alloy," *Encyclopedia Britannica Online*, retrieved Jan. 23, 2005.
[3] Yang J.B., Yelon W.B. and James W. J., Cai Q., Roy S. and Ali N. 2002 *J. Appl. Phys.* **91** 10
[4] Suryanarayana C and Grant N M 1998 *X-Ray diffraction: a practical approach* (Springer).
[5] Reynold R C 1986 *Clays Clay Miner* **34** 359.

Effect of chemical treatment on electrical properties of coir fibre reinforced epoxy composites

A Khan[1] and S Joshi[2]
[1] Motilal Vigyan Mahavidyalaya, Bhopal (M.P), India
[2] Institute for Excellence In Higher Education, Bhopal (M.P), India

E-mail: khanalveeera@gmail.com

Abstract: In this paper, the effect of chemical treatment on electrical properties of coir fibre-reinforced epoxy composites has been reported. For this purpose, epoxy composites reinforced with chemically treated coir fibre prepared by hand molding method. Samples were characterized for their electrical properties, such as dielectric constant (e_0), and AC conductivity (σ_{ac}), at different temperatures and frequencies. It was observed that dielectric constant increases with increase in temperature and decreases with increase in frequency from 5 KHz to 30 kHz. The peak height at the transition temperature decreases with increasing frequency. Electrical characterization of the samples has been done by impedance analyzer. Morphology of the samples has been done by SEM technique. Crystalline nature of the sample has been done by XRD analysis.

1. Introduction

Over the previous few years, natural fibers have been used as the main reinforcement for composite materials. Nowadays investigation on properties of natural fibre reinforced polymer composite is the most attractive area for researchers. Study of the properties and applications of fibre-reinforced polymer composite materials is a very fast growing area of research nowadays. The attention arises in natural fibre arises owing to their excellent performance in mechanical and electrical properties, low cost and considerable advantages of the composite material [1-3]. As natural fibre are renewable, are of no cost or of very low cost, light in weight due to which they find applications in different areas like in aerospace, electrostatic shielding, electromagnetic shielding etc.

Compared to other natural fibres coir fibre is of particular interest due to its remarkable mechanical and electrical properties. Coir fibre is one of the most widespread natural fibres found abundant in nature, mostly grown in tropical countries. Coir is one of the cheapest fibre among all natural fiber even cheaper than jute and sisal fibre [4]. The use of coir fibre as reinforcement or filler in polymer composites is important as it is inexpensive material when compared to artificial material like glass, aramid etc. The use of coir fibre as reinforcement in many polymers composite has been studied [5].

Research work has been done on the electrical properties of natural fibre reinforced epoxy composite but they never studied the effect of chemical treatment of coir fibre done by annealing method at higher temperatures. The main motive of this work is to analyze the electrical properties of chemically treated coir fibre reinforced epoxy composite at different temperatures and frequencies.

2. Experimental

2.1. *Chemical treatment of coir fibre*:

The treatment of processed coir fiber was done with nitro compounds. Ferric nitrate (Nanohydrate Extra pure, Fe $(NO_3)_3$. $9H_2O$) and ammonium chloride (NH_4Cl) was taken in the ratio 10:4 in 500 ml distilled water. Stir the mixture till homogeneous solution obtained. Submerged 100gm of processed coir fiber to it and pour 100 drops of liquid ammonia to it and left the solution for one hour. Again the mixed is dried and then fired it in a muffle furnace at 1000^0C and kept it at that temperature for 15 min. The fired sample was then powdered for their further study

2.2. *Fabrication of composite*:

Weighted amount of the treated fiber was mechanically mixed with epoxy polymer in the ratio of 10:8: .1(resin: hardener: treated fiber) until a homogenous solution was obtained. Pallets of 10 mm diameter were prepared by pouring the homogenous mixture into the mold cavity of desired diameter. The curing was done at room temperature for one day. Cylindrical rod of diameter 10mm was obtained. The rod was cut in pallets of 2mm thickness

3. Characterization of the sample

3.1. *X ray diffraction (XRD) analysis*

The XRD measurements were carried out using Bruker D8 X-ray diffractometer. The x-rays were produced using a sealed tube and the wavelength of x-ray was 0.154 nm (Cu K-alpha).

3.2. *Scanning electron microscope (SEM) analysis*

SEM images of the prepared samples were taken by JSM 6390A (JEOL Japan) at dissimilar magnification. Before examination the prepared samples were layered with gold in a vacuum coating unit

3.3. *Electrical measurements*

Capacitance (C) and tan d values were measured using a Wayne Kerr 6500B Impedance Analyzer in the temperature range from 30 to 180 ^{0}C at frequencies (5 to 30) KHz keeping the heating rate constant at 2 ^{0}C/min. Dielectric constant (e_0) of the sample has been calculated using the following relation

$$\varepsilon' = \frac{C}{C_o}$$

where C and C_O are the capacitance with and without dielectric, respectively. Value of C_o in pF is given by the following relation

$$C_0 = \frac{(0.08854)A}{d} pF$$

where A (cm^2) is the area of the electrodes and d (cm) the thickness of the sample. Ac conductivity of the material is given by the following relation

$$\sigma_{ac} = \varepsilon_o \omega \varepsilon' \tan\delta$$

where ε_o is the permittivity of free space, tan δ the dielectric dissipation factor and ω the angular frequency of the applied electric field.

4. Results and discussions

The XRD analysis of the two samples are given below 1 (a) pure epoxy composite 1 (b) chemically treated coir fibre reinforced epoxy composite

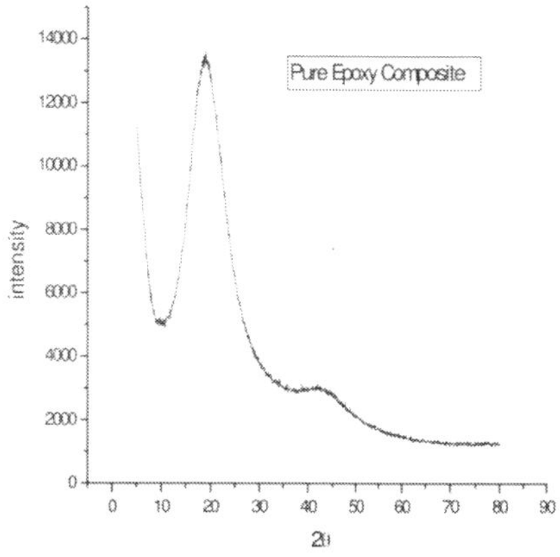

Figure 1(a). XRD of pure epoxy composite.

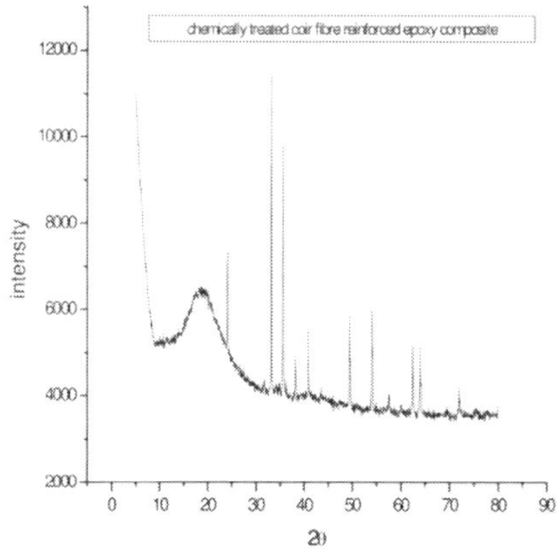

Figure 1(b). XRD of chemically treated coir fibre reinforced epoxy composite

Fig 1(a) and 1(b) shows the XRD structure of pure epoxy composite and chemically treated coir fibre reinforced epoxy composite.]. XRD structure of the composite shows their crystalline nature as it is reported by Khan et al. [6].

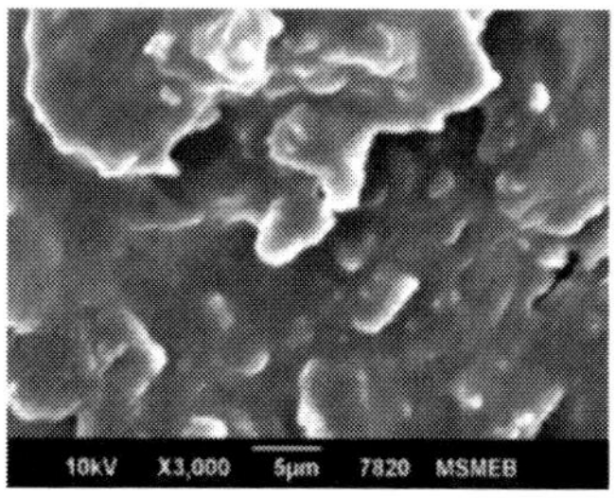

Fig.2 (a) Pure epoxy composite

Fig.2 (b) Chemically treated coir fibre reinforced epoxy composite

Figure 2(a) and 2(b) show the pictorial view of pure epoxy composite and chemically treated coir fibre reinforced epoxy composite. It has been found that chemically treated fibre is not fully bonded but is in poor contact with the matrix. Chemically treated fibre could not adhere with the epoxy matrix and therefore interfacial bonding is poor. In addition to this, the homogeneity of the sample is found to be lost, as the composite appears segregate into dissimilar phases. However, the hardness appears on physical observation. The surface of the composite is not smooth representing that the compatibility between fibers and epoxy matrices is poor.

Figure 3(a) shows that dielectric constant increases with increase in temperature from 30°C to 180°C and there is a decrease in dielectric constant with increase in frequency from 5 to 30 KHz. the dielectric constant of pure epoxy have merged at lower temperatures. Fig 3(b) shows there is a increase in dielectric constant on increasing the temperature from 30°C to 180°C and there is a decrease in dielectric constant on increasing the frequency from 5 to 30 KHz. The dielectric constant of chemically treated coir fibre reinforced epoxy composite is found to be greater than pure epoxy composite. This increase in dielectric constant is due to the agglomeration of chemically treated coir fibre in the epoxy composite. Same work has been reported by some researchers [7].

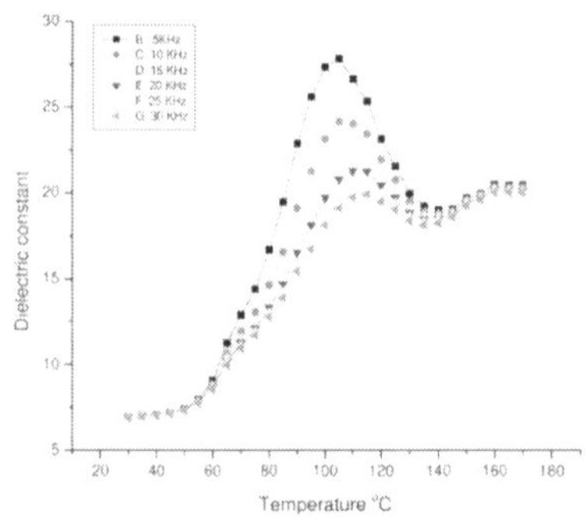

Figure 3(a). The variation of dielectric constant (ε_o) with temperature (T) for pure epoxy, measured at 5 to 30KHz.

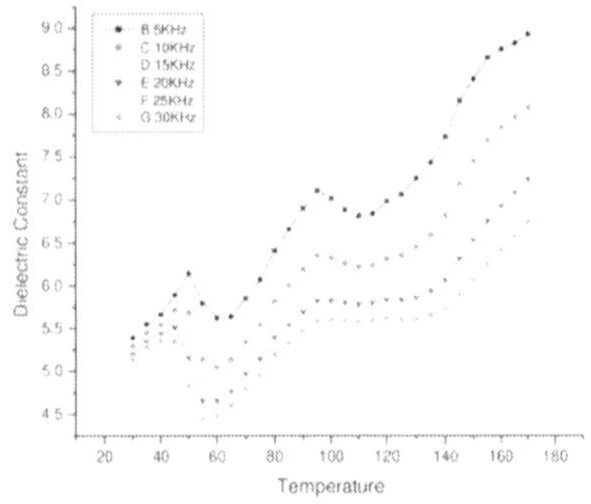

Figure 3(b). Variation of dielectric constant (ε_o) with temperature (T) for chemically treated coir fibre reinforced epoxy composites measured at 5to30KHz.

From figures 4(a) and 4(b) it has been found that the AC conductivity of chemically treated coir fibre reinforced epoxy composite is more than the pure epoxy composite at all frequencies due to the presence of chemically treated coir fibre in the pure epoxy composite

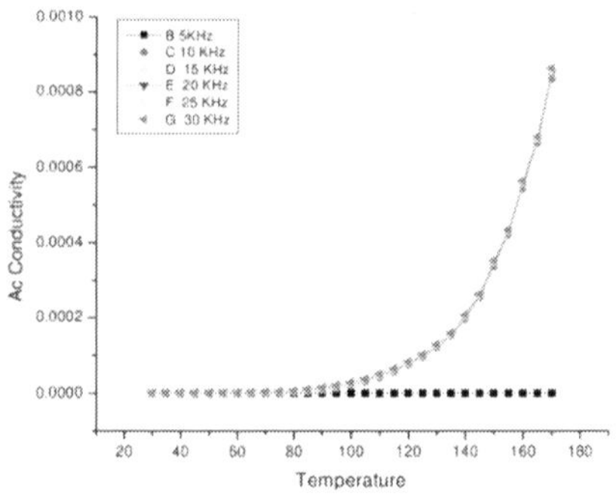

Figure 4(a). Variation of AC conductivity (σ_{ac}) with Temperature (T) for pure epoxy , measured at 5to30KHz

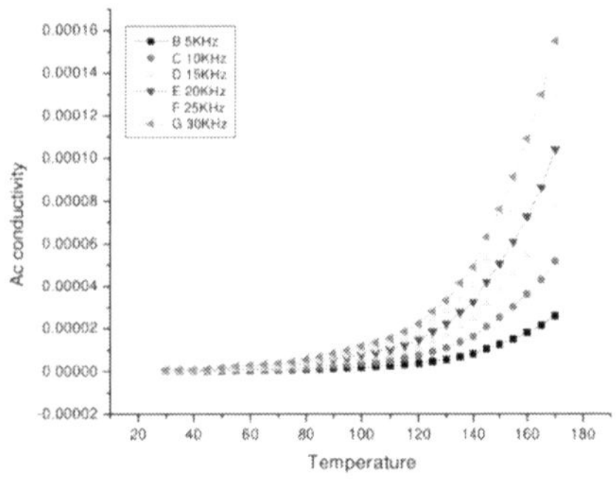

Figure 4(b). Variation of AC conductivity (σ_{ac}) with Temperature (T) for chemically treated coir fibre reinforced epoxy composite measured at 5to30KHz

References

[1] Sreekumar P A, Saiter J M, Joseph K, Unnikrishnan G and Thomas S 2012 *Compos A* **43** 507
[2] Pathania D and Singh D 2009 *Int J Theor Appl Sci* **1** 34
[3] Ronga M Z, Zhang M Q, Liu Y, Yang G C and Zeng H M 2001 *Compos Sci Technol* **61** 1437
[4] Devi L U, Bhagawan S S and Thomas S 1997 *J Appl Polym Sci* **64** 17395
[5] Joseph S, Sreekala MS, Oommen Z, Koshy P and Thomas S 2002 *Compos Sci Technol* **62**1857
[6] Khan A, Ahmed M, Joshi S and Said A F, 2014 *American J of Mech. Eng. and Automation.* **1** 1
[7] Trihotri M, Jain D, Dwivedi U K, Khan F, Malik M M and Qureshi M S 2013 *Polym. Bull.* **70** 3501

Extended X- ray absorption fine structure study at the K-edge of copper in mixed ligand complexes having benzimidazole as one of the ligands

V K Hinge[1], S K Joshi[2], N Nitin Nair[3], Vikram Singh Verma[3], B D Shrivastava[3], J Prasad[4] and K Srivastava[4]
[1]Physics Department, Ujjain Engineering College, Ujjain 456010, India.
[2]S. V. Govt. Commerce College, Ratlam 457001, India.
[3]School of Studies in Physics, Vikram University, Ujjain 456010, India.
[4]Chemistry Department, University of Allahabad, Allahabad 211002, India.

E-mail: vijayhinge1@gmail.com

Abstract. Extended X-ray absorption fine structure (EXAFS) spectra have been studied at the K-edge of copper in some of its biologically important complexes, viz., $[Cu(BzImH)_4X_2]$ and $[Cu(BzIm)_2]$, where X= Cl, Br, $1/2SO_4$, ClO_4, NO_3, and BzIm = Benzimidazolato anion. The spectra have been recorded using a bent crystal 0.4 m Cauchois-type transmission spectrograph. The positions of EXAFS maxima and minima have been used to determine the bond lengths in the complexes with the help of three different methods, namely, Levy's, Lytle's and Lytle, Sayers and Stern's (L.S.S.) methods. The phase uncorrected bond lengths have also been determined from Fourier transforms of the experimental spectra. The results obtained from these methods have been discussed and it has been found that the results obtained by L.S.S. method are comparable with the results obtained by Fourier transformation method and that these two methods give phase uncorrected bond lengths.

1. Introduction

The X-ray absorption spectra at the K-edge of copper have been studied in some of its biologically important complexes, viz., $[Cu(BzImH)_4X_2]$ and $[Cu(BzIm)_2]$, where X= Cl, Br, $1/2SO_4$, ClO_4, NO_3, and BzIm = Benzimidazolato anion. All the complexes studied have benzimidazole as primary ligand. Benzimidazole is extensively used in industrial processes as corrosion inhibitor [1] for metal and alloy surfaces particularly that of copper. Earlier [2] we have reported the measurements on X-ray absorption near edge structure (XANES), e.g., chemical shift, energy of the principal absorption maximum and edge-width in all the complexes. The present paper reports the results of the studies of extended X-ray absorption fine structure (EXAFS) of these complexes.

2. Experimental

All the complexes were prepared according to the standard methods reported in literature [3] and their purity was checked. A sealed Machlett X-ray tube with tungsten target was employed as X-ray radiation source and was operated at 17 kV and 15 mA. The X-ray absorption spectra at the K-edge of copper were recorded on a bent crystal 0.4 m Cauchois-type transmission spectrograph employing a mica crystal, oriented to reflect from (100) planes, for dispersion. The digital records of the K absorption spectra were obtained using a Carl-Zeiss GII microphotometer. Computer programs *Origin* and *Athena* [4] have been used for data analysis. The normalized spectra for all the copper complexes under consideration are given in Fig. 1. EXAFS have been observed on the high energy side of the K-absorption edge in all the copper complexes studied. In Fig.1 the maxima have been labeled as A, B, C, ... while minima by Greek letters α, β, γ, ... The values of energy (E) of maxima and minima and the corresponding k values (where $k= 0.263(E - E_0)^{1/2}$) are given in Table 1. Other experimental details are the same as the reported earlier [2].

Table 1. EXAFS parameters for Cu(II) complexes

Structure	Q	n	$Cu(BzImH)_4(NO_3)_2$		$Cu(BzImH)_4(ClO_4)_2$		$Cu(BzImH)_4SO_4$		$Cu(BzImH)_4Cl_2$		$Cu(BzImH)_4Br_2$		$Cu(BzIm)_2$	
			E	K	E	k	E	K	E	k	E	k	E	k
A	2.04	0	12.	1.80	15.8	2.03	13.	1.89	12.	1.77	13.0	1.84	17.	2.1
α	-	1	40	3.24	40	3.24	40	3.24	40	3.24	35	3.03	42	3.3
B	6.04	2	71	4.32	68	4.22	63	4.07	78	4.52	65	4.13	78	4.5
β	-	3	115	5.49	124	5.71	10	5.32	12	5.71	109	5.35	120	5.6
C	12.0	4	157	6.42	163	6.54	14	6.21	16	6.58	158	6.44	163	6.5
γ	-	5	201	7.27	232	7.81	20	7.36	21	7.55	217	7.55	226	7.7

3. Results and Discussion

Bond lengths have been determined from EXAFS by four different methods as given below.

3.1 Levy's method:

Levy [5] has given a method for obtaining the bond lengths from the analysis of EXAFS. According to Levy, the maximum, labeled A should correspond to a transition from the 1s to a vacant 4p level, and the maximum, labeled B, to a kind of ionization or escape energy for the 1s electron. Furthermore, the distance ΔE, from the maximum at B to the minimum at β should be a measure of the radius R_1 of the first coordination sphere through Bragg relation $R_1= (151 /\Delta E)^{1/2}$ Å . The values of R_1 determined by this method are given in Table 2. R_1 calculated in this way represents the average bond length.

3.2 Lytle's method:

Lytle [6] has proposed a theory in which a nearly spherical polyhedron is constructed in the lattice and approximated by a sphere of equivalent volume with radius R_s .In this method photoelectron ejected as a result of absorption of X-rays are thought of as if they are in a particle in a box. Solving the Schrodinger equation, Lytle has found that an absorption maximum will occur whenever E satisfies the equation= $(h^2/ mR^2_s)Q$, where h is Planck's constant, m is the electron mass, and Q represents the zero roots of the half-order Bessel function which appear in the radial part of the solution of the wave equation. The Q values for the K-edge are 2.04, 6.04, 12.00 and 20.00. A simple data analysis scheme is obtained, when E is plotted against Q, where the slope of the straight line contains R_s. If M is the slope, the following relation holds $R_s= (37.60/M)^{1/2}$Å . The plot of E vs Q for the complexes studied are found to be linear and are shown in Fig 2. The values of R_s determined from these plots are given in Table 2. It may be remarked here that the interatomic distance obtained by this method has to be multiplied by a factor greater than unity and appropriate to the geometry of the system to obtain the correct bond length.

3.3 Lytle, Sayers and Stern's (L.S.S.) method:

In Lytle, Sayers and Stern's (L.S.S.) method [7], plot of n vs k for the different maxima (n= 0, 2,4, ...) and minima (n= 1, 3, 5, ...) should results in a straight line. n vs k graph for the complexes studied are linear as shown in Fig. 3. The slope (M') of this line is given by $M'= 2(R_1- \alpha_1)/ \pi$. The value of $(R_1 - \alpha_1)$ can be obtained from the slope of the n vs k plot. Once α_1 is evaluated for a standard,

R_1 for unknown materials can be determined. It is found that for chemically similar systems, the value of α_1 remains more or less the same. In the present work, the value of α_1 has been taken as 0.43 [8]. The values of R_1 thus obtained by this method are given in Table 2. The distance R_1 obtained from this method is phase uncorrected bond length.

3.4 Fourier transform method:

The fine structure amplitude function, $\chi(E)$, of an EXAFS spectrum has been determined using the relation $\chi(E)=[\mu(E)- \mu_0(E)]/\Delta\mu_0(E)$, where $\mu(E)$ is the experimental absorption coefficient, $\mu_0(E)$ is the atomic contribution to the absorption coefficient and $\Delta\mu_0(E)$ is the measured absorption jump. This $\chi(E)$ has been first converted to $\chi(k)$, and then $\chi(k)$ has been Fourier transformed (Fig. 4). The position of the first peak in the Fourier transform corresponds to the distance between the absorbing atom and its nearest neighboring atoms, i.e., the radius of first coordination sphere. However, the bond length thus obtained is less than the actual bond length, because it is phase uncorrected. The values for the phase uncorrected bond lengths for all the complexes are given in Table 2.

Table 2. Values of bond lengths (in Å ± 0.1)

Complex No.	Complex	Levy's method R_1	Lytle's method R_s	LSS method		Fourier Transform method (R_1 phase uncorrected) $R_1 - \alpha_1$
				$R_1 - \alpha_1$	R_1	
1	$Cu(BzImH)_4(NO_3)_2$	1.85	1.61	1.42	1.85	1.48
2	$Cu(BzImH)_4(ClO_4)_2$	1.64	1.59	1.35	1.78	1.28
3	$Cu(BzImH)_4SO_4$	1.83	1.67	1.46	1.89	1.42
4	$Cu(BzImH)_4Cl_2$	1.81	1.56	1.35	1.78	1.25
5	$Cu(BzImH)_4 Br_2$	1.85	1.60	1.36	1.79	1.23
6	$Cu(BzIm)_2$	1.89	1.60	1.41	1.84	1.39

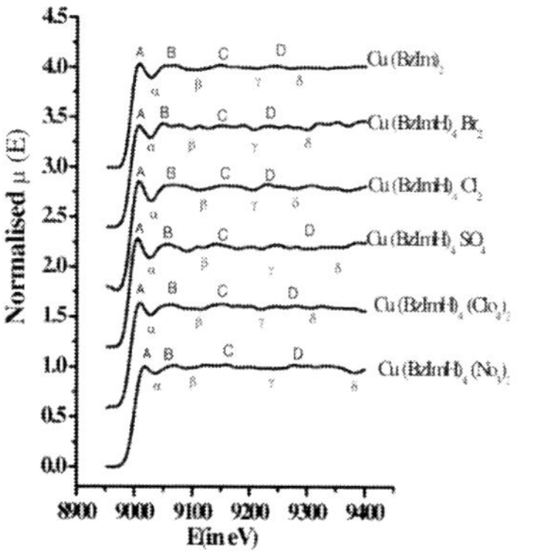

Figure 1. The K-absorption edge of copper along with the fine structure for the copper complexes. The different spectra have been shifted vertically for better presentation.

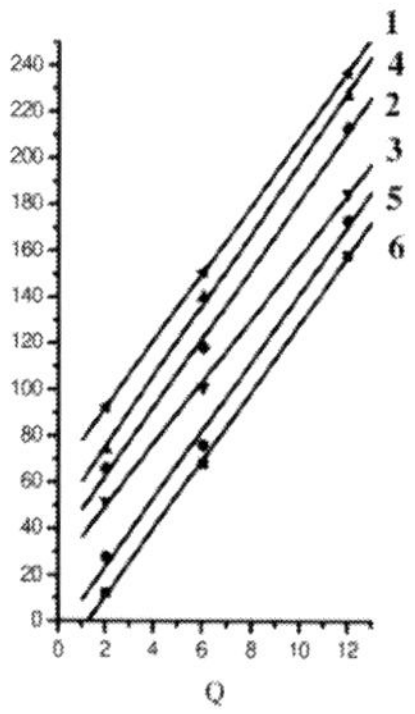

Figure 2. E vs Q plot (Lytle plot) for the copper complexes. The different spectra have been shifted vertically for better presentation.

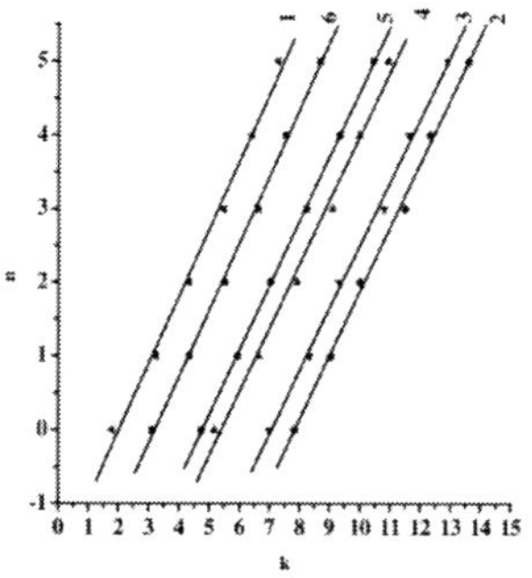

Figure 3. n vs k (in $Å^{-1}$) plot (LSS plot) for the copper complexes. The different spectra have been shifted horizontally for better presentation.

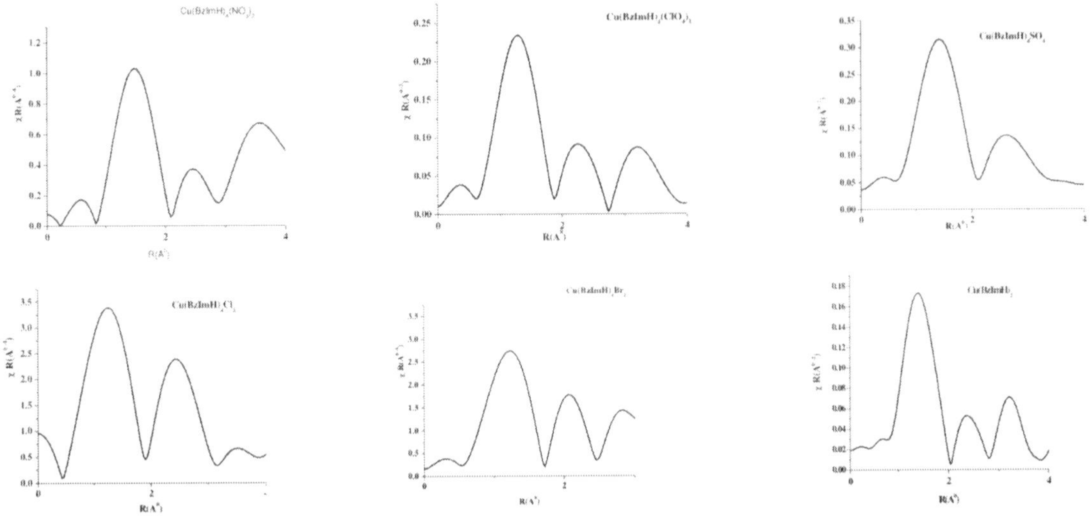

Figure 4. Fourier transformed EXAFS spectra for the complexes

4. Conclusions

A comparison of the values of R_s are found to be slightly lesser than the corresponding bond lengths obtained by Levy's method. This may be ascribed to the fact that Levy's method gives the radius of coordination sphere directly where as Lytle's method does not. In the later method, the interatomic spacing is obtained by multiplying R_s with a factor appropriate to the geometry of the system. The values of the phase uncorrected bond length obtained from LSS method are comparable with the values of phase uncorrected bond length obtained from Fourier transformation method with an error of ±0.05 Å.

References:

[1] Materazzi S, Curini R and D'Ascenzo G 1996 *Thermochimica Acta* **286** 1
[2] Hinge V K, Joshi S K, Shrivastava B D, Prasad J and Srivastava K 2011 *Ind. J. Pure Appl. Phys.* **49** 168
[3] Goodgame M and Haines L I B 1966 *J.Chem. Soc. A* 174
[4] Ravel B and Newville M 2005 *J. Synchrotron Rad.* **12** 537
[5] Levy R M 1965 *J. Chem. Phys.* **43** 1846
[6] Lytle F W 1966 *Adv. X-ray Annl.* **9** 398
[7] Lytle F W, Sayers D E and Stern E A 1975 *Phys. Rev. B* **11** 4825
[8] Martens G, Rabe P, Schwentner N and Werner A 1978 *Phys. Rev. B* **17** 1481

Zn doping induced ferromagnetism in NiO

Ridhi Master[1], Mayura Karandikar[2], Amruta Jambhale[2], Tarachand[1], D. K. Shukla[1], R. J. Choudhary[1], and D.M. Phase[1*]

[1]UGC- DAE Consortium for Scientific Research, University Campus, Khandwa Road, Indore - 452 017, India.
[2]Department of physics, Fergusson College, Pune - 411004, India

E-mail: dmphase@csr.res.in

Abstract. Here we report the effect of Zn doping in NiO, in thin film form. Thin film with the composition $Ni_{0.98}$ Zn $_{0.02}$O has been prepared on [001] oriented Al_2O_3 substrate using pulsed laser deposition technique. X-ray diffraction, SEM, and magnetic measurements were performed to investigate the properties of this material. XRD study revealed oriented growth of the film along the [111] direction. To characterize the magnetic behavior of the film, magnetization vs temperature (M-T) and magnetization vs field (M-H) measurements were performed. Anomalous behavior in M-T measurement has been observed in the low temperature region. At the temperature ~ 70 K, Zn doped NiO thin film is showing antiferro to ferro type of transition. This is in contrast to the pure NiO, which remains antiferromagnetic at low temperatures. Observation of isothermal magnetization hysteresis behavior with finite moment confirms ferromagnetism in Zn doped NiO at low temperatures.

1. Introduction

The NiO, a transition metal oxide, has attracted considerable interest because of its special electrical, optical and magnetic properties. NiO adopts NaCl structure and is an antiferromagnetic insulator at room temperature [1]. Stoichiometric NiO is a Mott insulator, but one can increase its p type conductivity by introducing Ni^{2+} vacancies or by doping with cations[2]. Electrical and magnetic properties of the NiO in thin film form strongly depend upon the deposition parameters like substrate temperature and oxygen content during the deposition. Therefore, selection of the proper substrate, growth condition and deposition technique play crucial role in the synthesis of NiO in thin film form. In the past NiO thin films have been fabricated on a variety of substrates by various deposition techniques like pulsed laser deposition, sputtering, chemical methods, spray pyrolysis methods etc [3-5]. It has been reported that as compared to other substrates, film grown on insulating substrates like TiO, $SrTiO_3$ and Al_2O_3 is more likely to be epitaxial in nature and contain less surface defects [6].

The NiO thin films find various scientific and technological applications. It can also be used as an exchange biased layer. One of its most important uses could be its applicability for spintronic application by doping with a suitable impurity. There has been various reports on the effort to develop NiO based diluted magnetic semiconductor (DMS) system by adding magnetic impurity/atom but there are very few reports on the doping effect of non magnetic impurity in NiO.

In the present work we have used pulsed laser deposition (PLD) technique to grow the thin film with the composition $Ni_{0.98}$ Zn $_{0.02}$O. In PLD one can easily tune the parameters to grow the stoichiometric single phase compounds. In this paper we report effect of the Zn doping on the structural and magnetic properties of the Zn doped NiO thin film grown on Al_2O_3 substrate.

2. Experimental details

Thin film with the composition $Ni_{0.98}Zn_{0.02}O$ has been prepared using pulsed laser deposition technique using a single phased composite target made out of high purity NiO and ZnO powders. The single phased target was prepared by the standard solid state reaction technique. Final sintering of target material was performed at 1000°C in air at atmospheric pressure for 24 hrs. [0001] oriented single crystal Al_2O_3 was used as a substrate. A KrF excimer laser source (λ = 248nm, pulse width = 20 ns) was used to ablate the target. During deposition, the laser pulse repetition rate was set at 10Hz. Substrate to target distance was kept at 5 cm. Deposition was carried out under oxygen partial pressure of $1*10^{-4}$ Torr at an optimized substrate temperature (Ts= 450 °C). Nominal thickness of the grown film is ~ 80 nm

To study the structural properties of the grown thin film X-ray diffraction (XRD) was performed in Θ-2Θ geometry using Cu Kα source. Magnetic behavior of the thin film was recorded using SQID vibration magnetometer from Quantum Design.

3. Results and discussions

Prior to deposition of the thin film, crystal structure of bulk sample was examined by XRD. Figure 1 presents the XRD pattern of the bulk sample which has been used as a target material. From the XRD pattern it is clear that sample is single phase and do not contain any impurity phases. XRD peak positions matches very well with the values given in JCPDS database (78-0643) for the cubic NiO. Inset of the figure shows the scanning electron microscopy (SEM) image of the bulk target. Quality of the SEM image ensures that prepared target is quite homogenous and very well sintered, suitable for PLD growth process.

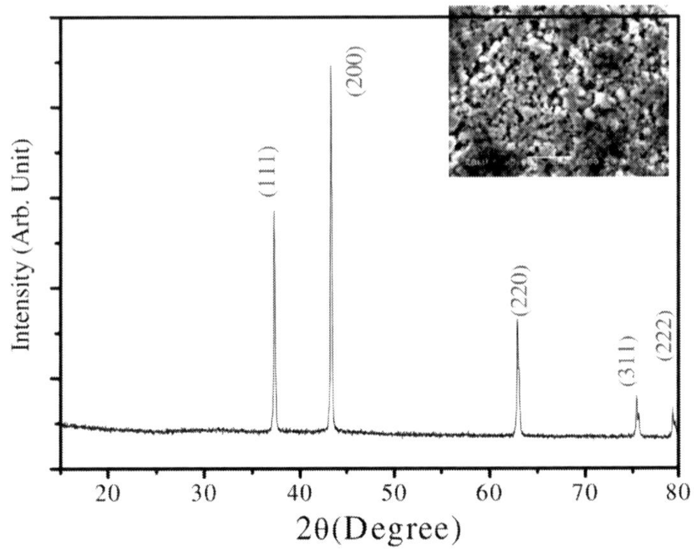

Figure 1. X-ray diffraction pattern of the 2% Zn doped NiO bulk target sample sintered at 1000 °C. Inset shows the SEM image of same sample

To study the crystal structure of deposited thin film, XRD pattern has been recorded, shown in Figure 2. XRD pattern of the film shows the peaks corresponding to (111) and (222) planes of NiO along with substrate peaks. No extra peaks corresponding to any other phase or orientation were observed in the spectrum. This confirms the single phase nature and [111] oriented growth of the film. We have calculated particle size of the film using Scherrer's formula, which is ~ 20nm. Calculate value of lattice parameter is a= 4.15 Å, which is slightly lower than pure NiO.

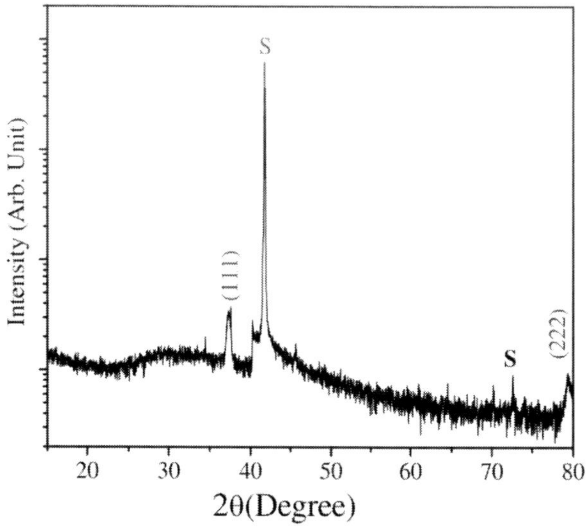

Figure 2. X-ray diffraction pattern of the $Ni_{0.98}Zn_{0.02}O$ thin film grown on Al_2O_3 substrate.

Figures 3 and 4 illustrate the magnetic property measurements of the grown thin film. Temperature dependent magnetization behavior of $Ni_{0.98}Zn_{0.02}O$ thin film taken at 500 Oe measuring magnetic field under zero field cooled (ZFC) and field cooled (FC) protocol is shown in figure 3. In FC data, value of the magnetization is seen decreasing with the decreasing temperature but at temperatures below 70 K reverse behavior was observed. This anti- ferromagnetic to ferromagnetic type of transition is prominent in FC data. Observed bifurcation in FC and ZFC data is a feature which is common to the systems having magnetic anisotropy and grain boundary effects.

Figure 4 shows low temperature (10K) M-H behavior of the film after subtracting the substrate contribution. Observed hysteresis, with the finite coercivity of ~ 180 Oe, in film confirms the ferromagnetic like behavior at low temperatures. It has been reported in literature that pure NiO thin films can show week ferromagnetism even at room temperature. The presence of Ni clusters into NiO lattice, uncompensated spins, and lattice distortion are some of the reasons reported to explain magnetic hysteresis in NiO [7]. However, observation of magnetic hysteresis here in Zn doped NiO film at low temperature may be attributed to the presence of Zn which causes the distortion in NiO as evident from the lattice parameter from XRD pattern. Such distortion can break the anti ferromagnetic interaction in NiO and lead to canted anti ferromagnetic interaction. Similar behavior has been reported for Zn doped NiO nano clusters [8]

4. Summary

Pulsed laser deposition technique was used to grow the $Ni_{0.98}$ $Zn_{0.02}O$ thin film on single crystal [0001] oriented sapphire substrate. XRD confirms that the prepared sample is single phase in bulk as well in thin film form and do not contain any impurity. Grown film is oriented along the [111] direction. Magnetic measurements revealed a antiferro to ferro type of transition at ~ 70 K. Low temperature ferro magnetic behavior were further confirm by isothermal hysteresis measurement performed at 10

K. Present results indicate that presence of Zn in NiO helps in field induced spin canting of the antiferromagnetic sub-lattices which is giving rise to observed magnetic hysteresis. For the better understanding of observed anti ferro to ferro magnetic transition at low temperatures more measurements are under process.

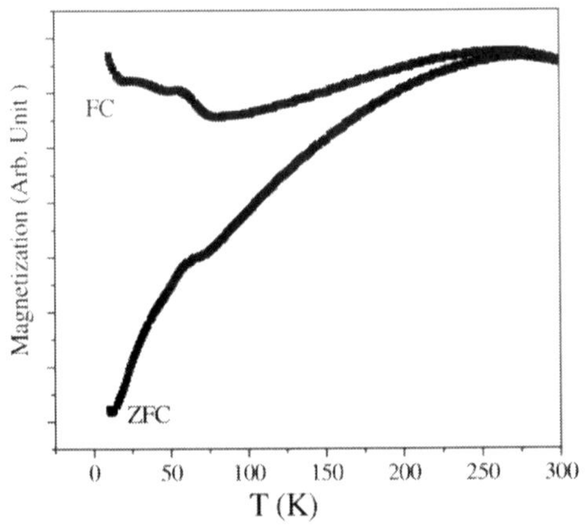

Figure 3. Zero field cooled and field cooled magnetization vs temperature behaviour of the $Ni_{0.98}Zn_{0.02}O$ thin film. Measuring field was 500 Oe.

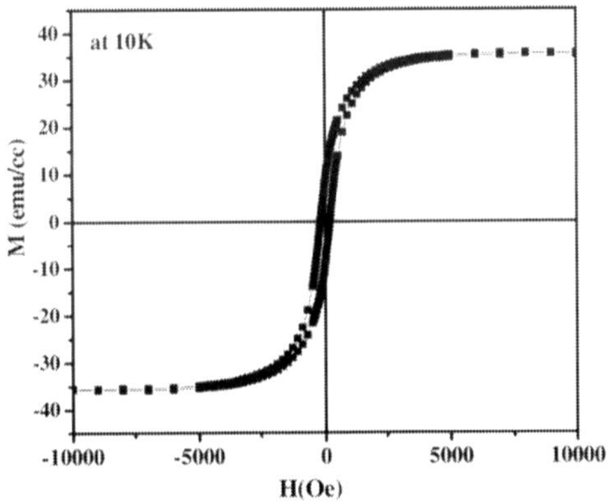

Figure 4. Isothermal magnetic hysteresis loop of the $Ni_{0.98}Zn_{0.02}O$ thin film at 10K.

Acknowledgments
Authors are thankful to Dr M. Gupta for XRD measurement. One of us RM would like to acknowledge CSIR, India for financial support

References

[1] Fujii E, Tomozawa A, Torii H and Takayama R 1996 *Japan. J. Appl. Phys* **35** L328.
[2] Wu J, Nan C W, Lin Y and Deng Y 2002 *Phys. Rev. Lett.* **89** 217601
[3] Nandy S, Goswami K S and Chattopadhyay K 2010 *Appl Surf Sci* **256** 3142
[4] Wang H, Wang Y and Wang X 2012 *Electrochem Commun* **18** 92
[5] Xia X, Tu J, Zahang J, Wang X, Zahang W and Huang H 2008 *Sol Energy Mater Sol Cells* **92** 628
[6] Lee Ju Ho, Kwon Yong Hun, Kong Bo Hyun, Lee Jeong Yong and Cho Hyung Koun 2012 *Cryst Growth Des.* **12** 2495
[7] Wongsaprom K and Maensiri S 2013 *Chiang Mai J. Sci.* **40** 99
[8] Peck M A , Huh Y, Skomski R, Zhang R, Kharel P, Allison M D, Sellmyer D J and Langell M A 2011 *J. Appl. Phys.* **109** 07B518

X-ray near edge studies on copper (II) carbamide complexes

Ashutosh Mishra, Pradeep Sharma, Pramod Malviya and J. Bhale
School of Physics, Devi Ahilya University, Indore-452001 India

Abstract: Synthesis of metal copper (II) complexes have been carried out by the chemical root method. The X-ray K- absorption near edge measurements were carried out and the spectra have been recorded at the K-edge of Cu using the dispersive beam line at 2.5GeV Indus-2 synchrotron radiation source RRCAT (Raja Ramanna Center for Advance Technology), Indore, India.

1. Introduction

The present paper includes the X-ray absorption of copper (II) micro cyclic carbamide complexes. There has been a lot of interest in synthesis structure and properties of nitrogen donor ligands particularly carbamide, thiocarbamide and its derivatives due to their wide application as pharmaceuticals [1] and in wood protection [2-5].Metal complex with ligand systems containing oxygen and nitrogen donor atoms are very important class of coordination compound. It is known that the carbamide complex of copper plays an important role [6].In this paper describes X-ray studies of the five copper (II) Carbamide complexes : Copper (II) Carbamide Bromide {Cu (II) UBr}[Cu (NH_2CONH_2)]Br_2, Copper (II) Carbamide Chloride {Cu (II) UCl} [Cu (NH_2CONH_2)] Cl_2, Copper (II) Carbamide Nitrate {Cu (II) UN} [Cu (NH_2CONH_2)] $(NO_3)_2$, Copper (II) Carbamide Sulphate {Cu (II) US} [Cu(NH_2CONH_2)] $(SO_4)_2$ and Copper (II) Carbamide Acetate {Cu (II) UA}[Cu(NH_2CONH_2)] $(CH_3COO)_2$.

2. Experimental

All complexes are prepared by chemical root method. X-ray absorption spectra of carbamide ligand copper complexes at the K-edge of copper have been recorded at the recently developed EXAFS beam line set-up at the Indus-2 synchrotron source at RRCAT, Indore, India.

3. Results and discussion

Fig. 1 shows that copper K –absorption edges and for five complexes studied. The energy of the K-absorption edge E_K and the energies of the principal absorption maximum E_A of copper in the various complexes are given in Table 1. The chemical shifts (in eV) of the K-absorption edge of copper in these complexes with respect to the metal edge are also given in Table 1 and it has been found that. K-absorption edge of copper is shifted towards the high energy side in the five complexes.

3.1 Chemical shifts

The shift of the X-ray absorption edge (i=K, L, M…) [7] of an element in a compound /complex with respect to that element is written as: $\Delta E_K = E_{K\,(complex)} - E_{kK\,(metal)}$.

The chemical shift values of the present work are ranging from 8.3 to12.3 eV. Shift for the nitro complex is greater than the shift for the sulphate complex .This has been attributed to the large polarizability of nitro as compared to the sulphate. Thus; the contribution due to such a bonding towards chemical shift in the sulphate complex is more than that of nitro complex. It has been found that shift in copper complexes under investigation is towards high energy side. The chemical shift in the studied samples is of the order- [Copper(II)(Carbamide)]Dinitrate > [Copper(II)(Carbamide)]Dibromide > [Copper(II)(Carbamide)] Dichloride > [Copper(II)(Carbamide)] Diacetate > [Copper (II) (Carbamide)] Disulphate. Earlier researchers V.K. Hinge et al [8] have reported the chemical shift value of various copper (II) complexes between 8.3 to 12.3 eV. Hence on comparing values of chemical shifts of all complexes are found to have copper in oxidation state +2.

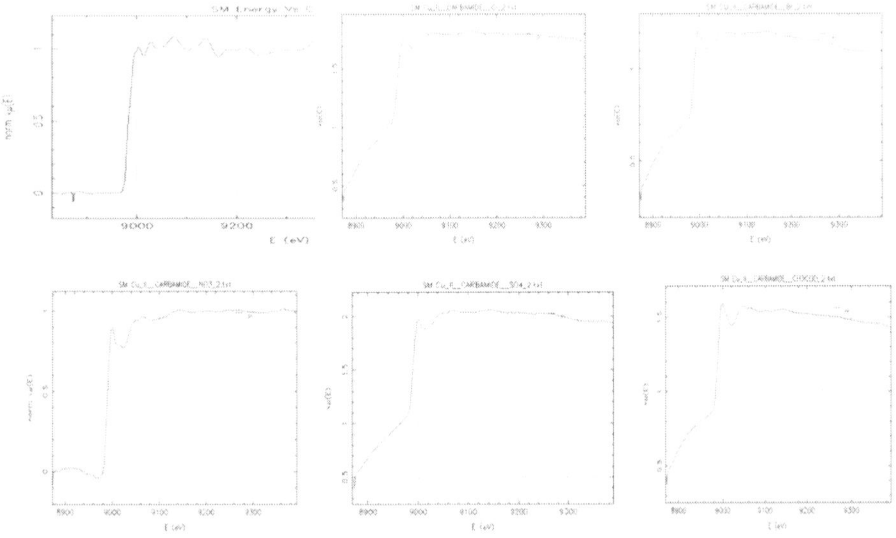

Fig.1 (a, b, c, d, e, f). X-ray K-absorption edge of copper metal and in its complexes

3.2 Shift of the Principal Absorption Maximum and Edge Width

The shift of principal absorption maximum depends upon the type of overlap between metal atom and the ligand orbital's. The greater the overlaps of the metal and the ligand orbital, the more stable are bonding molecular orbitals and hence the corresponding antibonding molecular orbitals are more unstable. Since the principal absorption maxima occurs due to the transitions from 1s orbital to the unoccupied antibonding molecular ($1s \rightarrow 5t_{1u^*}$ in octahedral), the principal absorption maxima, therefore, shifts to the higher energy side is towards the high energy side of the edge [9]. The edge widths are shown in Table 1 and related to the electro negativity differences between the central metal ion and its surrounding neighbours in the co-ordination sphere. In the present work edge width of Cu (II) complexes in Table 1 are ranging from 5 to 12 eV. Further, it may be seen that as the co-ordination stoicheiometey varies as a results of gradual replacement of metal –oxygen by metal –halogen bonds in the complexes, the edge width shows a marked decrease .In fact; edge width has decreases with decreasing covalent character of the metal –ligand bond.

3.3 Effective Nuclear Charge and Percentage Covalency

Effective charge can be defined as a total charge within a specified volume around the nucleus. Various theoretical and phenomenological methods have been proposed for the estimation of the effective nuclear charge .In the present work effective nuclear charge has been obtained from the measured chemical shift by using the semi- experimental methods by employing the procedure suggested by Nigam and Gutpa

[10]. ENC on the copper in the complexes under present study varies between 0.90 to 1.53 electrons /atom.

Chemical shift is also a measure of degree of covalency [11]. Clementi's [12] results of 1s orbital energies for different oxidation states of copper are available. Shift in 1s orbital energy has been calculated [13]. The percentage covalency of metal ligand bonding in copper complexes is the ranging from 36.34 to 49.29 and they are reported in table 1.

4. Conclusion

From these, the shift of the K-edge (chemical shift), shift of the principal absorption maximum and edge-width has been obtained. The order of chemical shift may also be taken as representative of the relative ionic character of the bonding in these complexes. The values of the chemical shifts suggest that copper is in oxidation state +2 in all of the complexes.

Table-1 -Chemical shift and other parameters of copper (II) carbamide complexes

Complexes	E_K-Edge (eV)	E_A (eV)	Chemical Shift (eV)	Shift of principal absorption maxima (eV)	Edge width (eV)	ENC Electron/ atom	Percentage covalency (%)
Copper metal	8980.5	9003.7	-	-	-	-	-
[Cu(II)Carbamide)](NO_3)	8992.8	8999.2	12.3	18	6	1.53	36.34
[Cu(II)Carbamide)] (Br)	8990.8	8996.8	10.3	16	5	1.03	42.75
[Cu(II)Carbamide)](Cl)	8989.8	9002.5	9.3	22	12	0.96	45.00
[Cu(II)Carbamide)](CH_3COO)	8988.8	8999.7	8.3	19	10	0.93	47.25
[Cu(II)Carbamide)](SO_4)	8988.5	8998.5	8.3	18	9	0.90	49.29

References

[1] Sorenson R J 1982 *In Metal Ions in Biological Systems*. Ed. Sigel H.and Dekker M. **14** 77.
[2] Bojankozlevkar Ninalahi, Simeon Makuc, Primoz Segedin 2000 *Acta Chim Slov.* **47** 421
[3] Richardson B.A 1993 *Wood Preservation* (London: Chapman & Hall).
[4] Prisakar V.I, Tsapkor V.I, Buracheva S.A, Bryke M.S and Gulya A. P 2005 *Pharm. Chem. J.* **39** 6.
[5] Premkumar T. and Govindarajan S. 2005 *World Journal of Microbiology*. **21** 479.
[6] Stojceva Radovanovic B.C. and Premovic P.I. 1992 *J. of Thermal Analysis* **38** 715
[7] Agarwal B. K.and L. P. Verma 1970 *J. Phys. C:Solid State Phys* **3** 533.
[8] Hinge V.K., Joshi S. K.and Shrivastava B. D., Prasad J and Srivastava K 2012 *J. Phys. Conf. Ser.* **365** 012029 .
[9] Rajput O. P., Nigam A. N. and Shrivastava B. D. 1984 *X-ray Spectrom.* **13** 156
[10] Nigam A. K. and Gupta M.K. 1974 *J. Phys. F: Metal Physics* **4** 1084.
[11] Agrawal B. K. and Verma L. P. 1970 *J. Phys. C* **3** 535.
[12] Clementi E. 1965 *IBM J. Res. Dev. Suppl.* **9** 2.
[13] Khan A.N. and Leydekar A. J. 1964 *Phys. Rev. A* **135** 1321.

Journal of Physics: Conference Series **534** (2014) 012027 doi:10.1088/1742-6596/534/1/012027

Synthesis and X-ray diffraction study of new copper (II) complexes of α-aminonitrile derived from P- methoxybenzaldehyde with aromatic amine.

Pradeep Sharma[1], Jaishree Bhale[2,5], Ashutosh Mishra[3] and Pramod Malviya[4]
[1]Govt. Holkar Science College, Indore(M.P), India
[2]Shree Cloth Market Institute of Professional Studies, Indore (M.P), India .
[3]School of Physics, Devi Ahilya University, Indore (M.P),India
[4] Govt.College, Nagda (M.P),India

Email: bhalejaishree@gmail.com

Abstract: New metal complexes of transition metal ion Cu (II) of the ligand HL = (P-methoxy anilino)-P- methoxy phenyl acetonitrile were synthesized. The ligand was prepared by chemical root procedure which included the reaction of P-methoxybenzaldehyde with P-anisidine. The ligand HL act as primary ligand and sulphate, chloride, nitrate and acetate act as secondary ligand. The data obtained has been preceding using XRD data analysis program Origin 6.0 Professional. From the experimental measurements, various parameters- lattice parameter and particle size have been estimated. Particle size is found to be in the range of nanometer. The XRD analysis revealed the crystalline nature of all the complexes.

1. Introduction:

The chemistry of nitrile and α-aminonitrile compounds and their derivatives has received special attention because of their application as potential ligands for a large number of metal ions (1).Nitriles and α-aminonitrile derivatives had a biological activities (2) as herbicides (3) , pharmacological agents (4) and biological synthesis of chemical compounds by it's microbial metabolism in some organisms (5). Complexes containing more than a metal centre represent the synthetic models of ferromagnetic interaction between the metal centers which can explain oxidation- reduction processes in biological systems in addition to their catalytic and biological activities (6). Beside that, some aminonitriles were used to prepare racemic compounds (7).A search through literature reveals that there is no work has been done on the XRD of transition metal complexes of the HL. Keeping this in view, the present paper describes the results of the synthesis and the characterization- XRD of new metal complexes of some transition metal ions Cu(II) of HL [(P-methoxy anilino) –p- methoxy phenyl aceto nitrile] .

2. Experimental

2.1 Materials

All reagent used were analytical grade purity .All metal salts of Cu(II) 99% (Merck), p-methoxy benzaldehyde 99%(Merck), glacial acetic acid 99%(Merck), concentrated H2SO4 99%(Merck), ethanol absolute 99% (Fluka) were used as received from the suppliers. P-anisidine and p-anisidine sulfonic acid were supplied by Prashant group of industries. All chemicals are prepared by chemical root method.

2.2 Preparation of α-aminonitriles (HL) :

The aldehyde p-methoxy benzaldehyde 0.05 mole was dissolved in 50 ml of glacial acetic acid , p-anisidine sulfonic acid was added in small portions to bring the pH to 2 , followed by the additions of 0.05 mole of the amine. The pH was adjusted to 3-4 by adding concentrated H2SO4 drop wise . KSCN 0.05 mole was added to the mixture which was kept stirring . The end of the reaction was checked by the disappearance of the starting material (the amine) and development of a higher spot on T.L.C . The reaction mixture was poured on ice and made slightly alkaline with ammonia . The solid product was filtered , washed with water and dried .

2.3 Preparation of Cu(II) complexes of HL:

A solution of a salt of Cu(II) in absolute ethanol was added to ethanolic solution of the ligand with a continuous stirring. The molar ratio of the reactants was 2:2. Precipitation of complex took place immediately. The product was filtered off , washed several times with ethanol and dried under vacuum.

2.4 X-ray diffraction

The sample is irradiated with a beam of monochromatic x-rays over a variable incident angle range.The X-rays were produced using a sealed tube and the wavelength of X-ray as 0.154nm (Cu K-alpa). The X-rays were detected using a fast counting detector based on silicon strip technology (Bruker LynxEye detector). Interaction with atoms in the sample results in diffracted x-rays when the Bragg equation is satisfied. The X-rays were detected using a fast counting detector based on silicon strip technology (Bruker LynxEye detector).X-ray diffraction pattern have been recorded by Bruker D8 advance diffractometer at IUC, Indore.

3. Results and discussion:

XRD patterns are shown in 'figures 1'. All the samples are characterized at room temperature by X-ray diffraction using Cu Kα radiation. The diffraction pattern of complexes are recorded between 2θ ranging from 10° to 80°. The particle size of the samples is estimated using the Scherrer's formula. According to Scherrer's equation , the particle size is given by $t = 0.9\ \lambda/B\cos\theta$, where t is the crystal thickness (in nm), B is half width (in radians),θ is the Bragg angle and λ is the wavelength. The particle size corresponding to each diffraction maxima are determined from the measurement of the half width of the diffraction peak. Lattice parameter for simple cubic crystal structure is determined by $a^2= \lambda^2(h^2+k^2+l^2)\ /4\sin2\theta$. The value of Lattice parameter and the particle size are shown in 'table 1' for all the four complexes. The particle size was found to be within in the range 11.43- 51.58 nm for all the complexes.

Table-1. Lattice parameter and particle size

Complexes	Lattice parameter (Å)	Particle size (nm)
$[Cu_2(HL)_2\ (H_2O)_8](SO_4)_2$	9.36	11.43
$[Cu_2(HL)_2(H_2O)_8](Cl)_2$	9.33	26.92
$[Cu_2(HL)_2(H_2O)_8](NO_3)_2$	9.22	30.24
$[Cu_2(HL)_2(H_2O)_8](CH_3COO)_2$	10.08	51.58

4. Conclusion

The XRD pattern is indicative of their crystalline in nature which is confirmed by the main peaks positioned .The X-ray analysis reveals that the sample is cubic in phase as seen from the presence of extra peaks in XRD pattern. All the peaks match with the soft wear JCPDF.

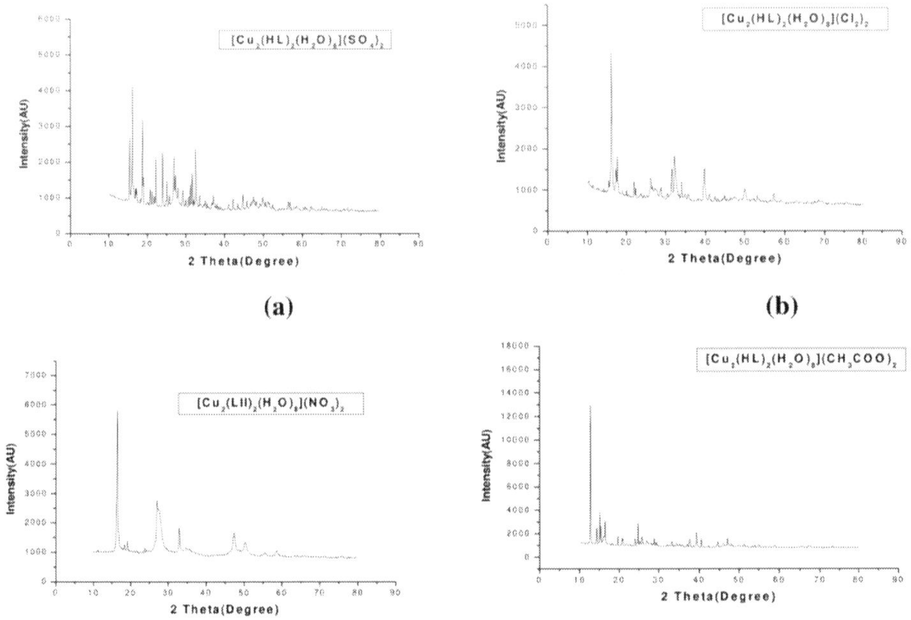

(a) (b)

(c) (d)

Figure 1: XRD of Cu(II) complexes of HL

Acknowledgment

We are grateful to P.Sharma and Dr. Mukul Gupta ,IUC, Indore for XRD studies of prepared complexes.We are also grateful to Mr.Nihar Patel (Prashant group of industries) for his valuable support.Thanks are also due to Dr.V.Mishra, SCMIPS, Indore for his valuable support during this research work.

References

[1] Nlbertin G A., Antoniutti S, Lanfrnchi M, Pelizzi G and Bordignon E 1986 *Inorg. Chem* **25** 950.

[2] Webb L J and Boxer S G 2008 *Biochemistry* **47** 1588.

[3] Helga S, Woldemer W, Juergen H, Kramer and Wilfried D D 1985 Ger. (East). 223, 624 (Cl.101N57/20), 19 Jun (1985) appl, 256,649, 14 Nov (1983); 8pp. C.F. C.A. 104 (1986) P16541d.

[4] Kalir A, Ederey H, Pelah Z,Baldormon D and Porath G 1969 M*ed J. Chem.* **12** 474.

[5] Wyatt J M and Linton E A 2007 *Cyanide Compounds in Biology* (John Wiley).

[6] a) Meyer F, H-Kryspin I, Kaifer E and Kircher P 2000 *Eur. J Inorg. Chem.* **4** 771
b) Incarvito C, Rheingold AL, Gavrilova Qin,C and Bosnich B 2001 *Inorg. Chem.* **40** 4101 .

[7] Beaufort – Droal V, Pereira E, Thery V and Aitken D J 2006 *Tetrahedron* **62** 11498

Comparative study of experimental and theoretical analysis of EXAFS data of copper complexes using FT method

A Mishra[1], S Mishra[1], P. Kekre[1] and Pankaj Choudhary[1]
[1]School of physics, Devi Ahilya University, Khandwa Road, Indore 452001
Email: amishra1960@yahoo.co.in

Abstract.
The X-ray absorption spectra at the K-edge for a series of copper mixed ligand, having hydroxypyridine as one of the ligands, have been investigated in the laboratory X-ray spectroscopic set-up. In the series only X-ray absorption near edge structure (XANES) parameters and bond length using modified Lytle, Levy's and LSS methods were calculated. In the present study the bond lengths are calculated by Fourier Transform method theoretically using IFEFFIT software and compared with experimental results.

Introduction

The transition metal Schiff base complexes of copper have been playing an important role in the development of coordination chemistry. Schiff base metal complexes have been widely studied, because of their antifungal and biological applications. [1]. In the present studies we have taken Schiff base compounds of copper (II) complexes with hydroxypyridine as ligand.

The structure extending from 40 to 1000eV from the edge is called extended X-ray absorption fine structure (EXAFS) and it gives information about the local structure of the complexes. The aim of the present investigation is to measure the spectra of copper complex, and to compare first shell inter atomic distance using the programmed called IFEFFIT.

2. MATERIAL SYNTHESIS

Preparation of Schiff Base Ligand

Standard method was adopted for the preparation of the complexes. The series of the copper (II) complexes with hydroxypyridine as ligand is shown in table 1 below.

S.N.	Name Of Complexes	Abbreviations	Mol. Formulae
1	Copper (II) sulphathiazole salicylaldimine	CU(SUTSA)	$(C_{16}H_{12}N_3O_3S_2)_2Cu$
2	Copper(II) Sulphadiazine salicylaldimine	CU(SUDSA)	$(C_{17}H_{13}N_4O_3S)_2Cu$
3	Copper(II) sulphamethoxy salicylaldimine	Cu(SUMSA)	$(C_{17}H_{14}N_3O_4S)_2Cu$
4	Copper(2,3-dihydroxypyridine)	Cu(DHP)	$(C_5H_5NO_2)_2Cu$
5	Copper(2-amino,3-hydroxypyridine)	Cu(AHP)	$(C_5H_6N_2O)_2Cu$
6	Copper(3-hydroxypyridine 2-carboxylic acid)	Cu(HPA)	$(C_6H_5NO_3)_2Cu$

Table 1: Series of copper (II) complexes hydroxypyridine as ligand.

3. EXPERIMENTAL TECHNIQUE

X-ray absorption fine structure studies were carried out using a conventional Siefert sealed X-ray Tube with Tungsten target operating 20 kV and 40 mA. After this process, the scanning of the X-ray films was completed on Carl-Ziess microdensitometer coupled with computer to convert the data into FT format. The FEFF series of programs for the calculation of x-ray absorption spectra has had a transformative impact on EXAFS analysis because of its accuracy, flexibility, and portability. Its primary use has been in supporting a path by path analysis of experimental data using auxiliary programs such as IFEFFIT, Artemis, Six Pack, etc. In this paper alternative strategies for XAFS analysis that combine FEFF are described [2]. IFEFFIT is an interactive program for XAFS Data Analysis. It runs like a command-line 'shell' in which commands are entered to process and manipulate data. IFEFFIT has a fairly high-level command language so that one can do the complex manipulation needed for XAFS analysis (such as background subtraction and Fourier transforms) with simple commands [3,4].

The EXAFS data usually analyzed on following steps.

(a) Pre edge background removal from raw data.

(b) Isolation of EXAS oscillations from the post edge data.

(c) Conversion of EXAFS oscillations from energy space to photoelectron wave vector space.

(d) Fourier Transform of the EXAFS oscillations into r-space.

(e) Extraction of individual shell contributions to EXAFS by inverse Fourier Transform.

4. RESULTS AND DISCUSSION

EXAFS analysis

The bond lengths of copper complexes were calculated using IFEFFIT method and compared with LSS, Levy's, Lytle's methods [5, 6, 7]. According to LSS method, we have determined the bond length using the slope of n vs k plot. The phase parameter $\alpha_1$and β_1, the metal ligand bond length R_1 have been estimated with expression

$$(1/2+n)\,\pi=2k\,(R_1-\alpha_1)+2\,\beta_1-\pi$$

The EXAFS curves are shown in figure1. By using the Levy's method, the bond length has been obtained from the expression

$$r=(151/\Delta E)^{1/2}$$

where, ΔE is the energy difference between first maxima and first minima [7]. Lytle method is modification of LSS method and the bond length is given by the expression

$$R=(37.60/M)1/2$$

where M is slope between E vs Q plots. In FEFFIT programming the EXAFS data were analyzed by the standard procedure. The preedge background absorption was subtracted to yield the atomic absorption spectrum of the atom of interest background above the edge was removed from the spectrum using sp line fit. The data converted to energy space and then to k space. A Fourier transform is applied to these data to convert into the r-space. The bond lengths of copper complexes calculated by IFEFFIT programming is compared with LSS, Lytle and Levy's methods as depicted in table 2.

The theoretical analysis of EXAFS of copper complexes having hydroxypyridine as ligand has been done and compared with experimental values .Following table shows the results.

S.N.	Complexes	R^{a}_{Levy}	R^{a}_{Lytle}	R^{a}_{LSS}	$R^{d}_{IFEFFIT}$
1	Cu(SUTSA)	2.4	2.2	2.5	1.2
2	Cu(SUDSA)	2.1	2	2.6	1.7
3	Cu(SUMSA)	1.8	1.8	2.5	1.5
4	Cu(DHP)	2	2	2.1	1.7
5	Cu(AHP)	2.2	2.4	2.6	2.5
6	Cu(HPA)	2.2	2.1	2.4	2.6

$R_{Levy's}$: Bond Length by Levy's method R_{Lss} : Bond Length by LSS method
R_{Lytel} : Bond Length by Lytel method $R_{IFEFFIT}$: Bond Length by FT method

Table 2: The average values of metal-ligand bond length in ($\overset{0}{A}$)$^{-1}$ for copper (II) complexes, hydroxypyridine as ligand.

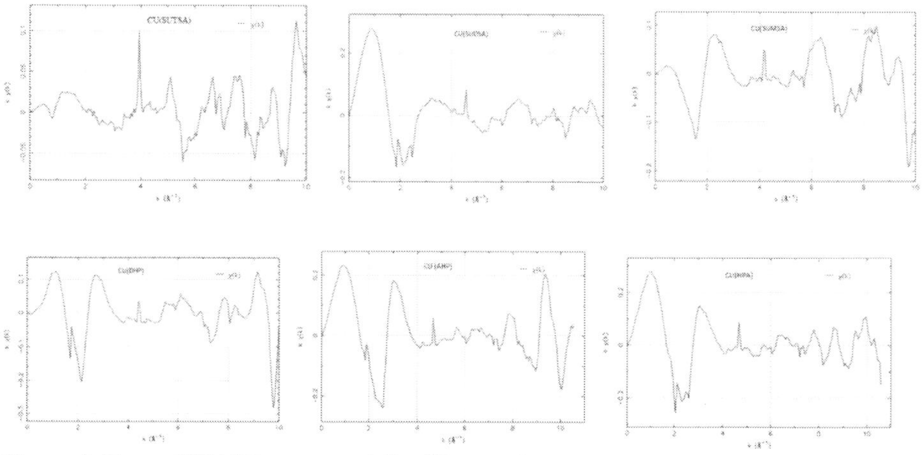

Figure 1 Shows EXAFS spectra of Cu (II) complexes

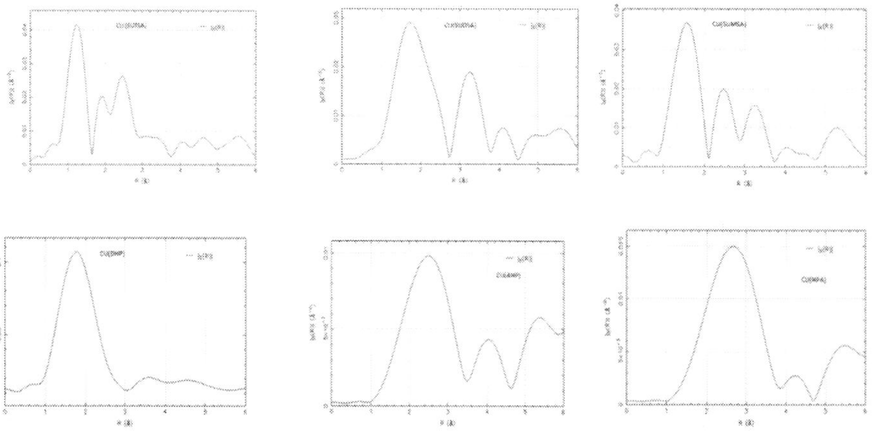

Figure 2 Magnitude of F.T. of $k^2\chi(k)$ for six copper (II) complexes hydroxypyridine as ligand

5. CONCLUSION

As is evident from the analysis that the bond lengths determined by Fourier transforming the FEFFIT programming are comparable with the bond length obtained by LSS, Levy's and Lytle methods. The theoretical and experimental values agree well witheach other. This means that the parameterized theoretical calculation of the EXAFS spectra of copper complexes described here is in good agreement with physical reality.

6. REFERENCES

[1] Kekre P A, Joshi K P and Mishra A, 1996 Indian Journal of Pure and applied Physics 34306.

[2] Alternative Strategies for Xafs Data Analysis using Feffand Mathematica 7JCPDS International Centre for Diffraction Data, 2011, ISSN 1097-0002

[3] Konningsberger D C, R. Prins (Eds.), 1998 , "X-ray Absorption Principle, Technique of EXAFS, SEXAFS, XANES" Wiley, New York,.

[4] Cullity B D, 1978, Element of X-ray Diffraction, Second Edition, Addision- Wesely Publishing Company,

[5] F. W. Lytle, D. E. Sayers, E. A. Stern, 1975, Phys. Rev. B 11, 4825.

[6] F. W. Lytle, 1966, Adv. X-Ray Anal. 9, 398.

[7] R. M. Levy, 1965, J. Chem. Phys., 43, 1846.

Journal of Physics: Conference Series **534** (2014) 012029 doi:10.1088/1742-6596/534/1/012029

EXAFS study of copper thiosemicarbazide complexes

Ashutosh Mishra, S Mishra, S Ninama and A Pritam
School of physics, Devi Ahilya University, Khandwa Road, Indore 452001, India

Email: amishra1960@yahoo.co.in

Abstract. The present paper deals with the synthesis of transition metal Schiff base complexes of copper by chemical root method. The synthesized metal complexes have been studied by extended X-ray absorption fine structure (EXAFS) technique. The metal ligand bond length have been determined using LSS, Levy's and Lytle's methods. Bond-lengths have also been determined from the Fourier transform of the EXAFS data. Bond lengths determined from these methods are comparable to each other.

1. Introduction

In the present investigation, the X-ray absorption spectra of Copper(II) complexes of thio-semicarbezide mixed ligands is done using the synchrotron radiation, i.e., on beamline BL-8 at RRCAT, Indore [1 – 3]. In organic chemistry, **semicarbazide** is a derivative of urea, where NH_2 on one side has been replaced with H_2NNH hydrazine, yielding $H_2NNHC(=O)NH_2$. A **thiosemicarbazide** is an analog with a sulfur atom in place of the oxygen atom. Semicarbazone is a ketone form of semicarbazide which is derived by the condensation reaction between a ketone (or aldehyde) and a semicarbazide.Semicarbazide products (semicarbazones and thiosemicarbazones) are known to have an activity of antiviral, antiinfective and antineoplastic through binding to copper or iron in cells.

X-ray absorption fine structure (EXAFS) refers to small amplitude oscillations in the X-ray absorption coefficient that can extend hundreds of electron volts above a core-level absorption edge. These oscillations arise from final state interference effects of backscattered photoelectrons from neighboring atoms. In contrast, an isolated atom exhibits a smooth and essentially structure less absorption background corresponding to the photoionization process induced by a core-level excitation. EXAFS features on the high-energy side of the k absorption edge, up to several hundred eV have been observed in the copper and cobalt metal complexes. Different theories developed to explain the EXAFS have already been studied.

2. Material synthesis

The copper(II) compexes were prepared by chemical root method. The procedure is described below in brief.

(a) Synthesis of ligand:

The thiosemicarbazide (0.01M) was dissolved in 10 ml of methanol in a 100 ml round bottom flask, a solution of 0.01M substituted aromatic aldehyde in methanol was added drop wise over a 10 min. period with continues stirring, after addition the reaction mixture was stirred for 3 hours at room temperature, reaction was monitored by TLC. After completion solvent was evaporated and residue was washed with cold methanol and dried at room temperature.

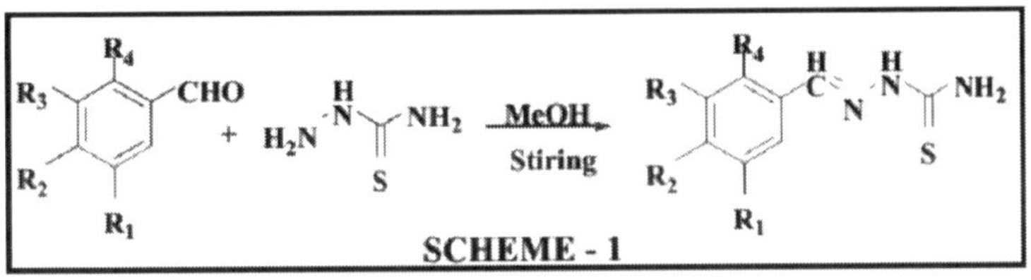

In the ligands, substituted methoxy group and anilines are at different positions.

(b) Preparation of metal complexes:

Corresponding metal acetate (0.01mol) was dissolved in min. quantity of water and then was added to the hot solution of ligand (0.02mol) in methanol (50-60 ml) .The reaction mixture was heated on 80 – 90 C for 1 hr. with constant stirring and then the reaction mixture stirred for 3 days, until a colored solid mass separated out. The ppt was filtered, washed with methanol and finally with diethyl ether and dried in vaccum.

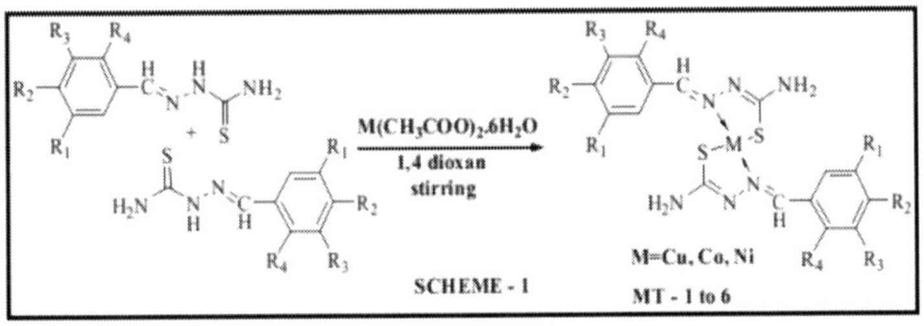

Here the series of copper(II) compounds are mentioned in table 1 which are under study.

Table 1 Copper(II) complexes with name ,mol. formulae and abbreviations.

S. No.	Complex	Molecular formula	Abbreviation
1	Cu[(2,5-dimethoxy-2-nitrobenzylidene)] thiosemicarbazide	$C_{20}H_{22}Cl_2$ $CuN_8O_8S_2$	Cu[2,5-dimethoxy]
2	Cu[(3,5-dimethoxy-2-nitrobenzylidene)]thiosemicarbazide	$C_{20}H_{22}Cl_2$ $CuN_8O_8S_2$	Cu[3,5- dimethoxy]
3	Cu[(4,5-dimethoxy-2-nitrobenzylidene)]thiosemicarbazide	$C_{20}H_{22}Cl_2$ $CuN_8O_8S_2$	Cu[4,5- dimethoxy]
4	Cu[(2-Chloroquinolin-3-yl) Methylene]thiosemicarbezide	$C_{22}H_{16}Cl_2CuN_8S_2$	Cu[2-chloroquinolin]
5	Cu[(2-chloro-8-methylquinolin-3-methylene)]thiosemicarbazide	$C_{24}H_{20}Cl_2CuN_8S_2$	Cu[3-methylene]
6	Cu[(2-chloro-8-methylquinolin-4-methylene)]thiosemicarbazide	$C_{24}H_{20}Cl_2CuN_8S_2$	Cu[4- methylene]

3. Experimental

X-ray absorption fine structure studies were carried out on beamline BL-8 coupled with computer to convert the data into IFEFFIT. The FEFF series of programs for the calculation of x-ray absorption spectra has had a transformative impact on EXAFS analysis because of its accuracy, flexibility, and portability. Its primary use has been in supporting a path by path analysis of experimental data using auxiliary programs such as IFEFFIT, Artemis, Six Pack, etc. In this paper alternative strategies for XAFS analysis that combine FEFF are described[4].

4. Results and discussion

The EXAFS curves are shown in figure 1 and the Fourier transforms are given in figure 2. The bond lengths of copper complexes were calculated using Fourier transform method and compared with LSS, Levy's, lytle's methods [5, 6, 7].

According to LSS method, we have determined the bond length using the slope of n vs k plot. The phase parameter $\alpha 1$ and $\beta 1$, the metal ligand bond length R1 have been estimated with expression

$$(n+1/2)\pi = 2k\,(R_1 - \alpha_1) + 2\,\beta_1 - \pi$$

Using the Levy's method, the bond length has been obtained from the expression

$$r = (151/\Delta E)^{1/2}$$

where, ΔE is the energy difference between first maxima and first minim [8].

Lytle method is modification of LSS method and the bond length is given by the expression

$$R = (37.60/M)^{1/2}$$

where M is slope between E vs Q plots.

In FEFFIT programming the EXAFS data were analyzed by the standard procedure. The pre-edge background absorption was subtracted to yield the atomic absorption spectrum of the atom of interest background above the edge was removed from the spectrum using spline fit. The data converted to energy space and then to k space. A Fourier transform is applied to these data to convert into the r-space . The bond lengths of copper complexes calculated by Fourier transform is compared with LSS, Lytle and Levy's methods as depicted in table 2 given below.

Table 2: The average values of metal-ligand bond length in (Å) for copper complexes

S.N.	Abbreviations	R_{Levy}	R_{Lss}	R_{Lytle}	R_{FT}
1	Cu[2,5-dimethoxy]	1.2	1.2	1.4	1.5
2	Cu[3,5- dimethoxy]	1.8	1.7	1.9	1.1
3	Cu[4,5- dimethoxy]	1.5	1.2	1.4	1.6
4	Cu[2-chloroquinolin]	2.1	2	1.9	1.9
5	Cu[3-methylene]	1.8	1.8	1.6	1.4
6	Cu[4- methylene]	1.8	1.7	1.8	1.3

$R_{Levy's}$: Bond length by Levy's method, R_{Lss} : Bond length by LSS method
R_{Lytle} : Bond length by Lytle method and R_{FT} : Bond length by Fourier transform

5. Conclusion

As is evident from the analysis that the bond lengths determined by Fourier transforming the FEFFIT programming are comparable with the bond length obtained by LSS, Levy's and Lytle methods. The theoretical and experimental values agree well with each other. This means that the parameterized theoretical calculation of the EXAFS spectra of copper complexes described here is in good agreement with physical reality.

References

[1] Bhattacharya D., Poswal A. K., Jha S. N., Sangeeta and Sabharwal S. C., 2009 *Bull. Mater. Sci.*,**32** 103.

[2] Poswal A. K., Jha S. N., Bhattacharya D., Kumar S., Jain A., Tomar B. S., Manchanda V. K. and Sabharwal S. C., 2009 *Proceeding of Workshop on Energy Dispersive X-ray Absorption Spectroscopy Scientific Opportunities and Technical Challenges*, Grenoble, France.

[3] Kumar S., Jain A., Poswal A. K., Jha S. N., Bhattacharya D., Tomar B. S., Manchanda V. K. and Sabharwal S. C., 2009, *Proceeding of Nuclear and Radiochemistry (NUCAR) Symposium*, Mumbai, p. 309.

[4] Lytle F.W., Sayers D.E. and Stern E.A. 1975 *Phys.Rev. B* **11** 4825; Teo B. K. and Lee P. A., 1979 *J. Am. Chem. Soc.* **101** 2815.

[5] Shulman R.G., Eisenberger P., Blumberg W.E, and Stombauch N A 1975 *Proc. Nat. Acad. Sci.,*

USA **72** 4003.
[6] Lytle F.W. 1966 *Advances in X-ray Analysis* **9** 398
[7] Mishra A, Awate R, Soni N, Johar K K and Shrivastava Bhakht Darshan *Ultra Science*, 2008
[8] Mishra Ashutosh, Parsai Neetu, Soni Namrata and Awate Ruchita 2010 *Ind. J. Pure Appl. Phys.* **48** 81.
[9] Mishra A., Mishra Shweta, Ninama Samrath, Sura Kamaljeet 2010 *Search and Research* **3** 46

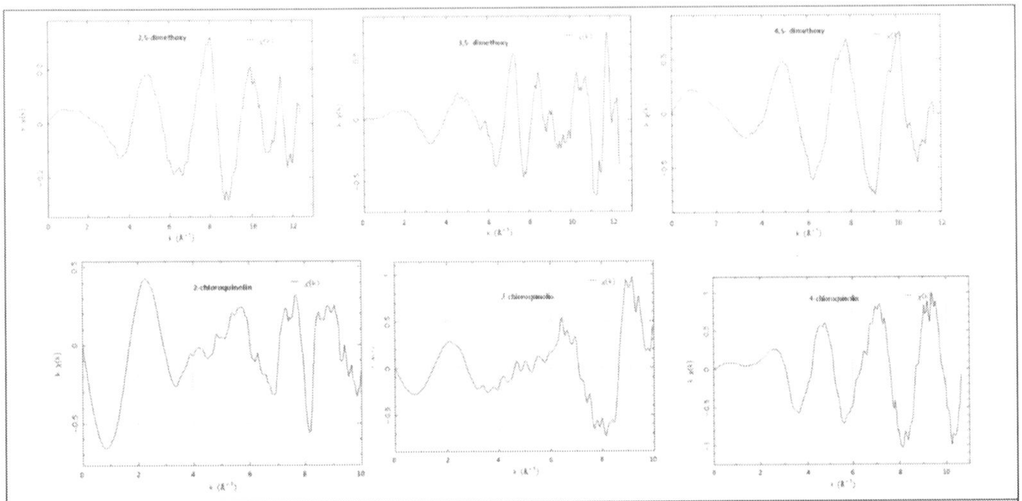

Figure 2 Magnitude of Fourier transform of k2χ(k) for Cu(II) complexes.

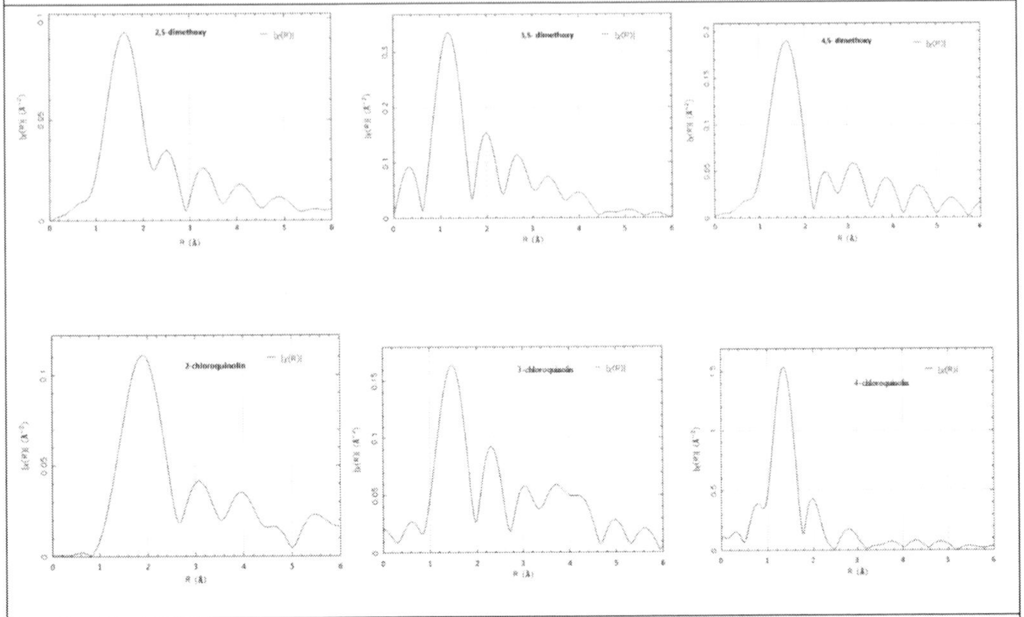

Figure 1 Shows EXAFS spectra of Cu(II) complexes.

Structural and optical properties of silver doped TiO_2 thin films for solar cell applications

V.V.S. Murty and Vinay Dashore
Physics Department, Govt. Autonomous Holkar Science College, Indore-452017, India

e-mail: drvvsmurty@yahoo.co.uk

Abstract. Electron collection efficiency and hence the device performance of a dye sensitized solar cell can be increased by localized surface plasmons (LSP) produced by the thin films of 2% silver doped TiO_2 photo electrode. These films of silver doped TiO_2 on glass substrates are prepared by pulsed laser deposition (PLD) method and have a lowest grain size of 8-31 nm. XRD, AFM and raman studies are made to characterize the surface at various annealing temperatures and annealing time durations. The lowest grain size of 8 nm of these films developed.

1. Introduction

The widespread use of solar photovoltaic technology for terrestrial and domestic applications is limited due to its high cost per unit watt. Thin film and hence nano film based solar cell is one of the best options. Nano particles (Ag or Au, with particles sizes in the range of 8 to 100 nm) are deposited on the solar cell surface, due to surface plasmon resonance excites and enhances the electromagnetic field near the metal surface where the absorption of photon increases[1]. The maximum absorption wavelength of plasmonic particles has been found to dependent on factors such as size, shape, and dielectric environment [2]. This property is exploited in photo voltaic cell to improve photo current through its spectral response. Silver doped composite thin films show that they are nano crystalline in nature, highly transparent, homogeneous with no visible pinholes and suitable for solar cell applications [3]. In the present study, silver doped TiO_2 thin films are prepared by Pulsed Laser Deposition (PLD) method on glass substrate. Dependence of various parameters is studied at different annealing temperatures suitable for solar cell applications.

2. Experimental

In our experiment thin films are prepared by PLD method. It has four different stages. It include the ablation process of the target material by the laser irradiation, the development of a plasma plume with high energetic ions, electron as well as neutrals and the crystalline growth of the fim itself [4-5]on the heated substrate.

TiO_2 Powder (Molecular Weight 79.9, Purity 99.9%) was mixed with Ag_2O(Molecular Weight 231.749, Purity 98.9%) and ground for 6 hours by wet ball mixing using acetone/alcohol as solvent and pressed to target at hand pressure. The target having dimensions of thickness 2mm and diameter 24mm were prepared after sintering 1200° C for 8 hours. Target was mounted on pellet with silver epoxy by heating for 2 hours. Target along with pellet was fixed in PLD chamber. Si substrate was arranged on sample holder in PLD chamber. Film of TiO_2 mixed with Ag_2O was deposited by using rotating arrangements in PLD chamber by PLD technique. The base pressure was bought down to $3x10^{-6}$ torr and thin film a deposited at a working pressure of $1x10^{-4}$ in an ultra pure oxygen atmosphere. The target was ablated with KrF exclaimer laser pulse (248 nm) that was synchronized with rotational motion. Films deposited under Laser power of 230 mJ with repetition rate of 10 Hz and duration of 10 minutes. During the deposition, the distance between target and substrate is maintained at 5 cm. Post annealing was performed in oxygen atmosphere at 435°C for 105 minutes with pressure of $1x10^{-4}$. Substrate was cooled for 90 minutes gradually by decreasing the temperature in presence of oxygen atmosphere by reducing the heater current of 0.5 ampere step by step interval of 10 minutes.

3. Results and discussion

To evaluate the performance of plasmon resonance effect and to enhance the efficiency of photochemical cells, thin films of TiO_2 is doped with 2% Ag (By molecular weight) particles on glass. We found that the photo response in the visible region is increased. These results suggest that plasmon resonance effect contributes to the enhancement of photocurrent, and indicates the possibility of improving the energy conversion efficiency with silver doping. The characterization of the film is to study 1. surface morphology 2. optical Properties .

3.1 XRD analysis

The XRD peaks slightly different in silicon substrate is shown in fig 1. These differences can be attributed to the differences in particle size between the samples. In order to quantify the particle size effects, we can estimate the mean grain size (*L*) of these samples by applying the scherrer equation

$$\beta(2\theta) = \frac{K\lambda}{L\cos\theta}$$

where K=0.94 Constant and $\beta(2\theta)$ Full width half maximum L= grain size, and λ is the wavelength of X-rays, $\beta(2\theta)$ is the pure full width of the diffraction line at half the maximum intensity, and θ is the bragg's angle.

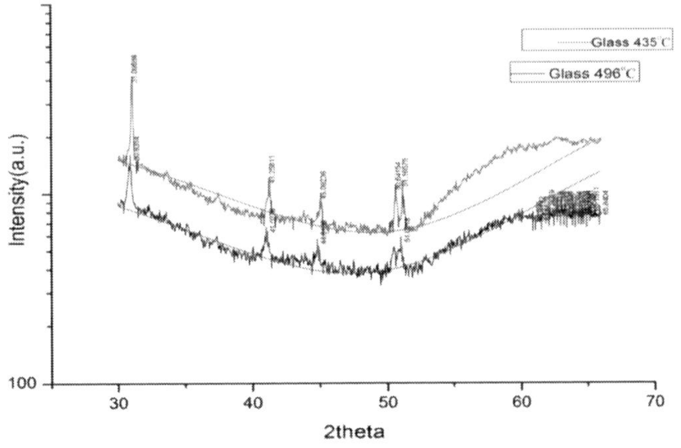

Figure 1. XRD pattern of the Ag covered TiO_2 on glass substrate with annihilated with different temperture A (496°C) b B(435°C)

Table 1: XRD Data of silver doped thin films at various substrate temperatures

Sample	2θ	Grain size
Glass 435°C	51.1695	18nm
Glass 496°C	30.9354	8 nm

The mean grain sizes of samples A and B are 18 and 8 nm, respectively.

3.2 AFM Analysis

Figure 2 shows the AFM images of the samples obtained after heating at 435°C and 496°C . It is obvious from these images that these samples of Ag covered TiO_2 particles are disperse, with an average particle size of 25 nm-35 nm for sample A and 5 nm-21nm for sample B. The particle size of samples A and B are determined from the AFM images are in good agreement with the values determined from the XRD analysis. XRD and TEM results (table 1 and 2) show that the size-selected Ag covered TiO_2 nanoparticles were prepared from the SSR method and that both samples A and B have an anatase structure.

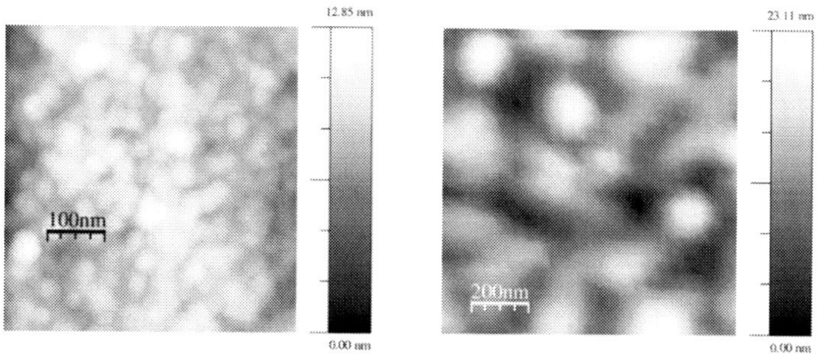

(a) (b)

Figure 2 Ag covered TiO_2 on glass substrate (a) 435°C (500nm×500nm) sample A and (b) 496°C (976.6nm×976.6nm) sample B

Table 2. AFM results of Ag covered TiO_2 on glass substrate 435°C and Ag covered TiO_2 on glass substrate 496°C

S.No	Sample	Hole	Hill	Intervals
1	Glass 435°C(A)	29.01%	70.97%	70.96%
2	Glass 496°C(B)	42.33%	57.69%	57.53%

3.3 Raman Analysis

The ultimate goal in light-harvesting technologies such as solar cells is to collect as much sunlight as possible and to efficiently convert this light into useful electricity. As the presence of non-crystalline silicon leads to reduced conversion efficiency, Raman spectroscopy is one such tool that is widely used in the solar industry to monitor the highly crystalline, defect-free and strain-free silicon films/layers.

The ability to quantify both thermal and interfacial stress in solar cell materials is critical. Stress can dramatically impact conversion efficiencies in PV cells. It is critical to understand the location of the stress and their impact on the cell performance and improvement for process yields [7].

Raman spectroscopy provides a direct measurement of stress (or strain) in silicon based solar cells by monitoring the spectral position of the $520 cm^{-1}$ peak on the sample, from which a stress map can be generated with sub-micrometer spatial resolution. As result, Raman spectroscopy provides direct insight into solar cell processing and prevents less efficient cells from leaving production. High-performance Raman spectroscopy using filters with steep edges and small transition widths allows one to detect low-energy vibration modes. From fig. 3 Raman peak at 558.06 cm^{-1} for the sample annealed at 496°C (1) and Raman peak at 557.014 cm^{-1} for the sample at 435°C (2).

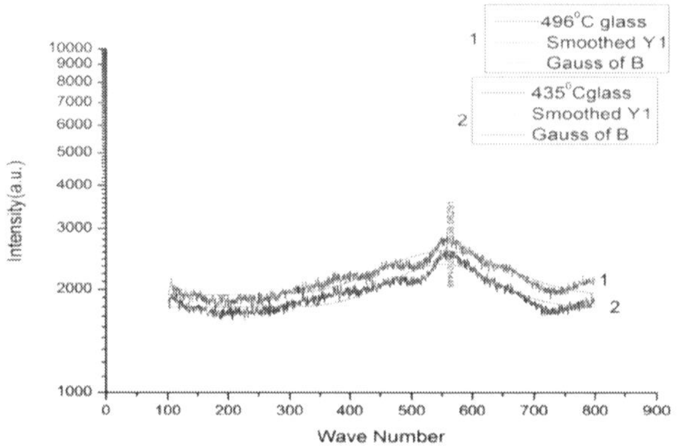

Figure 3. 2% silver doped TiO_2 films at 496°C (1) and 435°C (2) on glass substrate exhibit green photonics for PV Technology detect low-energy vibration modes

4. Conclusion

Understanding the temperature dependent behaviour of Ag covered TiO_2 on glass substrate, for calculating grain size and porosity is key behaviour for exhibiting the various properties of a solar active device. In this study, it is observed that particle size and porosity depends on temperature. The size and structure of 2% Ag covered TiO_2 on glass nano particles are calculated by using XRD,AFM . The XRD patterns and AFM images showed that the prepared samples of 2% Ag covered TiO_2 on glass substrate produces nano particles with an average particle size of 18 nm-35nm (sample A), and 5 nm-21 nm(sample B). The raman spectra revealed that samples A and B have detected low-energy vibration modes of green photonics useful for PV technology.

Acknowledgement

The authors are very grateful to Ajay Gupta, Ram Janay Choudhary, V. Ganeshan, Mukul Gupta and Vasanth Sathe UGC –DAE-CSR, Indore, India. The authors are also very much thankful to R.K. Tugnawat, Govt. Autonomous Model Holkar science college, Indore, India.

References

[1] Schaadt D M, Feng B and Yub E T 2005 *Appl. Phys. Lett.* **86** 063106 .
[2] Kelly K L, Coronado E, Zhao Lin Lin, and Schatz George C. 2003 *J. Phys. Chem. B* **107** 668.
[3] Saikia D., Saikia P. K., Gogoi P. K., Saikia P. 2011 *Digest Journal of Nanomaterials and Biostructures* **6** 589.
[4] *Handbook of Analytical Methods for Materials* 2001 (Materials Evaluation and Engineering Inc.) p.7-10.
[5] Shaw D. G. and Langlois M. C 1993 *Proc. Soc. Vac. Coaters* **36** 348.
[6] Schrader B. (Ed.) 1995 *Infrared and Raman Spectroscopy: Methods and Applications* (Chichester: John Wiley)
[7] Grasselli, J. G., Bulkin, B. J. (Eds.) 1991 *Analytical Raman Spectroscopy*; (Chichester: John Wiley)

On the structural and magnetic study of $Mg_{1-x}Zn_xFe_2O_4$

S Raghuvanshi[a], **M Satalkar**[a], **P Tapkir**[a] , **N Ghodke**[c] **S N Kane**[a,1]
[a]School of Physics, D. A. University, Khandwa road Campus, Indore – 452001, India
[b]UGC-DAE CSR, University Campus, Khandwa Road, Indore – 452 001, India

[1]Corresponding author, E-mail: kane_sn@yahoo.com (S. N. Kane)

Abstract. $Mg_{1-x}Zn_xFe_2O_4$ (with x = 0.0, 0.2, 0.4, 0.6, 0.8 and 1.0) spinel ferrite was synthesized via sol-gel auto combustion method. XRD confirms the formation of spinel phase with a secondary phase of Fe_2O_3 (in case of x = 1.0). Structural parameters changes as the Zn concentration increases which can be ascribed to the replacement of an ion with lower ionic radius (Mg^{2+}) with that of an ion with higher ionic radius (Zn^{2+}). Variation in the magnetic properties with Zn content are interpreted in terms of migration of Fe^{3+} between tetrahedral and octahedral sites.

1. Introduction

Study of soft magnetic nanocrystalline ferrites is growing, owing to their use in various applications [1]. In spinel ferrites cations are dispersed among A (Tetrahedral) and B (Octahedral) sites, exhibiting different structural and magnetic environments. In spinel ferrites, a specific synthesis process (e. g. - e. g.- solid state reaction, co-precipitation, sol gel auto-combustion method etc.), substitution and post-preparation thermal treatments can be effectively used to introduce changes in the spinel structure, thus allowing one to be able to selectively change their properties. By using a specific preparation technique e. g. - sol gel auto-combustion method, it is possible to synthesize ferrites without subsequent heat treatment [1]. It is known that heat treatment of as-prepared powder (also called as '*dry gel*') leads to an increase of grain diameter and, magnetic properties strongly depend on grain diameter. Mg-Zn based ferrites have attracted researchers, due to their applications in high frequency applications [2]. Magnesium ferrite ($MgFe_2O_4$) has predominantly an inverse spinel, whereas Zinc ferrite ($ZnFe_2O_4$) possesses a normal spinel structure: $(Zn^{2+})_A\,[Fe_2^{3+}]_B O_4^{2-}$ (where Zn^{2+} ions occupy the teteahedral 'A' site and, Fe^{3+} ions occupy octahedral (B site). Substitution of Mg by Zn in $Mg_{1-x}Zn_xFe_2O_4$ would modify magnetic properties [3]. Therefore, in the present work, we report the synthesis, structural and, magnetic characterization of $Mg_{1-x}Zn_xFe_2O_4$ (with x = 0.0, 0.2, 0.4, 0.6, 0.8 and 1.0).

2. Experimental details

Magnesium – Zinc based ferrite of nominal composition $Mg_{1-x}Zn_xFe_2O_4$ (with x = 0.0, 0.2, 0.4, 0.6, 0.8 and 1.0) in powders form were prepared using the sol-gel auto-combustion method. Schematic diagram of the synthesis process is shown in figure 1. For synthesizing Magnesium – Zinc based ferrite, citrate-nitrate precursors [Zinc Nitrate – $Zn(NO_3)_2.6H_2O$, Magnesium Nitrate – $Mg(NO_3)_2.6H_2O$, Ferric Nitrate – $Fe(NO_3)_3.9H_2O$ and Citric acid - $C_6H_8O_7$] were mixed in stoichmetric ratio with citric acid as fuel. Metal salt to fuel ratio was taken as 1:1. Citric acid primarily

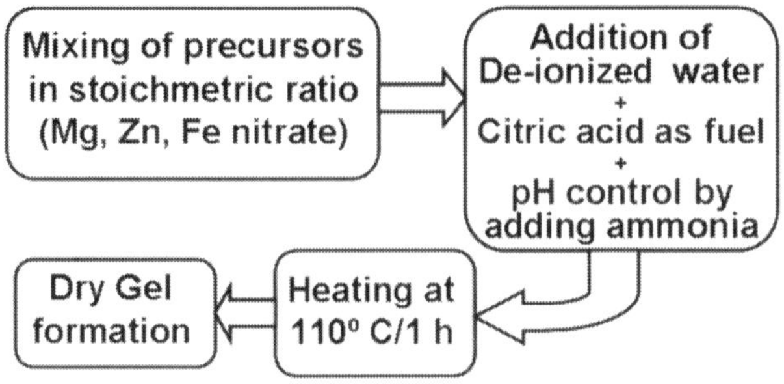

Figure 1. Schematic diagram of the synthesis process of $Mg_{1-x}Zn_xFe_2O_4$ spinel ferrite

acts as a chelating agent for metal ions of varying ionic sizes, which helps in preventing their selective precipitation to maintain compositional homogeneity among the constituents. Afterwards it also serves as a fuel in the combustion reaction [4]. De-ionized water was used to dissolve all the precursors and pH was maintained at 7 by adding ammonia solution. Then the solution was heated at 120 °C in air till the fluffy powder was formed, called as '*dry gel or as-burnt powder*' which was used for structural and magnetic measurements. It is of value to note that formation of cubic spinel ferrite phase in *dry gel* form without any heat treatment was reported earlier [5-8]. Structural characterization of the synthesized samples was done at room temperature by x-ray diffraction (XRD) measurements, done in θ-2θ configuration by Bruker D8 advance diffractometer utilizing CuK_α (λ= 1.54056 A°) radiation. The x-rays were detected by using a fast counting detector based on Silicon strip technology (Bruker LynxEye detector). The lattice parameter (***a***) corresponding to [311] reflection was obtained by:

$$a = d\sqrt{h^2 + k^2 + l^2}$$

where d – inter-planer spacing and, (h, k, l) - miller indices. X-ray density (ρ_{XRD}) of the studied samples was obtained by:

$$\rho_{XRD} = 8M_w/N(a)^3$$

where M_w – Molecular weight, N – Avagrado's Number, a - Lattice parameter.
Grain diameter is obtained by the line width of [311] reflection, using Scherrer's formula:

$$D = 0.9\lambda / \beta \, Cos\theta$$

where, λ - wavelength of the x-ray used, β - line width, θ - peak position (in 2θ scale).
Hoping length for site A and B were calculated by :

$$L_a = \frac{a\sqrt{3}}{4} \quad and, \quad L_b = \frac{a\sqrt{2}}{4}$$

where L_a and L_b are respectively hoping length for site A and B, a – lattice parameter.
Magnetic moment (n_B) in Bohr magneton (μ_B), can be calculated by using the relation:

$$n_B = [M_w \times M_s] / 5585$$

where M_w is the molecular weight of a particular ferrite composition and M_s is saturation magnetization (emu/g) and, 5585 is the magnetic factor. Room temperature hysteresis loops of the synthesized samples were measured at 50 Hz by conventional induction technique, applying maximum field – $H_{max} \approx \pm$ 119 kA/m (1500 Oe). Loops were analyzed by a Matlab® based program [9] to obtain coercivity (H_c).
Satutation magnetization (M_s) values were obtained by plotting magnetization against inverse of magnetic field (1/H) and, extrapolating the data to 1/H = 0. Anisotropy constant (K_1) was calculated by:

$$K_1 = [H_c \times M_s] / 0.96$$

where H_c – coercivity and M_s – saturation magnetization.

3. Results and discussion

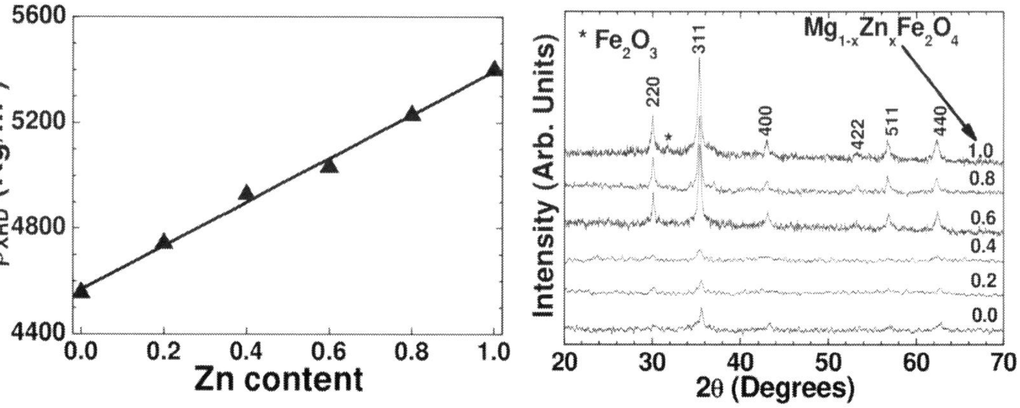

Figure 2. Variation of ρ_{xrd} with Zn content (left panel) and, XRD patterns (right panel) of the studied specimens. Zn content (x) is shown on right side in the figure.

Figure 2 (right panel) depicts the XRD patterns of the studied samples. XRD confirms the formation of spinel phase in the studied samples. Fig. 2 (left panel) shows the variation of ρ_{xrd} with Zn content.Table 1 depicts the lattice parameter values *(a)*, Scherrer's grain diameter (*D*), x-ray density (ρ_{xrd}) cell volume (a^3), hoping length for site A (L_a) and, hoping length for site B (L_b) obtained by analyzing XRD patterns. Although XRD data shows the formation of spinel phase, presence of Fe_2O_3 is also detected which is ascribable to the increase of Fe/Zn ratio in the samples due to loss of Zn during combustion process. Perusal of table 1 shows that the grain diameter varies between 17.1 to 22.9 nm, exhibits the formation of nano-crystalline ferrites. With increase of Zn content (x), the x-ray density 'ρ_{xrd}' for the studied samples increases. Changes of lattice parameter (***a***) with increase of Zn-content can be understood by the fact that an ion with lower ionic radius (ionic radius of Mg is 0.066 nm) is being replaced by an ion with higher ionic radius (ionic radius of Zn is 0.074 nm). Changes of ***a*** are also reflected in unit cell volume. ρ_{xrd} increases linearly with increase of Zn-content. As ρ_{xrd} depends on the molecular weight of the sample therefore, with increase of Zn content there is an increase in x-ray density of the sample because atomic mass of Zn is higher than that of Mg. Changes in L_a and L_b follows the same trend as the variation of ***a*** with Zn content The changes can be ascribed to the fact that the smaller Mg^{2+} ions are being replaced by larger Zn^{2+}.

Figure 3 shows the representative hysteresis loops of the studied samples with varying Zn content. Changes in the hysteresis loops can be observed as the zinc content increases which can be ascribable to the changes in cation distribution on tetrahedral (***A***) site and octahedral (***B***) site. Fig 3 inset depicts the variation of H_c with Zn content H_c first decreases with Zn content then increases as Zn concentration increases.

Table 1: Zn-content dependence of lattice parameter (***a***), grain diameter (***D***), x-ray density (ρ_{XRD}), cell volume (***V***), hoping length for site A (L_a) and hoping length for site B (L_b) for $Mg_{1-x}Zn_xFe_2O_4$.

x	a (nm)	D (nm)	ρ_{XRD} (Kg/m^3)	Cell Volume (V) (nm^3)	L_a (nm)	L_b (nm)
0.0	0.8354	17.1	4556.2	0.5830	0.3617	0.2953
0.2	0.8354	18.4	4743.3	0.5830	0.3617	0.2953
0.4	0.8354	14.5	4930.4	0.5830	0.3617	0.2953
0.6	0.8401	22.8	5032.1	0.5929	0.3637	0.2970
0.8	0.8384	22.9	5228.5	0.5893	0.3630	0.2964
1.0	0.8401	22.8	5400.1	0.5929	0.3637	0.2970

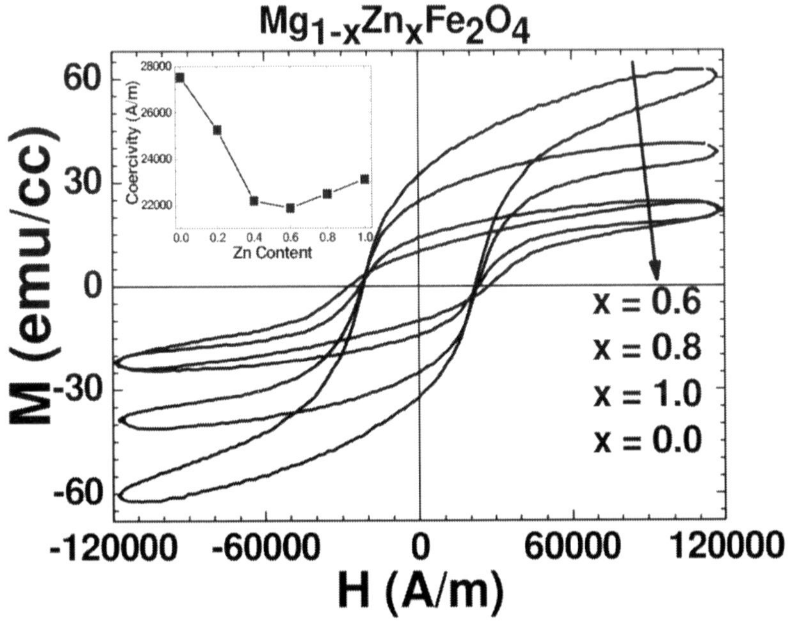

Figure 3. Representative hysteresis loops of the studied samples with varying Zn content (x). Inset: Variation of coercivity with Zn content.

Figure 4 (left panel) depicts the variation of $\boldsymbol{M_s}$ and, $\boldsymbol{K_1}$ with Zn content and fig. 4 whereas inset shows the changes in $\boldsymbol{n_B}$ with Zn concentration. $\boldsymbol{M_s}$ values are obtained by plotting M versus 1/H and extrapolating the data to 1/H =0 (Fig. 4 right panel). $\boldsymbol{M_s}$ and $\boldsymbol{n_B}$ follow the same trend as the doping content (Zn) increases. Magnetic moment of the $Mg_{1-x}Zn_xFe_2O_4$ sample depends on the migration of Fe^{3+} (5 μ_B) ions between ***A*** and ***B*** site as Zn^{2+} (0 μ_B) and Mg^{2+} (0 μ_B) ions are diamagnetic in nature. $\boldsymbol{M_s}$ and $\boldsymbol{n_B}$ increases from x= 0.0 to 0.6 due to the increase in the magnetization on B site (Fe^{3+} ions on B site increases) and magnetization on A site decreases (Fe^{3+} ions on A site decreases) and then decreases from x= 0.6 to 1 due to the decrease of magnetization on B site and increase in magnetization on A site and also due to the spin canting in one of the sub-lattice due to the substitution of non-magnetic cation on the other sub-lattice [10]. $\boldsymbol{K_1}$ also shows the same trend as $\boldsymbol{M_s}$ due to the fact that $\boldsymbol{K_1}$ depends on $\boldsymbol{M_s}$.

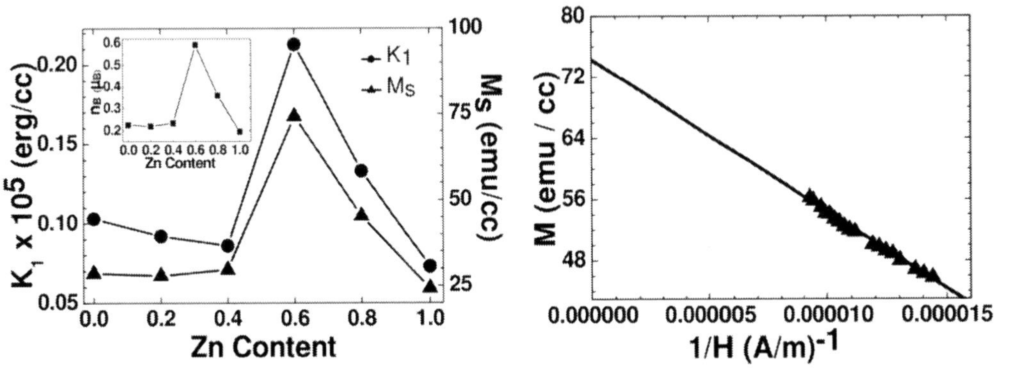

Figure 4. Variation of M_s and, K_1 with Zn content (left panel). Inset: n_B variation with Zn content. M versus 1/H curve showing the extrapolation of the data to 1/H =0 for sample with Zn-

To summarize, $Mg_{1-x}Zn_xFe_2O_4$ spinel ferrite was synthesized via sol-gel auto combustion method XRD confirms the formation of spinel phase with a secondary phase of Fe_2O_3 (in case of x = 1.0). Replacing an ion with lower ionic radius (Mg^{2+}) by an ion with higher ionic radius (Zn^{2+}) leads to changes in structural parameters obtained from XRD measurements. Observed changes in the magnetic properties with Zn content are explained on the basis of migration of Fe^{3+} between tetrahedral and octahedral sites.

Acknowledgments
This work is supported by MPCST project No. 1783/CST/R & D/Phy & Engg Sc/2013. Research fellowship to MS form UGC-DAE CSR project No. CSR-I/CSR51/2011-12 is gratefully acknowledged. Authors are grateful to Dr. M. Gupta for doing XRD measurements.

References

[1] Ming-Ru Syue, Fu-Jin Wei, Chan-Shin Chou, and Chao-Ming Fu, 2011 *J. Appl. Phys.* **109** 07A324-3; Satalkar M, Kane S N, Ghosh A, Ghodke N, Barrera G, Celegato F, Coisson M, Tiberto P, Vinai F, 2014 *J. Alloys Comp.* http://dx.doi.org/10.1016/j.jallcom.2014.01.248; Batra A S, Satalkar M, Kane S N, Ghosh A and Ghodke N, 2014 *AIP Conf. Proc.* **1591** 537
[2] Kassabova-Zhetcheva V D, 2009 *Central European J. Chem.* **7** 415
[3] Joshi H H, Kulkarni R. G. 1986 *J. Mater. Sci.* **21** 2138
[4] Kapse D, Ghosh S A, Raghuwanshi F C, Kapse S D, 2009 *Mater. Chem. Phys.* **113** 638
[5] Syue M R, Wei F J, Chou C H, Fu C M, 2011 *J. Appl. Phys.* **109** 07A324–3
[6] Castro S, Gayoso M, Rivas J, Greneche J M, Mira J, Rodriguez C, 1996 *J. Magn. Magn. Mater.* **152** 61
[7] Khot V M, Salunkhe A B, Phadatare M R, Thorat N D, Pawar S H, 2013 *J. Phys. D: Appl. Phys.* **46** 055303.
[8] Ghosh A, Satalkar M, Rathod S, Nag S P, Vyas P, Kane S N, Ghodke N, Prasad R, and Dwivedi R 2013 *Int. J. Mod. Phys.: Conf. Ser.* **22** 28
[9] Kane S N, Gercsi Z, Ghodke N, Khinchi S S and Gupta A, "A digital hysteresis loop tracer for studying soft magnetic materials", in Proc. of the DAE Solid State Physics Symposium vol. 50, edited by Aswal V K, Bhushan K G and Yakhmi J V, Mumbai, Prime Time Education, Mumbai, India, 2005, pp 287-288
[10] Haque M M, Huq M, Hakim M A, 2009 *Physica B* **404** 3915-3921

Extended X-ray absorption fine structure data analysis of copper (II) hydroxamic acid mixed ligand complexes

N Parsai[a], A Mishra[a], B D Shrivastava[b]

[a]School of Physics, Devi Ahilya University, Indore 452001
[b]School of Studies in Physics, Vikram University, Ujjain 456010

* E-mail: parsai.neetu@gmail.com

Abstract. The X-ray absorption spectra of copper mixed ligand complexes, having hydroxamic acid as one of the ligands, have been recorded at the K-edge of copper at BL-8 Dispersive EXAFS beamline at the 2.5 GeV INDUS-2 Synchrotron, RRCAT, Indore, India. For the analysis of EXAFS data, crystallographic data of the complex or of its analog is required, which is not available. Hence, for the analysis of EXAFS data, theoretical EXAFS data of the studied complexes have been generated using the EXAFS equation employing computer software program *Mathcad*. Firstly, the experimental data has been processed using the computer program *Athena* to obtain the normalized absorption versus energy data. From the experimental EXAFS data, the phase shift parameter (an energy independent constant δ) has been computed using Lytle, Sayers and Stern's (LSS) method. The backscattering amplitude has been taken from the available theoretical tabulations and other parameters have been taken from crystallographic data of the copper metal. Fourier transforms of both the experimental and theoretical data have been computed, and the two Fourier transforms are found to agree with each other for all the complexes. The position of the first peak in the Fourier transform gives the value of the first shell bond length, which is shorter than the actual bond length as a result of energy dependence of the phase factor (δ(k)) in the sine function of the EXAFS equation. Since, the Fourier transform method and LSS method both are uncorrected for phase and other parameters of the EXAFS equation, the present method gives phase uncorrected bond length of the first coordination shell.

1. Introduction

For the analysis of extended X-ray absorption fine structure (EXAFS) data, the crystal structure of either the complex studied or of its analogous complex is required. The crystallographic information is used to generate theoretical model by using software programs like *Artemis*. In the present work some copper complexes have been studied of which the crystallographic data is not available. Hence, for the analysis of their EXAFS data, the method outlined by us in our earlier communications has been used [1, 2]. The method is simple as it generates theoretical EXAFS data employing computer software program *Mathcad*.. The copper complexes studied have been listed in table 1. The experimental EXAFS data of the copper complexes have been recorded at the K-edge of copper at BL-8 Dispersive EXAFS beamline at the 2.5 GeV INDUS-2 Synchrotron, RRCAT, Indore, India.

2. Analysis of experimental EXAFS data

Firstly, the experimental data has been processed using the computer program *Athena* to obtain the normalized absorption (μx) vs. energy (E) data and then the EXAFS oscillations χ(k) vs. wave vector (k) data is obtained, which are shown in fig. 1. Fourier transforms of the experimental EXAFS data, given fig. 1, have been computed, which are given in fig. 2.

3. Computation of theoretical EXAFS data

The theoretical EXAFS data has been computed for the complexes using the standard form of EXAFS equation [3].

$$\chi(k)=\sum_j \frac{N_j S_0^2 F_j(k)}{kR_j^2}\times \exp(-2\sigma_j^2k^2)\times\exp\left(-2R_j/\lambda(k)\right)\times\sin[2kR_j+\delta_j(k)] \quad \ldots (1)$$

The subscript j denotes the various scattering paths, N_j is the number of atoms in the jth coordination shell, R_j is the average radial distance to the jth atom and σ_j^2 is mean square deviation about the bond length. S_0^2 is the passive electron reduction factor and accounts for the slight relaxation of the remaining electrons in the presence of the core-hole vacated by the photoelectron. S_0^2 usually has a value between 0.7 and 1.0 and is different for different elements. The function $F_j(k)$ is the photoelectron backscattering amplitude and $\delta_j(k)$ is the scattering phase shift for each scattering path. $\lambda(k)$ is the photoelectron mean free path which is dominated by lifetime of the excited state.

In the our earlier works [1, 2], we have developed a simple analysis procedure for extracting bond length for the first coordination shell from the experimental EXAFS data. The method is simple and novel because calculations are done using the computer software *Mathcad*. The theoretical model is generated through the EXAFS equation given in eqn. (1) for the first shell. The values of N_1, σ_1^2 and λ have been taken from the crystallographic data of copper metal [4, 5]. The backscattering amplitude $F_j(k)$ has been taken from the table of Mckale [6]. The phase shift δ_1 has been extracted from the experimental data in the following manner.

The analysis depends on the argument of the sine term in eqn. (1), i.e., $\sin[2kR_j + 2\delta_j(k)]$ where δ_j is assumed as linear in k, i.e., $\delta_j = -\alpha_jk + \beta_j$. Substituting into the argument of sine and rearranging for the first coordination shell (j = 1), it is convenient to define n by [7]:

$$(n + ½)\pi = 2k(R_1 - \alpha_1) + 2\beta_1 \quad \ldots (2)$$

where n = 0, 2, 4... for maxima and n = 1, 3, 5.... for minima. Using the graphical method outlined by Lytle, Sayers and Stern (LSS method) [4, 5], a plot of n versus k for the maxima and minima of the measured EXAFS data determines $(R_1 - \alpha_1)$ from slope and β_1 from intercept.

4. Results and Discussion

From the experimental χ(k) versus k plots, given in fig. 1, the values of k corresponding to maxima and minima of χ(k) have been obtained and are given in table 3. The n versus k curves are given in fig. 3 for the complexes. The values of $(R_1 - \alpha_1)$ and β_1 determined respectively from the slope and the intercept of these plots. Taking α_1 to be 0.31 Å, R_1 and δ_1 have been calculated and used in eqn. (1) and the theoretical EXAFS χ(k) data have been obtained for all the complexes and are given in fig 4. This theoretical EXAFS data has been Fourier transformed which are given in fig. 5. Fourier transforms of both the experimental and theoretical data, given in figs. 2 and 5 respectively, have been found to agree with each other for all the complexes. The position of the first peak in this Fourier transform gives the value of the bond length, i.e., the average distance between the absorbing atom and the neighboring atoms. However, the position of the first peak in these Fourier transforms gives the value of the first shell bond length, which is shorter than the actual bond length as a result of energy dependence of the phase factor (δ(k)) in the sine function of the EXAFS equation [8]. The phase uncorrected bond lengths thus determined from the theoretical EXAFS data are given in table 2.

Thus the present method gives phase uncorrected bond length of the first coordination shell. It can be seen from Figs. 2 to 5 that the theoretical χ(k) curves as well as their Fourier transforms are in good agreement with their experimental counterparts, showing that the simplified method presented in this work is quite satisfactory. The values of bond length for the first shell in all copper complexes obtained from the present analysis of the experimental and theoretical EXAFS data, given in table 2 are in good agreement with each other.

Table 1 Copper (II) hydroxamic mixed ligand complexes with their abbreviations and molecular formulae.

S.No.	Complex	Abbreviation	Molecular formula
1.	Cu(II) 5 chloro-benzohydroxamic acid	Cu 5CBHA	$CuC_7H_6O_2NCl$
2.	Cu(II) malanohydroxamic acid	Cu MHA	$CuC_2H_4O_4$
3.	Cu(II) succenohydroxamic acid	Cu SuHA	$CuC_4H_8O_4N_2$

Table 2 Values of the first shell bond lengths R(Å) obtained from the Fourier transforms of the experimental EXAFS data and from the theoretical data.

S.No.	Complex	R (from FT of experimental data)	R (from FT of theoretical data)
1	Cu 5CBHA	1.81	1.87
2	Cu MHA	1.38	1.44
3	Cu SuHA	2.07	2.11

Table 3 Wave vector k ($Å^{-1}$) for EXAFS maxima and minima at the K-absorption edge of copper in the complexes and their corresponding values of n for LSS plots.

n	Wave vector k ($^{-1}$)		
	Cu 5CBHA	Cu MHA	Cu SuHA
0	1.94	1.90	1.85
1	3.45	3.40	4.05
2	5.00	4.80	4.95
3	5.75	5.60	5.55
4	6.45	6.95	6.45
5	7.80		7.15

5. Conclusions

A simplified technique has been presented for the analysis of extended X-ray absorption fine structure (EXAFS) data and determining the first shell bond length. The method employed here is simple and straight forward, thereby providing a physical picture of the X-ray absorption fine structure. The analysis uses easily available computer program *Mathcad*. The method can be applied for any pure metal and is quite suitable for complex systems also. In the present case, analysis has been carried out only for the first shell, but it can be extended to other shells also.

References

[1] Mishra A, Parsai N & Shrivastava B D 2011 *Indian J Pure & Appl Phys* **49** 25.
[2] Mishra A, Parsai N, & Shrivastava B D 2012 *X-Ray Spectrom.* **41** 219.
[3] Rehr J J & Albers R C 2000 *Rev Mod Phys* **72** 621.
[4] Lytle F W, Sayers D E & Stern E A 1975 *Phys Rev B* **11** 4825.
[5] Stern E A, Sayers D E & Lytle F W 1975 *Phys Rev B* **11** 4836.
[6] McKale A G, Knapp G S & Chan S K 1986 *Phys Rev B* **33** 841.
[7] Deshpandey A P, Sapre V B & Mande C 1988 *Phys Stat Sol (b)* **145** 77.
[8] Stern, E A & Heald, S M 1983 *Handbook on Synchrotron Radiation* Vol. 1 edited by Koch E E Elsevier p. 955.

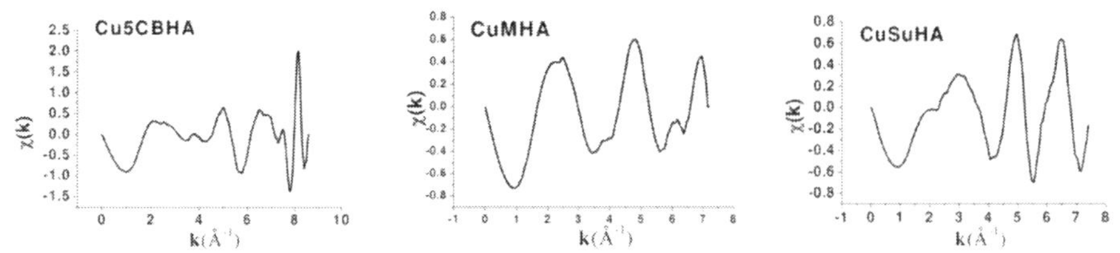

Figure 1 - Experimental χ (k) versus k curves for copper complexes.

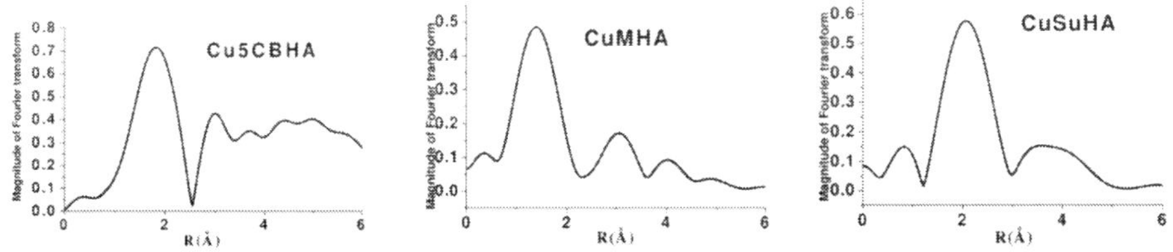

Figure 2 - Magnitude of Fourier transform of χ (k) versus k curves of the experimental EXAFS data, given in figure 2, for copper complexes.

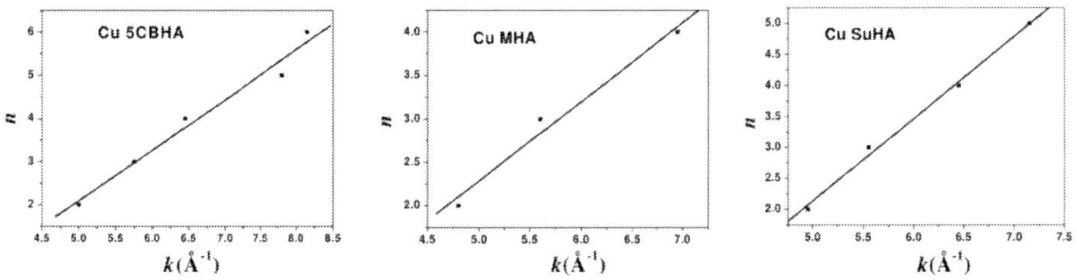

Figure 3 - n versus k curves for the copper complexes.

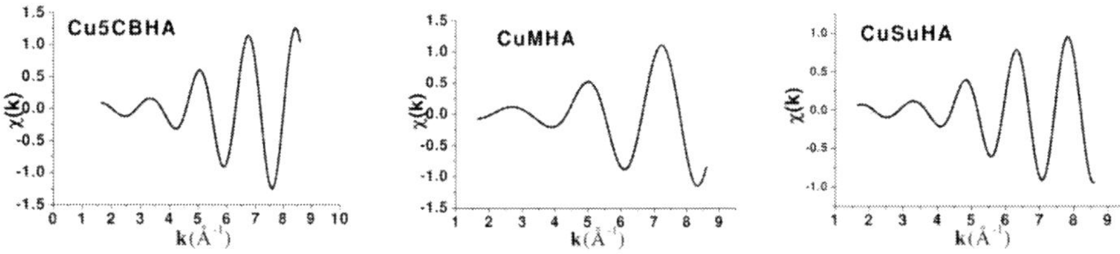

Figure 4 - Theoretical χ (k) versus k curves for copper complexes. For calculations, backscattering amplitude has been taken from Mckale [6] (values for $R = 2.5$ Å)

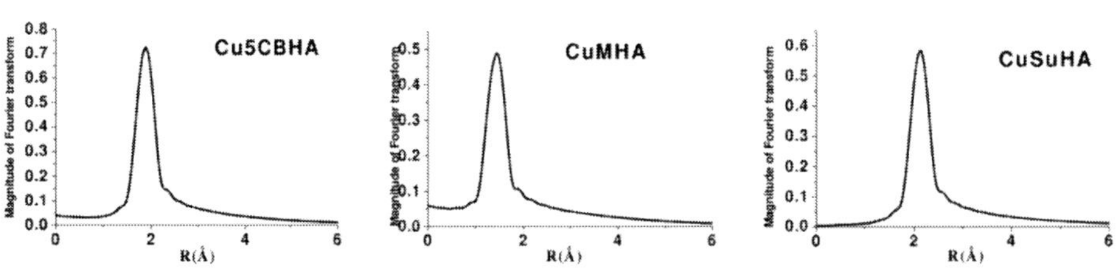

Figure 5 - Magnitude of Fourier transform of χ (k) versus k curves of the theoretical EXAFS data, given in figure 3, for copper complexes.

XRD, IR and XAFS studies of cobalt complexes having amino pyrazole dicarboxylate (APD) as ligand

Ashutosh Mishra, Garima Jain, H. Patil
School of Physics, Devi Ahilya University, Khandwa Road, Indore, 452001

E-mail: garimasop@gmail.com, amishra1960@yahoo.co.in

Abstract. X-ray absorption fine structure spectroscopic (XAFS) studies have been done on two cobalt complexes using APD (diethyl 4-amino-1-phenyl-1H-pyrazole-3,5 dicarboxylate) as ligand. The X-ray absorption spectra of the complexes have been recorded on beam line of synchrotron at Raja Ramanna Centre for Advanced Technology (RRCAT), Indore (India). The X-ray diffraction of the samples has also been carried out. FTIR studies of two samples were also reported in the present communication.

1. Introduction

X-ray Absorption Fine Structure (XAFS) spectroscopy is a unique tool for studying, at the atomic and molecular scale, the local structure around selected elements that are contained within a material. XAFS can be applied not only to crystals, but also to materials. XAFS is often divided into X-ray absorption near edge structure (XANES), which lies within the first 40 eV of the edge position and extended X-ray absorption fine structure (EXAFS), which lies beyond 40 eV above the absorption edge. We have studied the X-ray absorption spectra at the K-edge of cobalt of a series of three cobalt complexes have been investigated.

X-ray diffraction (XRD) is a tool to identify the crystalline nature of the samples. Infrared spectroscopy is one of the most important analytical techniques. FT-IR provides information about identify unknown materials, determine the quality or consistency of a sample and determine the amount of components in a mixture.

The present study deals with analysis and structural information of samples using X-ray diffraction, FTIR and X-ray K-absorption spectra of two cobalt (II) complexes i.e.

1. Bis-(diethyl 4-amino-1-(P-nitrophenyl) -1H-pyrazole-3,5 dicarboxylate) cobalt (II) chloride **(Co(D4A1(P-n)1HP35D)**.
2. Bis-(diethyl 4-amino-1-(4-bromophenyl) -1H-pyrazole-3,5 dicarboxylate) cobalt (II) chloride **(Co(D4A1(4-br)1HP35D).**

2. Experimental

All the studied complexes have been prepared by chemical root method. The X-ray K-absorption spectra of the cobalt complexes have been recorded using synchrotron radiation. The X-ray spectroscopy setup is available at Raja Ramanna Centre for Advanced Technology (RRCAT), Indore and is called beamline. This beamline BL-8 has been recently commissioned at the 2.5 GeV INDUS-2 Synchrotron radiation source.

Journal of Physics: Conference Series **534** (2014) 012033 doi:10.1088/1742-6596/534/1/012033

The infrared absorption spectra of the complexes were measured at room temperature, in the wave number range 4000 to 400 cm^{-1} on Jasco FTIR-300 spectrometer using the KBr pellet technique.
The X-ray diffraction pattern of cobalt complexes have been recorded using Rigaku RINT-2000 X-ray diffractometer with a rotating anode with tube voltage of 40 kV and current of 100 mA. Copper target was used as the source of X-rays at wavelength λ=1.54Å.

3. Result and discussion

3.1 X-ray Diffraction

X ray diffraction patterns for two cobalt complexes (Co (D4A1(P-n)1HP35D) and (Co (D4A1(4-br)1HP35D) were recorded and calculated parameters are given in table 1. All the complexes had a monoclinic crystal lattice.

3.2 Fourier Transform Infrared spectroscopy (FTIR):

The IR spectra of samples were taken in the range between 4000–400 cm-1. IR spectrums of the synthesized samples are depicted in Figure 2.

3.3 X-ray absorption near edge structure:

The X-ray absorption near edge structure (XANES) parameters i.e., chemical shift, shift of principal absorption maxima and edge-width have been determined.

3.3.1 Chemical Shift

The shifts of the k-absorption edge of cobalt in the complexes with respect to that of cobalt metal have been determined according to the equation:

$$\Delta E = E_k\ (\text{Complex}) - E_k\ (\text{Metal}) \text{ where the cobalt metal K-edge } E_k = 7710.2\text{eV}$$

The K-absorption edge of cobalt has been found to be shifted towards the high energy side in all the complexes studied as compared to the k-absorption edge in the metal. The values of chemical shifts of the cobalt complexes is 8.19 eV for Co(D4A1(P-n)1HP35D) and 8.22 eV for Co(D4A1(4-br)1HP35D). These values of the chemical shifts suggest that cobalt is in oxidation state +2 in all of the complexes.

3.3.2 Principal absorption maximum

The distance (in eV) of the principal absorption maximum A with respect to the respective K-absorption edge of cobalt complexes are 15.43eV for Co(D4A1(P-n)1HP35D) and 15.26eV for Co(D4A1(4-br)1HP35D). The order of shift of principal absorption maximum for cobalt complexes is in reverse order of the chemical shift. The reverse order represents that the shift of the principal absorption maximum is inversely proportional to ionic character for the complexes.

3.3.3 Edge width

The edge-width of cobalt complexes is 7.24 eV and 7.04 eV for Co(D4A1(P-n)1HP35D) and Co(D4A1(4-br)1HP35D).The order of the edge-width is in the reverse order of chemical shift. This represents that the edge width is inversely proportional to ionic character for this series.

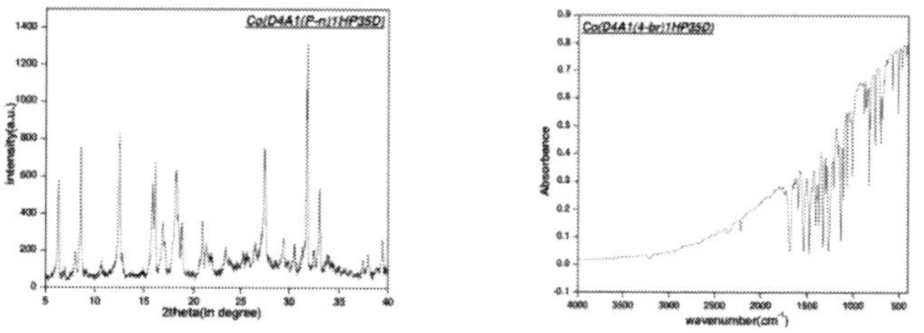

Figure 1.X-ray diffractometer records of complexes

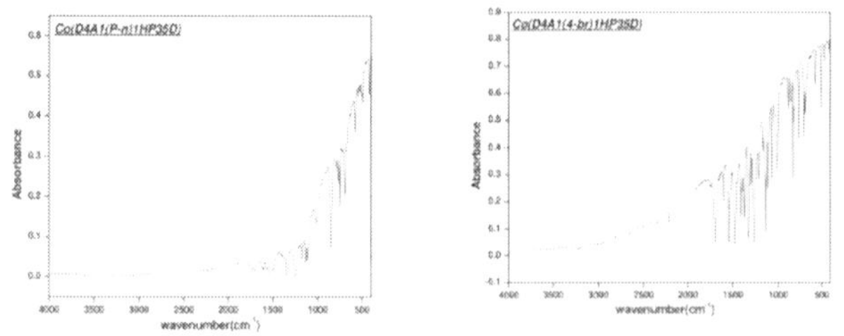

Figure 2. IR spectra of complexes.

Table 3. Results of X-ray diffraction and FTIR studies of complexes

Complex	Lattice parameter			Unit cell volume ($10^{-8}cm^3$)	IR ν_{max} (cm^{-1})
	a (nm)	b (nm)	c (nm)		
(Co(D4A1(P-n)1HP35D)	14.1	6.48	10.29	936.12	1739.54 cm^{-1}(C=O stretching),1522.79 cm^{-1}(No_2 stretching),1258.81 cm^{-1}(C-O stretching)
(Co(D4A1(4-br)1HP35D)	12.58	6.88	13.68	1137.22	3218.52 cm^{-1}(N-H stretching),1687.26 cm^{-1}(C=O stretching),1266.53 cm^{-1}(C-O stretching),509.63 cm^{-1} (C-Br stretching)

REFERENCES

[1] Hemchandran K and Chetal A R 1986 *Phys. Stat. Sol. (b)* **136** 181.
[2] Shrivastava U C and Nigam H L 1973 *Coord. Chem. Rev.* **9** 267.
[3] Cullity B D 1978 *Elements of X-ray Diffraction* (Addision- Wesely)
[4] Kekre P A, Joshi K P and Mishra A 1996 *Ind J Pure Appl Phys* **34** 306.

Surface, morphology and X-ray diffraction studies of Co (II) complexes of pyrazole ligands

A.Mishra[a], Garima Jain[a] and S. Ninama[a]

[a]School of Physics, D. A. University, Khandwa road Campus, Indore – 452001, India
Corresponding author, E-mail: garimasop@gmail.com, amishra1960@yahoo.co.in

Abstract. Pyrazole based complexes of the cobalt (II) Bis-(diethyl 4-amino-1-(P-nitrophenyl) 1H-pyrazole-3,5dicarboxylate) [Co (D4A1(P-N)1HP35D)] and cobalt (II) Bis-(diethyl 4-amino-1-(3-chlorophenyl) 1H-pyrazole-3,5dicarboxylate) [Co (D4A1(3-Cl)1HP35D)] were synthesized by chemical root method and characterized by different method viz. X-ray diffraction, Fourier transform infrared spectroscopy and Transmission electron microscopy studies. All these studies were in good agreement with the synthesized complexes.
Keywords: Cobalt (II), XRD, FTIR, TEM.

Introduction

Pyrazole-based complexes are important because of their attractive physicochemical, chemical, structural and biological properties [1-3]. The pyrazole derivatives have been widely used in many fields for their excellent biological properties like pharmaceutical evaluation, antitumor agents and chemical communication. Pyrazole derivatives exhibit important biological properties such as antitumour, anticoagulant, antihyperglycemic, analgesic, anti-inflammatory, anti-pyretic,antibacterial, hypoglycemic and sedative–hypnotic activity [4-6]. These derivatives have attracted significant attention because of the application in drug development.

2. Experimental details

Pertinent aniline derivatives were dissolving in distilled water and kept at freezing temperature in the refrigerator .To this, an aqueous solution of sodium nitrite in of distilled water was added with continuous stirring keeping the temperature in the vicinity of 0- 5°C. Meanwhile, in another beaker ethyl cyanoacetate, sodium acetate and distilled water were taken and cooled in an ice bath. The reaction mixture was kept for overnight period, filtered through suction, washed with water and dried in vacuum. Fine crystals of the compound were obtained, which were recrystallized from ethyl alcohol. To a solution of couple product, Triethylamine and Ethyl Bromo Acetate were added. The reaction mixture was the refluxed, and then poured onto ice-cold water and HCl, the solid product so formed was filtered off and crystallized from ethanol. The complex preparation is now mentioned here. The mixture of ethanolic solution of Ligand and $CoCl_2$ was refluxed for 1 hour. After reflux Solvent was evaporated to afford fine crystal of Bis-(diethyl 4-amino-1-(R-phenyl) 1H-pyrazole-3,5dicarboxylate) Co (II) chloride complexes. This was recrystallized from ethanol [7].
The XRD measurements were carried out with Cu Kα (λ=1.5406A°) radiation using a Rigaku powder diffractometer equipped with a rotating anode scanning (0.01 step in 2θ) over the angular range 0° to 60° using Ni-filtered at room temperature having voltage 40 KV and current 100 mA.

At Room temperature Fourier transform infrared (FTIR) spectra were recorded in the frequency range of 4000–400 cm−1 employing KBr disc technique using Bruker Germany make Jasco FTIR-300 spectrometer model vertex-70.
TEM images has been obtained on Tecnai (300 kV), FEI, Holland. It has a LaB6 electron gun, which can be operated between 50 to 300 kV.

3. Results and discussion

3.1 X-ray Diffraction (XRD):
The diffraction patterns have been successfully indexed and the values of different parameters are shown in table. X-ray powder diffraction patterns of the complexes are indicative of their crystalline nature and had a monoclinic crystal lattice. In some cases deviation between the calculated and observed d values was nearly equal to one, which indicates that the synthesized complexes are multiphase complexes [8]. Inter-planar d spacing and unit cell volume of the synthesized complexes were calculated by the formulae:

$$\frac{1}{d^2} = \frac{1}{sin^2\beta}\left[\frac{h^2}{a^2} + \frac{k^2 sin^2\beta}{b^2} + \frac{l^2}{c^2} - \frac{2hl cos\beta}{ac}\right]$$

And volume is given by V = abc sinβ [8].
As our synthesized samples are complexes, and due to the lone pair of electrons,distortion occurs due to which it undergoes lower symmetry, yet the characteristic and prominent peaks of complexes and the crystal system have been identified, which are equal and in some cases nearly match the standard diffraction card JCPDS also.
The X-ray diffraction patterns of the complexes are shown in figure 1 and the different results are reported in table 1 and 2.

Table 1. The experimental data and the calculated results for powder X-ray diffraction pattern of Co (D4A1(P-N)1HP35D). JCPDS CARD NO. (100754 - 441628)

2 theta (in degree)	d-spacing (in Å) observed	d-spacing (in Å) calculated	Relative Intensity (%)	h	k	l
6.33	13.97	14.04	43.77	1	0	0
8.58	10.29	10.00	57.34	0	0	1
12.54	7.05	7.02	62.22	2	0	0
16.16	4.48	5.01	50.45	1	1	1
18.29	4.84	5.01	47.87	1	1	1
20.85	4.25	4.01	27.31	0	1	2
27.44	3.24	3.24	57.08	0	2	0
31.85	2.80	2.86	100	-2	1	3
33.09	2.70	2.54	41.53	3	2	1

Lattice parameters and angle calculated are: a = 14.1 nm, b = 6.48 nm, c = 10.29 nm, β = 94.32◦ Unit cell volume of the complex: 936.12 × 10−8 cm3.

Table 2: The experimental and the calculated results for powder X ray diffraction pattern of the Co (D4A1(3-Cl)1HP35D), JCPDS CARD NO. (100754 - 441628)

2 theta (in degree)	d-spacing (in Å) observed	d-spacing (in Å) calculated	Relative intensity (%)	h	k	l
6.22	14.20	14.14	44.35	1	0	0
11.18	7.90	7.73	38.59	-1	0	1
16.08	4.50	4.87	100	1	1	1
18.42	4.81	4.71	79.19	3	0	0
20.8	4.26	3.60	49.64	-1	1	2
26.68	3.33	3.33	92.33	0	2	0
30.5	2.92	2.90	47.88	0	0	3
32.98	2.71	2.55	51.57	3	2	1
37.3	2.40	2.07	37.46	-1	1	4

Lattice parameters and angle calculated are: a = 14.20 nm, b = 6.66 nm, c = 8.76 nm, β = 95.07◦.Unit cell volume of the complex: 824.20 × 10−8 cm3.

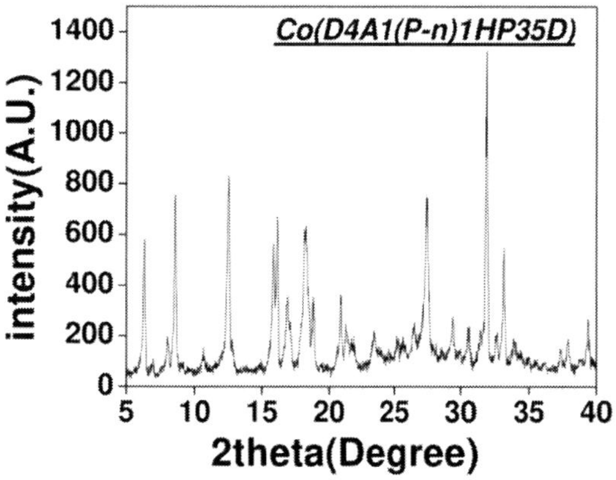

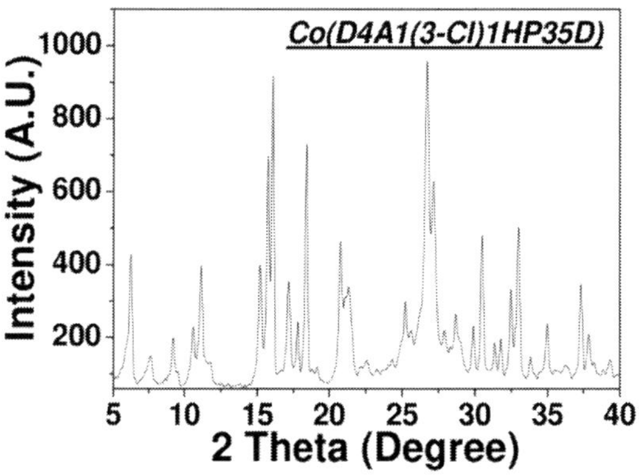

1(a) 1(b)

Figure 1: X-ray Diffraction pattern of Co(D4A1(P-N)1HP35D) **1(a)** and Co (D4A1(3-Cl)1HP35D) 1**(b)**

3.2 *Fourier Transform Infrafred Spectroscopy (FTIR)*

The IR spectra of samples were taken in the range between 4000–400 cm^{-1} in KBr pellets. IR spectrums of some of the synthesized samples are depicted in Figure. The IR spectrum of the samples shows that a band in the region from 3217-3230 cm^{-1} is due to –NH stretching. Similarly a strong band in the region between 1740 -1685 cm^{-1} is attributed to the carbonyl –C=O stretching. The other bands in the region between 1240- 1260 cm^{-1} (C-O), 1532-1516 cm^{-1} (-NO_2), 785-775 cm^{-1}(-Cl) further confirm the samples structure.[9]

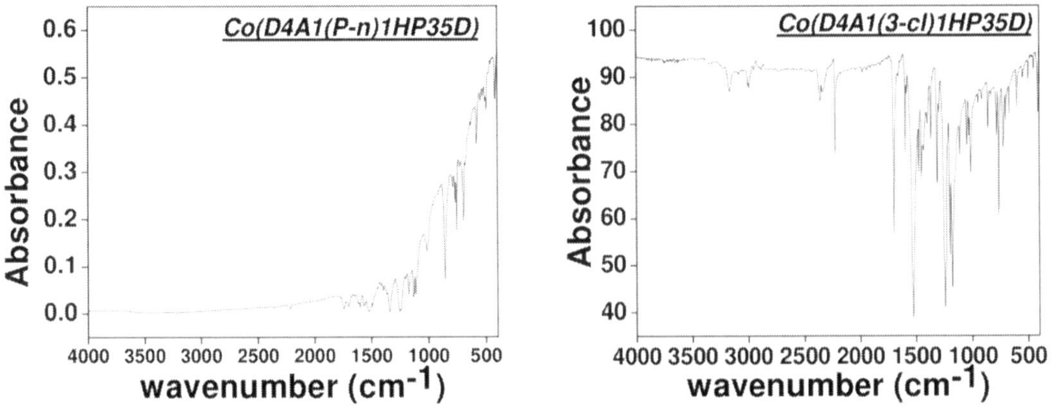

2(a)

2(b)

Figure 2: FTIR spectra of Co(D4A1(P-N)1HP35D) **2(a)** and Co (D4A1(3-Cl)1HP35D) **2(b)**

3.3 Transmission Electron Microscopy (TEM)

The values of particle size of the samples calculated from TEM images for Co (D4A1(3-Cl)1HP35D) is 43.59 nm and Co(D4A1(P-N)1HP35D) 149.55 nm respectively. TEM images are shown in figures 3.

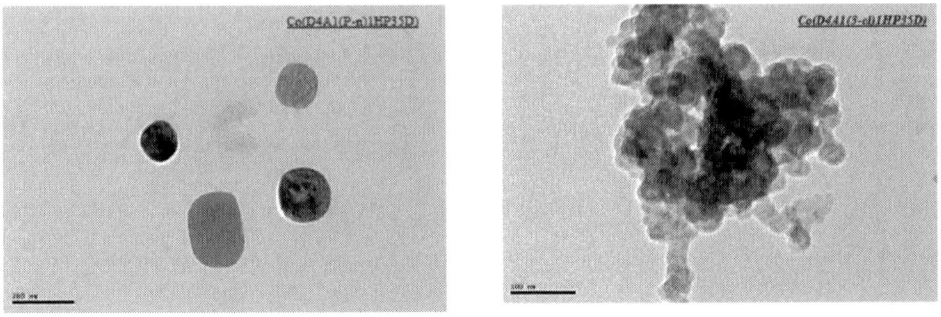

3(a) **3(b)**

Figure 3: TEM images Co(D4A1(P-N)1HP35D) **3(a)** and Co (D4A1(3-Cl)1HP35D) **3(b)**

Acknowledgement:

Thanks to Dr. T. Sripathy, Dr. N. P. Lalla for accessing the characterization facilities at UGC-DAE-CSR, Indore.

References

[1] Brundic A H, Kaitner B, Kamenar B, Leovac V M, Iveges E Z, Juranic N 1991 *Inorg. Chim. Actavol* **188** 151–158
[2] Kukushkin V Yu, Aleksandrova E.A, Leovac V M, Iveges E Z, Belsky V K, Konovalov V K 1992 *Polyhedron* **11** 2691–2696
[3] Leovac V M, Petkovic R, Kovacs A, Pokol G, MeszarosSzecsenyi K 2007 *J.Therm. Anal. Cal.* **89** 267–275
[4] Xia Y, Dong Z W, Zhao B X, Ge X, Meng N, Shin D S, Miao J Y 2007 *Bioorg.Med. Chem.* **15** 6893
[5] Cottineau B, Toto P, Marot C, Pipaud A, Chenault J 2002 *Bioorg. Med. Chem. Lett.* **12** 2105.
[6] Lee K Y, Kim J M, Kim J N 2003 Tetrahedron Lett. **44** 6737

[7] Ghozlan Said AS, Badahdah Khadija O and Abdelhamid Ismail A 2007 *Beilstein Journal of Organic Chemistry* **3** 15
[8] Chauhan H P S, Bakshi A 2011 *J Therm Anal Calorim*, **105** 937–946
[9] Niţu S V, Bercean V N, Badea V, Csunderlik C 2008 *Chem. Bull. "POLITEHNICA" Univ. (Timisoara)* **53(67)** 1-2

Journal of Physics: Conference Series **534** (2014) 012035 doi:10.1088/1742-6596/534/1/012035

Trioctylphosphine and oleylamine induced thermoelectric power of Ag nanoparticles

Jaiveer Singh[1], Ashvini Girothia[2], Kavita Mandre[2], N. Kaurav[2] and Gunadhor S. Okram[3]

[1]Department of Physics, ISLE, IPS Academy, Rajendra Nagar, Indore 452012, MP, India

[2]Department of Physics, Government Holkar Science College, A. B. Road, Indore 452017, MP, India

[3]UGC-DAE Consortium for Scientific Research, University Campus, Khandwa Road, Indore 452017, MP, India

E-mail: jaiveer24singh@gmail.com, netramkaurav@yahoo.co.uk

Abstract. In the present study Ag nanoparticles (NPs) of average particle size of 31-51 nm were prepared by the thermal decomposition method in the presence of appropriate concentration of Trioctylphosphine (TOP) and Oleylamine (OA). The X-ray diffraction measurements indicate the occurrence of face-centered cubic metallic Ag-NPs. The compacted pellets of these nanoparticles were investigated using thermoelectric power in the temperature range from 5 to 300 K. The observed thermoelectric power for all the samples are typical of a good metal and are found to be strongly dependent on grain size and the type of cappant used. The magnitude of thermo electric power increases as the particle size decreases. These results clearly indicate dominant role of disorder at grain boundaries and quantum confinement, which influence scattering of phonons and electrons in a significant way.

1. Introduction

Nano size materials exhibit remarkable structural and physical properties different from conventional macroscopic polycrystalline materials. Atoms at grain boundaries (GB) occupy a volume fraction significantly higher than that of conventional materials. Crystal structure, transport, magnetic, optical and mechanical properties can be influenced due to nanostructure [1], which serves as a bridge between molecule and condensed phase. Nano structured transition metal nanopartilces and their compounds are the most studied materials among other elements in the periodic table. These elements show variable valance states due to their low ionization potential. Over the past decades silver and its compounds have been used extensively in many applications. As an antimicrobial agent, the silver was widely applied in wound treatment, medical devices, water purification, air treatment, cosmetics, aqueous paint etc. Along with the rapid development of nanotechnology, special attention has been foused on the Ag nanoparticles because of their exhibited stronger microbial activity and wider range of applications. Due to high electrical conductivity, the Ag nanoparticles are also applied in conductive inks, adhesives and pastes for variety of advanced electronic devices [2]. In the present investigation the synthesis of Ag NPs by chemical route is discussed, which is an easy, simple and convenient route for preparing metal particles in nanometer range. The prepared silver nano particles were characterized using X-ray diffraction (XRD) and thermopower.

2. Experimental

Silver nanoparticles were prepared by thermal decomposition method [3]. Initially 1.0 gram of silver acetate was mixed with 5 ml of Oleylamine (OA) and 100 ml of ethylene glycol (EG) in a 100 ml 3-neck round bottle flask, equipped with condenser and thermometer. The chemical mixture was heated to 180°C. When temperature reached 180°C, the refluxing arrangements were made. The color of complex changed from off-white to light grey and then finally nearly black after an interval of about 2hr which indicates the initiation of formation of silver nanoparticles. The reaction was continued for another two hour. It was cooled to room temperature and sample was separated in the beaker with the help of 20ml hexane and 50ml ethanol. Then, the obtained precipitate was washed with ultrasonic processor and 3-4 times by ethanol. After washing the precipitate the sample was dried and finally it was obtained in the form of dark grey powder. This sample was coded as Ag-1. Similarly, after keeping all the parameters constant, OA was taken 10ml and 15ml to prepare second and third samples. They were coded as Ag-2 and Ag-3. Further Silver nanoparticles were prepared by Trioctylphosphine in place of Oleylamine and rest all experimental conditions were same. This samples was coded as Ag-4. Similarly, after keeping all the parameters constant, TOP was taken 10ml and 20ml to prepare second and third samples. They were coded as Ag-5 and Ag-6.

These NPs were examined by X-ray diffraction (XRD) using Bruker D8 advanced X-ray diffractometer with Cu Kα radiation in the 2θ range from 30° to 90°. Thermopower measurement down to the liquid helium temperatures (5-300k) were carried out using compacted pellet sandwiched between two oxygen free highly conducting blocks and absolute Thermopower was measured with reference to copper block.

3. Results and discussion

XRD data of samples denoted by Ag-1, Ag-2 and Ag-3 for various silver nanoparticles are shown in Fig. 1. Diffraction peaks were observed and matched with standard JCPDS data of two reports viz (87-0720), (75-0969) and lattice parameters were calculated using the matched (hkl) values. It is clear from Fig. 1 that all peaks correspond to face centered cubic structure of Ag-metal showing that the samples prepared were of pure silver metal nanoparticles. Further, XRD data of samples denoted by

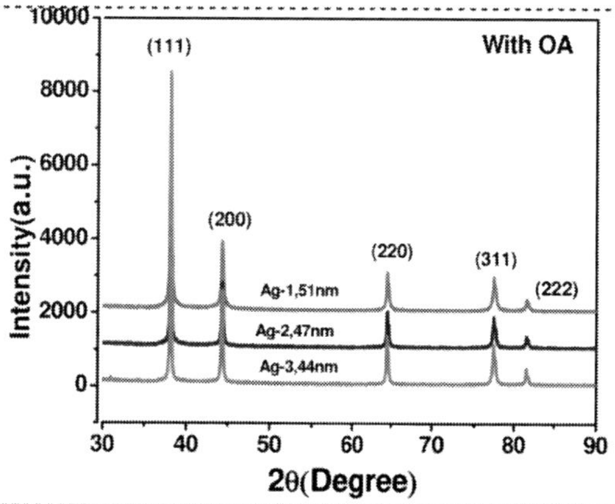

Figure 1. Shows the XRD patterns of Ag-1, Ag-2 and Ag-3 samples for Oleylamine content of 5ml, 10 ml and 15ml, respectively.

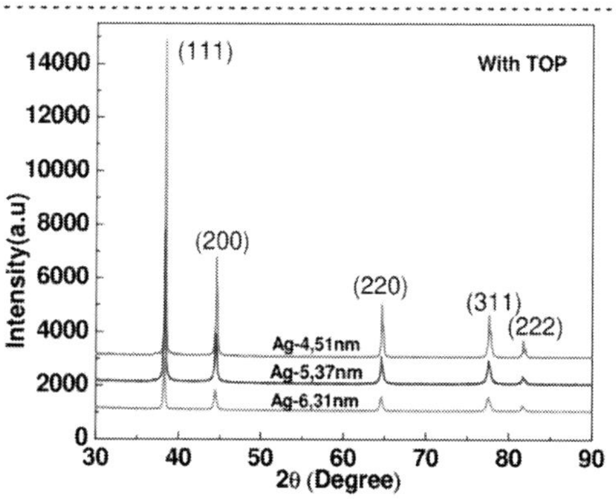

Figure 2. Shows the XRD patterns of Ag-4, Ag-5 and Ag-6 samples for TOP content of 5ml, 10 ml and 20ml, respectively.

Table 1: Particle size and lattice parameter of silver nanoparticles

Sample	Cappants (ml)	Angle (degree)	FWHM (β, degree)	Particle size (D) (nm)	Lattice parameter (nm)
Ag-1	OA(5ml)	38.162	0.19372	51	*0.4081*
Ag-2	OA(10ml)	38.163	0.17597	47	*0.4080*
Ag-3	OA(15ml)	38.149	0.17469	44	*0.4082*
Ag-4	TOP(5ml)	38.35	0.1798	51	0.4098
Ag-5	TOP(10ml)	38.29	0.2389	37	0.4061
Ag-6	TOP(20ml)	38.25	0.2561	31	0.4058

Ag-4, Ag-5 and Ag-6 for various silver nanoparticles are shown in Fig. 2. The observed diffraction peaks were matched with standard JCPDS data of two reports and lattice parameters were calculated using the matched (hkl) values. Fig. 2 shows that all the peaks correspond to face centered cubic structure of Ag-metal showing that the samples prepared were of pure silver metal nanoparticles. The calculated particle size and lattice parameter were reported in Table 1.

Figure 3 and 4 show the variation of thermopower (*S*) of silver nanoparticles with different particle sizes and cappants. For a comparison, the data for bulk Ag (Ag-bulk) is also included from literature [4]. It is clear that in the bulk Ag, the sign of *S* is positive throughout the temperature range (10-300K), indicating the holes as majority charge carriers as consequence of contribution from the necks of the Fermi surfaces [4]. In stark contrast, the *S* of the nanocrystalline silver show almost mirror-reflection of that of bulk Ag. Such an observation indicates that the transport behavior in the samples is completely modified as compared to that of the bulk. In the range ~85 K to 205 K, *S* exhibit positive values, indicating that the holes are majority charge carriers.

On the other hand, *S* shows negative values in the temperature range ~10 to 85 K and 205 to 300 K, indicating that the electrons are the majority charge carriers. Such behavior indicates that the overall contributions of the majority charge carriers are highly temperature-dependent. These results clearly indicate dominant role of disorder at grain boundaries, which influence scattering of phonons and

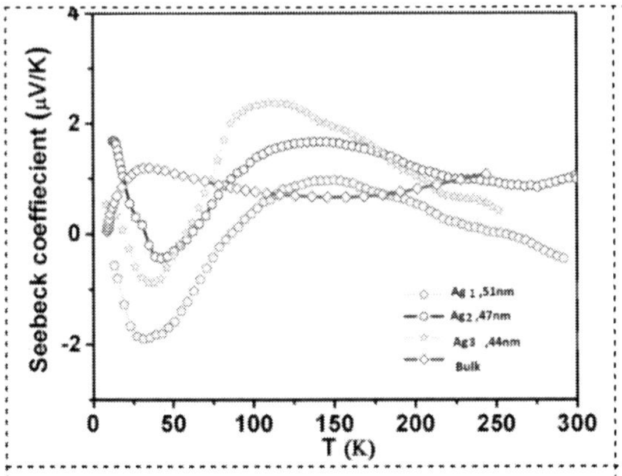

Figure 3. Variation of thermopower with temperature for Ag1, Ag2, Ag3 & Bulk samples.

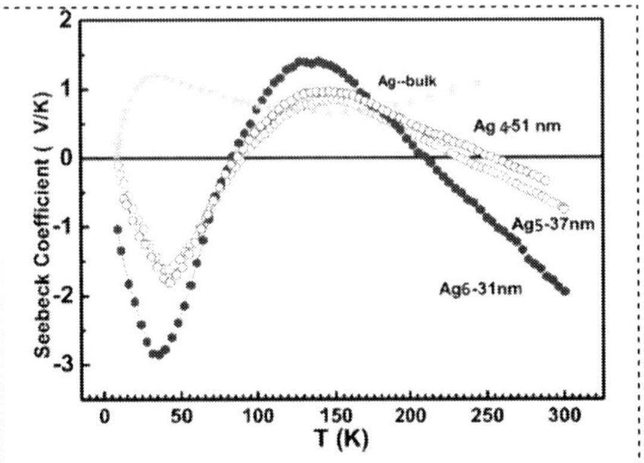

Figure 4. Variation of thermopower with temperature for Ag4, Ag5, Ag 6 & Bulk samples.

electrons in a significant way. Size effect on contributions from electrons and phonons is, in particular, attributed to their effective mean free path being reduced when D is comparable to or smaller than mean free path. In such situation, flow of phonon is restricted i.e. localized continually and transport of phonons is limited thereby decreasing phonon thermo power as D decreases.

4. Conclusion

Nanocrystalline samples were prepared by thermal decomposition method. The estimated particle size is found to be in the range of 31-51nm from XRD data. As we increase the amount of cappant in sample the particle size was found to be decreasing. The magnitude of thermopower decreases as the particle size decreases. These results clearly indicate dominant role of disorder at grain boundaries which influence scattering of phonons and electrons in a significant way.

Acknowledgements:

Dr. Mukul Gupta and Dr. V. Ganesan, Centre Director, UGC-DAE-Consortium for scientific research, Indore (M.P.) is gratefully acknowledged for providing XRD experimental faculties.

References

[1] Christopher P and Linic S 2008 *J. Am. Chem. Soc.* **130** 11264; Gleiter H 1982 *Mat. Sci. Engg.* **52** 91; Weins M J, Gleiter H and Chalmers B L 1971 *J. Appl. Phys.* **42**, 2639

[2] Tam L T, *et al.* 2013 *Appl Phys A* **113** 613

[3] Okram G S and Kaurav N 2011 *J. Appl. Phy.* **110** 023713; Soni A and Okram G S 2009 *Appl. Phys. Lett.* **95** 013101

[4] Mcdonald D K C 1962 *Thermoelectricity: An introduction to the principles* (New York: Wiley–Interscience); Barnard R D 1972 *Thermoelectricity in Metals and Alloys* (Taylor and Francis, UK)

Local structure studies of some cobalt (II) complexes using extended X-ray absorption fine structure

Ashutosh Mishra[1], Samrath Ninama[1], Apurva Trivedi[2]
[1]School of Physics,Vigyan Bhawan ,Devi Ahilya University, Indore, India
[2]Shri Vaishnav Institute of Management, Indore, India

E-mail:samrathn@gmail.com, mishraa1960@yahoo.co.in

Abstract: Extended X-ray Absorption Fine Structure (EXAFS) analysis of Cobalt (II) complex as a ligand of 2 –methyl-3-[(bis-aniline(R) phenyl]-3H-1,5 benzodiazepine for finding local structure using conventional method .The Co(II) complexes were prepared by chemical root method. The EXAFS spectra were recorded at Cobalt K-edge i.e.; 7709 eV using Dispersive EXFAS beam line at 2.5GeV Indus-2 Synchrotron Radiation Source(SRS) at RRCAT, Indore, India. The recorded EXAFS data were analysed using the computer software Athena for determine the nearest neighbouring distances (bond lengths) of these complexes with conventional methods and it compared with Fourier transform(FT) analysis. The Fourier Transform convert EXAFS data signal into r-space or k-space. This is useful for visualizing the major contributions to the EXAFS spectrum.

1. Introduction

In the present work, the X-ray absorption spectra have been recorded at the K-edge of cobalt in two macro-cyclic complexes. The coordination chemistry of macro-cyclic ligand is a fascinating area of research. The extended X-ray absorption fine structure (EXAFS) spectroscopy is an effective technique for selectively investigating the local coordination environment around the metal active site of complexes [1]. The EXAFS K-edge of Co (II) complexes were recorded at 7709 eV and also determined the bond length between metal-ligand of (a) Co1-Co(II)2-methyl-3[(bis-ortho nitophenyl)]-3H 1, 5 benzodiazepine chloride , (b) Co2-Co(II)2-methyl-3[(bis-meta nitrophenyl)]-3H 1, 5 benzodiazepine chloride.

A method for determining the metal-ligand bond length by an appropriate Fourier analysis of the K-edge data has been presented. This is illustrated by a detailed analysis of the data for a series of metal organic based on pyridine ligand systems, which model active centers of natural copper containing metal proteins [1]. The Fourier transform of the experimental EXAFS data is done using the IFEFFIT program [2] and metal-ligand bond length is determined. Applications of EXAFS can be found in literature on a variety of fields, namely, chemical reactions at the solid state [3], surface treatments [4] and structural studies of thin films [5]. A structural analysis of X-ray absorption fine structure (XAFS) spectroscopy has been used for a local structural refinement on an atom of interest [6]. In general, the small pre-edge peaks of the K-edge absorption spectra for transition metal compounds have been assigned to the transition from 1s to *n*d orbital, even though it is a dipole forbidden transition. Since the transition is very sensitive to chemical environment, the pre-edge feature in XAFS studies has been used to infer the local structure around the central atom. Some molecular orbital calculation studies have been reported that the pre-edge peak intensity is related to

3d and 4p orbital mixing by perturbation of site symmetry and increases gradually with the departure from a centro-symmetric environment [7]. In this context, the extended X-ray absorption fine structure (EXAFS) spectra have provided many quantitative structural characterizations such as inter-atomic distance, coordination number and Debye-Waller factor [8]. Since the EXAFS refinement does not depend on the long-range order of atomic arrangement and is very sensitive to atomic local sites, it is a powerful technique for local structural analysis of partially substituted inorganic compounds.

2. Material Synthesis

Preparation of ligand (L)

Cobalt Complexes have been prepared by chemical root method. The chemicals used are of pure grade. O-phenylenediemine (0.05 mol, 5.0g), substituted aniline(R) (0.05 mol, 6.9g) dissolved in ethanol (25 mL), and ethyl acetoacetate (0.05 mol, 6.7g) dissolved in ethanol (25 mL). The mixtures of both reactants were refluxed for 6 hour. The volume concentrated on the steam bath until white colored crystals began to form. The product was purified by crystallization from hot alcohol. Thus white crystalline pure macro-cyclic ligand was collected.

Preparation of [Co (II)-L-Cl_2]

Hot ethanol solution of ligand (L) (0.01 m. mol, 0.2 g), hot ethanol solution of cobalt chlorides [$CuCl_2.6H_2O$] with few drops of hydrochloric acid was added. The solution mixture was then refluxed for 4h. On cooling, colored precipitates were then filtered and washed with cold water and dried in vacuum [8].

3. Experimental

The complexes were finely powdered and then absorption screens were prepared by carefully spreading calculating amount of the powdered on 1 cm area of kapton tape. The cobalt K-edge EXAFS spectra of these absorption screens have been recorded at BL-8 dispersive EXAFS beamline at 2GeV Indus -2 synchrotron radiation source at Raja Ramanna Center for Advanced Technology (RRCAT), Indore, India [9-11]. The energy calibration of beam line for a particular setting of the polychromator can be done by recording the absorption spectra of two standards, whose absorption edge energies are well established. For example for recording the absorption spectra of Co sample, the two standards can be Co metal foil and Ho oxide powder. The absorption spectra of Co K edge and Ho_2O_3 L_3-edge have been recorded under the same setting of crystal bender and goniometer. In the present work we are taking the values of the energies of Co K-edge in metal and Ho L_3 –edge in Ho_2O_3 as 7709 eV and 8074 eV respectively [13].

4. Results and discussion

The X-ray absorption spectra, i.e., normalized μ (E) vs. E curve of Cobalt complexes is shown in figure 2. The bond lengths have been determined using the following four methods, i.e., Levy's, Lytle, Lytle, Sayers and Stern's (LSS) and Fourier transformation methods and the results obtained are given in table1.

4.1 Levy's method

The bond lengths According to Levy's method [14], are given by Bragg relation $R_1 = [151/\Delta E]^{1/2}$ Å, where ΔE is the difference in eV of the energies of the EXAFS maximum B and minimum β and R_1 is the radius of the first coordination sphere.

4.2 Lytle's method

According to Lytle [15] method the bond length (R_s) is evaluate by the radius R_s of equivalent polyhedron through the relation $R_s = [37.60 / M]^{1/2}$.where M is slope of E vs Q plot ,energy values (E) is EXAFS maxima and Q = 2.04, 6.04, 12.0, and 20.0 are constant . The values of R_s calculated with the help of this method are reported in table 1.

4.3 Lytle, Sayers and Stern's (LSS) method

In the LSS method [16], for determination of the nearest neighbor distances, gives the value of (1/2+n) $\pi = 2k(R_1 - \alpha_1) + 2\beta_1$,where R_1 is the bond length. $(R_1 - \alpha_1)$ is the phase uncorrected bond length shown in table 1. Although, Fourier transform of EXAFS data is a general and powerful technique to determine many physical parameters; however, it is possible to determine nearest-neighbur distances and phase shifts from a simple graphical technique. This is possible because the first shell scattering usually dominates the EXAFS curves; then, if only the major EXAFS peaks are used, the period will be that of the first coordination shell. In this method, the energy positions of the maxima and minima in EXAFS are measured with respect to the inflection point on the main absorption edge, which signifies the Fermi level in metals and the first unoccupied level of suitable symmetry available for the absorption of the ejected electron in the case of compounds.

4.4 Fourier transformation method

The magnitude of Fourier transform spectra are shown in figure 3. The Fourier transforms peaks at the radial distances of the neighboring atoms from the absorbing atom. The distances found in Fourier transform are, however, shorter by 0.2-0.5 Å than the actual distances due to energy dependence of the phase factors in sine function of the theoretical expression for EXAFS, which is known as the EXAFS equation [17]. The peaks in the Fourier transform are shifted towards the origin by an amount α_j and hence the peaks are at distances R_j-α_j. For the first peak j=1 and hence the position of the first peak determines the distance R_1-α_1. This distance is given in table 1. There are some subtleties associated with doing the FT of EXAFS data but techniques for dealing with these are well developed. Any signal that is a sum of sine waves can be Fourier transformed to separate each frequency component in distance R, i.e., a Fourier transform separates a signal into its Fourier components. Each Fourier component is defined by an amplitude and phase. The Fourier transform is a complex function containing both real and imaginary parts. The magnitude of the Fourier transform is an envelope of the real and imaginary parts of the Fourier transform.

Table 1 Bond length (Rs) of by Levy's, Lytle, LSS and F.T. method of Co (II) complexes.

Complex	Levy's	Lytle	LSS	F.T.
Co foil	2.10	2.12	1.75	2.02
Co1	2.14	1.97	1.86	2.01
Co2	2.18	1.95	1.72	1.99

5. Conclusion

Calibration of particular setting of the polychromator has to done and it has been shown by derivative method, and after this EXAFS spectra has been recorded and comparison of the values of bond length R_s is seen to have slightly lesser values than the corresponding bond lengths obtained by Levy's method .This may be described to the fact that Levy's method gives the radius of coordination sphere directly whereas Lytle method does not. In later method, the inter atomic spacing is obtained by multiplying R_s with a factor Appropriate to the geometry of the system .the value of the bond length obtained from LSS method are closer to obtained by Levy's method and Fourier transformation method.

Acknowledgement

The authors are thankful to Dr. Shambhu Nath Jha, RRCAT.,Indore for their help in recording the spectra at RRCAT, Indore, India and also thankful to Dr.Pratibha Sharma, school of chemical science, Devi Ahilya University, Indore - 452001, India for their help in preparation of Co(II) complexes.

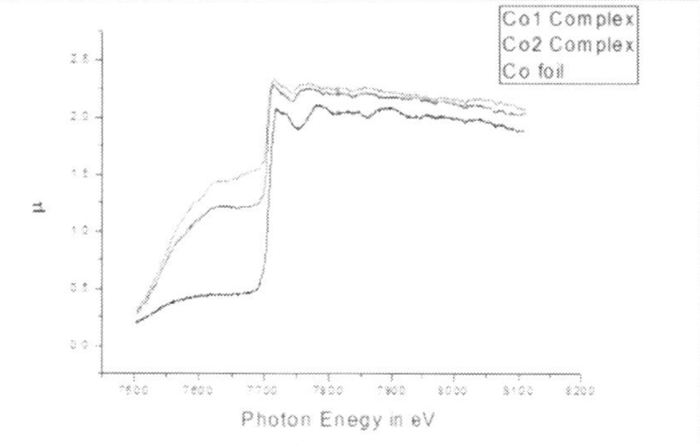

Figure 1- EXAFS spectra of Co (II) complexes.

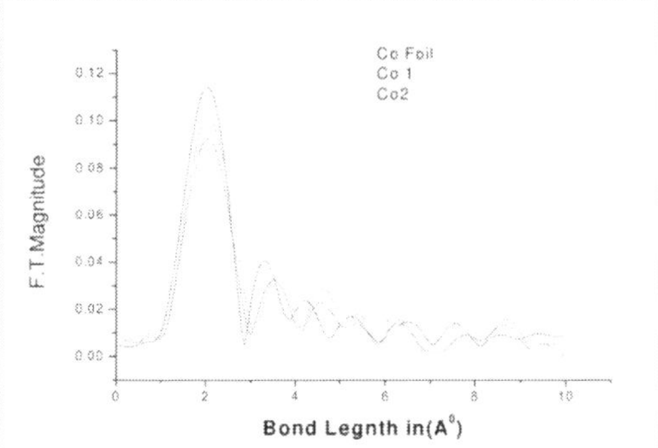

Figure 2- Magnitude of Fourier transformation in first shell maxima for Co (II) complexes.

References

[1] Vlasenko V G, Uraev A I, Zubavichus Ya V, Chernyshov A A, Garnovskii A D, & Mamin R K, 2008 *Bull. Russian Acad. Sci.: Phys.***72** 468.
[2] Newville M 2001 *J. Synch. Rad.* **8** 322.
[3] d'Acapito F, Ghigna P, Alessandri I, Cardelli A, & Davoli I,2003, *Nucl Instrum Methods: Phys Res B* **200** 421.
[4] Hecht Lutzenkirchen and Frahm R, 2005 *Physica B* **357** 213.
[5] d'Acapito F , Milita S, Satta A, and Colombo L 2007 *J Appl Phys* **102** 043524.
[6] Rehr J J, Leon J M, Zabinsky S I, Albers R C 1991, *J Am Chem Soc* **113** 5135.
[7] Westre T E, Kennepohl P, DeWitt JG, Hedman B, Hodgson K O, Solomon E I 1997 *J Am Chem Soc* **119** 6297.
[8] Mayura Panchbhai A et al. 2008 *E-Journal of Chemistry* **S1** 5.
[9] Bhattacharyya D., Poswal A. K., Jha S. N., Sangeeta , Sabharwal S. C. 2009, *Bull. Mater. Sci.* **32** 103.
[10] Bhattacharyya D, Poswal A K, Jha S N, Sangeeta , Sabharwal S C, 2009, *Nucl. Instrum. Meth. A*; **609** 286
[11] Das N C, Jha S N, Bhattacharyya D, Poswal A K, Sinha A K and Mishra V K 2004 *Sadhana* **29** 545.
[12] Lee P. L., Beno M. A., Jennings G., Ramanathan M., Knapp G. S., Huang K., Bai J., Montano P. A. 1994 *Rev. Sci. Instrum* **65**.
[13] Matthew A. Marcus, Andrew J. Westphal and Sirine C. Fakra, 2008 *J. Synchrotron Rad.* **15**, 463
[14] Levy R. M. 1965 *J. Chem. Phys.* **43** 1846
[15] Lytle F .W. 1966 *Advances in X-ray Analysis* **9** 398.
[16] Lytle F. W., Sayers D. E. and Stern E. A. 1975 *Phys. Rev. B.* **11** 4825.
[17] Stern E.A., Sayers D. E. and Lytle F. W. 1975 *Phys. Rev. B.* **11** 4836.

Origin of the satellites $L\alpha_3$, $L\alpha_4$ and $L\alpha_5$ in the elements from $_{40}$Zr to $_{50}$Sn

Renuka Kendurkar and B. D. Shrivastava
School of Studies in Physics, Vikram University, Ujjain -456010, India.

E-mail: renukakendurkar2511@gmail.com; rashmibasant@gmail.com

Abstract. The origin of the $L\alpha$ satellites $L\alpha_3$, $L\alpha_4$ and $L\alpha_5$ have been explained in the elements from $_{40}$Zr to $_{50}$Sn, on the basis of multiple ionization theory. The energies and intensities of the various transitions corresponding to the L_3M_x - $M_xM_{4,5}$ (where x = 1-5) transition array, which may give rise to these satellites, have been calculated theoretically. The energies of the transitions have been calculated using the available Hartree-Fock-Slater data for the energies of K-LM and L-MM Auger transitions. The intensities of the various transitions have been estimated by considering cross sections for L_1-L_3M_x Coster-Kronig transitions as well as for M-shell shake-off process occurring simultaneous to a L_3 hole creation. The total cross sections for initial two-hole states L_3M_x have then been distributed statistically amongst the various allowed transitions from these initial states to the final states $M_xM_{4,5}$. By assuming each transition as a Gaussian line, theoretical satellite spectrum has been computed as the sum of these Gaussian curves. The energies of the satellites, as obtained from the theoretical spectrum, have been found to be comparable with the measured energies of the satellites $L\alpha_3$, $L\alpha_4$ and $L\alpha_5$. Consequently, these satellites have been assigned the transitions.

1. Introduction

$L\alpha$ satellites $L\alpha_3$ (or $L\alpha'$), $L\alpha_4$ (or $L\alpha''$) and $L\alpha_5$ (or $L\alpha'''$) have been observed [1] in the elements from $_{40}$Zr to $_{50}$Sn. Poonia and Soni [2] have studied theoretically the $L\alpha$ satellites in some of these elements . In their calculations they have used the Coster-Kronig transition probabilities from the tables of McGuire [3] based on non-relativistic calculations. The tables of McGuire were superseded by the tables of Chen *et al.* [4], who have done relativistic calculations. Hence, we thought it proper to do fresh calculations using the tables of Chen *et al.* and to make a comparison of the present calculations with those of Poonia and Soni. Also, we have tried to explain the origin of the satellites in all the elements in which it has been observed.

2. Theoretical considerations

2.1. Initial and final states for the $L\alpha$ satellites

Poonia and Soni have assigned the origin of $L\alpha$ satellites to the transition array $2p^{-1}3x^{-1}$ - $3x^{-1}$ $3d_{3/2,5/2}^{-1}$ (x = s, p, d). Hence, we have also assumed that the origin of the $L\alpha$ satellites observed near the dipole lines $L\alpha_1(L_3$-$M_5)$ and $L\alpha_2$ $(L_3$-$M_4)$ can be ascribed to the transition scheme $2p^{-1}3x^{-1}$ - $3x^{-1}$ $3d_{3/2,5/2}^{-1}$ (x = s, p, d), i.e., L_3M_x - $M_xM_{4,5}$, (X = 1....5). The hole in M_x subshell remains a spectator during the transition, which gives rise to the satellite, and is called spectator vacancy. L_3M_x is initial state and $M_xM_{4,5}$ is final state. All the possible initial $2p^{-1}\,3x^{-1}$ (L_3M_x) states and final $3x^{-1}\,3d_{3/2,5/2}^{-1}$ ($M_xM_{4,5}$) states are calculated in LS coupling scheme. Considering the dipole selection rules, 49 transitions have been found out corresponding to the transition scheme L_3M_x - $M_xM_{4,5}$. In the present work, the energies and intensities of these 49 transitions have been calculated and then theoretical satellite spectra have been synthesized.

2.2. Calculation of energies of transitions in doubly ionized atoms

The formula for calculation of transition energy of say, $^3P_1L_3M_3 - {}^3P_0M_3M_5$ transition, has been written as

$$E\,({}^3P_1L_3M_3 - {}^3P_0M_3M_5) = E\,(K\alpha_1) + E\,(L_3 - {}^3P_0M_3M_5) - E\,(K - {}^3P_1L_3M_3)$$

The energy of $K\alpha_1$ line, i.e., E (K - L_3) has been taken from the tables of Cauchois and Senemaud [1]. The energies of the Auger transitions L_3 - $^3F_3M_3M_5$ and K - $^3D_2L_3M_3$ have been taken from the tables of Larkins [5].

2.3. Calculation of intensities of transitions in doubly ionized atoms

After the creation of a single-hole state, it is the probability of a particular subsequent process that leads to the formation of initial two-hole state, required for the emission of a satellite. The single-hole state can get converted into the two-hole states by two processes, i.e., Auger transition and shake-off process. To calculate the probability of creation of the initial double vacancy state by the process of Coster-Kronig transition, the cross section for creation of single vacancy state is first calculated by the formulae given by Moores *et al.* [6] and then this cross section is multiplied by the transition rates for the Coster-Kronig transition as given in the tables of Chen *et al.* [4] The singly ionized state may get converted to doubly ionized state also by an associated shake-off of an electron. Hence, to calculate the probability of creation of the initial double vacancy state by the shake-off process, the cross section for creation of single vacancy state, as calculated above, is multiplied by the shake-off probability as calculated from the reports of Carlson *et al.* [7]. Finally, the total probability of creation of doubly ionized initial state can be determined by adding both the cross sections calculated above, i.e., by Coster-Kronig transition and by shake-off process. This probability can then be distributed statistically among all the allowed transitions using the tables of White and Eliason as given by Condon and Shortley [8]. In this method, all the multiplets of supermultiplets from various levels of the set are considered.

2.4. Synthesis of satellite spectra

A composite spectrum formed by spectral lines emitted by the 49 transitions has been computed by taking each as a Gaussian line. A representative theoretical satellite spectrum for the element $_{48}$Cd is shown in Fig. 1. The three satellites under investigations, i.e., $L\alpha_3$, $L\alpha_4$ and $L\alpha_5$ along with the transitions giving rise to them are clearly visible in this figure. The observed and theoretical positions are also marked in the figure. The theoretical spectra have been compared with the available experimental results for energies of the satellites. The intense peaks have been identified as the observed satellite lines and the transition assignment to the satellites, based on the identification of the peaks to the particular transitions, has then been done.

3. Result and discussion

In the middle-Z elements from $_{40}$Zr to $_{50}$Sn, the theoretical spectra of $L\alpha$ satellites may have at least 8 peaks in each of these elements under consideration, In $_{50}$Sn the number of peaks is as large as 13. Some peaks are strong and should correspond to the observed satellites. Table 1 gives the theoretical energies and relative intensities of the satellites along with experimental values of the energy.

3.1. Satellite Lα' (or Lα₃)

In the present study, the second strong peak in the computed spectra has been identified as the Lα' (or $L\alpha_3$), satellite which has been reported in almost all the elements. The transition which gives rise to this peak and hence to the satellite Lα' (or $L\alpha_3$) is A, i.e., $^3F_4$ - $^3F_4$ (3d). Transition U, i.e., $^3P_1$ - $^3P_0$ (3d) also contributes to the intensity of the satellite.

3.2. Satellite Lα'' (or Lα₄)

In the present study, the third peak in the computed Lα satellite spectra has been identified as the Lα'' (or $L\alpha_4$) satellite. Considering the various transitions which give rise to this peak, we have found that the transitions D, V, C, W, G, g, i and u, i.e., $^1F_3$-$^1G_4$ (3d), $^3P_0$-$^3P_1$ (3d), $^3P_2$-$^3D_3$ (3s), $^1P_1$-$^1D_2$ (3p), $^3P_1$-$^3D_2$ (3s), $^3D_2$-$^3P_1$ (3p), $^3P_1$-$^3D_1$(3s) and $^3P_1$-$^3D_1$(3s) transitions are responsible for the origin of the satellite Lα'' (or $L\alpha_4$) in all the elements with Z = 40 to 50.

It may be remarked here that in the elements $_{40}$Zr, $_{41}$Nb, the satellite Lα'' (or $L\alpha_4$) appears as a clear peak, while the satellite Lα' (or $L\alpha_3$) appears as a shoulder on the lower energy side of this clear peak. On the contrary, in the elements $_{42}$Mo, $_{43}$Tc and $_{44}$Ru, the satellite Lα' (or $L\alpha_3$) appears as a clear peak, while the satellite Lα'' (or $L\alpha_4$) appears as a shoulder on the higher energy side of this clear peak. This is an important result of the present study.

Table 1(a) Assignments of transitions to Lα satellites $L\alpha_3$ and $L\alpha_4$ in the elements from $_{40}$Zr to $_{50}$Sn.

S. No	Z	Theoretically computed data for satellite $L\alpha_3$: Present work, Transitions A,U: Energy (eV)	Normalized relative intensity	Poonia 's work: Energy (eV)	Relative intensity	Measured energy (eV)	Theoretically computed data for satellite $L\alpha_4$: Present work, Transitions D,V,C,W,G,g, i,u: Energy (eV)	Normalized relative intensity	Poonia 's work: Energy (eV)	Normalized relative intensity	Measured energy (eV)
1	40	2048.8	115.50	2048.6 2048.6	311.15 311.50	2048.9	2049.2	142.57	-	-	2051.0
2	41	2172.9	137.83	-	-	2172.4	2174.3	84.10	-	-	2174.6
3	42	2300.1	124.80	2300.1	194.52	2298.3	2301.4	78.61	2301.2	112.41	2299.7
4	43	2432.9	136.36	-	-	-	2434.4	99.9	-	-	-
5	44	2566.1	142.40	2566.3	153.30	2565.4	2568.1	100	2567.4	94.95	2568.1
6	45	2705.1	131.50	-	-	2704.1	2706.9	113.81	-	-	2706.7
7	46	2847.1	137.20	2847.0	112.83	2846.2	2848.9	136.1	2848.4	74.99	2849.0
8	47	2993.1	112.50	-	-	2992.4	2995.9	134.66	-	-	2995.0
9	48	3143.0	111.17	3143.1	14.87	3142.5	3144.9	160.83	3144.9	9.95	3145.2
10	49	3296.2	107.56	-	-	3296.1	3298.8	145.84	-	-	3298.8
11	50	3454.1	108.20	-	-	3451.7	3456.1	134.01	-	-	3456.3

3.3. Satellite Lα''' (or Lα₅)

The fourth peak on the higher energy side of Lα'' (or $L\alpha_4$) satellite and well separated from it, has been identified as the satellite Lα''' (or $L\alpha_5$) in the theoretical Lα satellite spectra. The transitions which give rise to this peak and hence to the satellite Lα''' (or $L\alpha_5$) are E, F, S and T, i.e., transitions $^3D_3$-$^3F_4$ (3d), $^3D_2$-$^3F_3$ (3d), $^1F_3$-$^1D_2$ (3d) and $^1P_1$-$^1D_2$ (3d) along with some weak transitions. Poonia and Soni and Poonia have also assigned the same transitions to this satellite.

Table 1(b) Assignments of transitions to Lα satellite $L\alpha_5$ in the elements from $_{40}Zr$ to $_{50}Sn$.

S. No	Z	Theoretically computed data for satellite $L\alpha_5$				
		Present work Transitions E, F,S,T		Poonia's work		Measured energy (eV)
		Energy (eV)	Normalized relative intensity	Energy (eV)	Relative intensity	
1	40	2052.3	188.03	2052.1	309.74	2053.0
2	41	2176.3	152.42	-	-	2176.5
3	42	2304.0	147.26	2304.2	201.53	2302.0
4	43	2436.9	159.48	-	-	-
5	44	2570.9	156.08	2570.9	164.4	2570.5
6	45	2710.0	150.59	-	-	2709.4
7	46	2852.1	136.21	2852.2	122.75	2852.1
8	47	2999.1	101.52	-	-	2998.3
9	48	3148.9	151.6	3148.9	15.19	3148.5
10	49	3303.2	138.26	-	-	3304.1
11	50	3460.1	109.04	-	-	3460.1

Measured energies are from Cauchois and Senemaud [1].

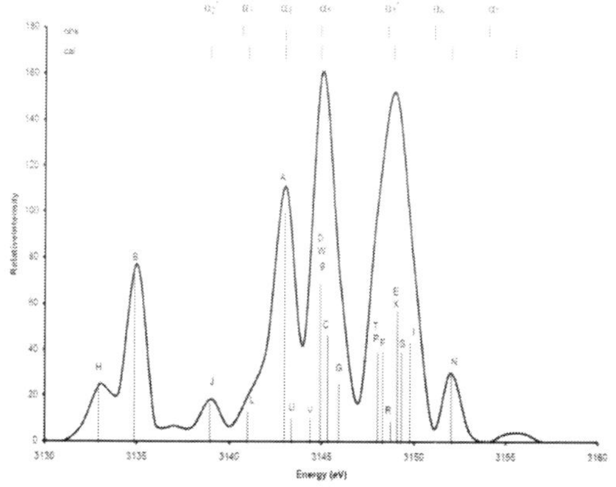

Fig.1 The computed $2p_{3/2}^{-1}$ $3x^{-1}$ - $3x^{-1}$ $3d^{-1}$ (L_3M_x - M_x $M_{4,5}$) spectrum of $_{48}Cd$. The satellite positions experimentally observed and theoretically calculated in the region of Lα satellites are shown at the top.

4. Conclusion

In the present study, the origin of the Lα satellites $L\alpha_3$ (or Lα'), $L\alpha_4$ (or Lα") and $L\alpha_5$ (or Lα"') have been explained in the elements from $_{40}Zr$ to $_{50}Sn$, i.e., in all the elements in which they have been observed. The present assignment of the transitions to these satellites is different from those of Poonia and Soni [2]. This is because, the present assignments are based on relativistic calculations while those of Poonia and Soni are based on non-relativistic calculations.

Reference

[1] Cauchois Y and Senemaud C 1978 *Tables Internationnal de constantes selectionnnees. 18.Longeurs d'onde des emissions X et des discontinuites d'absorption X* (Pergamon Press).
[2] Poonia S and Soni S N 2002 *Ind. J. Pure & Appl. Phys* **40** 786.
[3] McGuire E J 1971 *L-Shell Auger, Coster-Kronig and radiative matrix elements, and Auger and Coster-Kronig transition rates in j-j coupling* (Report No. Sc-RR-710075) (New Mexico: Sandia Laboratories).
[4] Chen M H, Crasmann B and Mark H 1979 *At. Data Nucl. Data Tables* **24** 13.
[5] Larkins F P 1977 *At. Data Nucl. Data Tables* **20** 311.
[6] Moores D L, Golden L B and Sampson D H 1980 *J. Phys.* B **13** 385.
[7] Carlson T A and Nestor C W Jr. 1973 *Phys. Rev.* A **8** 2887.
[8] Condon E U and Shortley G H 1967 *The theory of atomic spectra* (Cambridge) p.241

Study of structural and magnetic properties of $Fe_{73.5}Si_{3.8}C_{14}Mn_{0.7}B_4Al_4$ alloy

P Tapkir[1], M Satalkar[1], M Shah[1],N Ghodke[2], L K Varga[3] J P Araujo[4] and S N Kane[1]

[1] School of Physics, D. A. University, Khandwa Road, Indore-452001, India
[2]UGC-DAE Consortium for Scientific Research, University Campus, Khandwa Road, Indore-452001, India
[3] Research Institute for Solid State Physics and Optics, Hungarian Academy of Sciences, P.O. Box 49, 1525 Budapest, Hungary
[4]IFIMUP, Departamento de Fisica, Universidade de Porto, 4169-007 Porto, Portugal

E-mail: kane_sn@yahoo.com

Abstract. Present work, reports the influence of thermal annealing on magnetic and structural properties of $Ci_{92}B_4Al_4$ ($Fe_{73.5}Si_{3.8}C_{14}Mn_{0.7}$= Ci – Cast iron) alloy using, magnetic measurements, differential scanning calorimetry (DSC) and x-ray diffraction (XRD) to obtain information on structure, formed nano-crystalline phases and their influence on soft magnetic properties. Structural changes were achieved by annealing of the samples performed at 350, 370, 400, 425 and 450 °C for one hour. Studied specimen shows two-step crystallization, and the activation energy of crystallization, obtained using Kissinger's method, was 3.79 ± 0.6 eV (for main peak) and 3.02± 0.3 eV (for pre peak). Coercive field H_c of the studied samples varies between 22.06 – 838.67 A/m. Best coercivity (H_c) value of 22.06 A/m was obtained for the as cast sample. The measured saturation induction (B_{2000}) values range between 0.89 and 1.31 Tesla. Best saturation induction (B_{2000}) value of 1.31 Tesla was obtained for the as cast sample. XRD data shows that apart from α-Fe, phase (lattice parameter ~ 0.2859 nm) an additional Fe-Al phase is also formed, with Al ranging between 53 – 64 %, responsible for the reduction of magnetic induction as well as the increase of coercivity.

1. Introduction

Metallic glasses exhibit excellent soft magnetic properties, which can be further improved by choosing a specific composition and, also by post-preparation treatments. Materials exhibiting soft magnetic properties are very important for applications e.g. –as a core material for transformers, where lower core losses (~ 30% less compared to their crystalline counterparts). Soft magnetic materials showing low cores losses are important from energy saving applications point of view. For the same recently Ci-Al-B (Ci-cast iron) based amorphous alloy exhibiting excellent soft magnetic properties and also high saturation induction. As mentioned before, soft magnetic behaviour of Ci-Al-B based amorphous alloys can be further improved by thermal treatments [1,2]. Almost all the Fe-based bulk and bulk-type metallic glasses (BMGs) with high glass forming ability and good soft magnetic properties, high fracture were inevitably fabricated using high purity elements under high vacuum which unavoidably limits their industrial applications because of their high unit cost and low production efficiency. To reduce the production cost, development of BMG's based on the pig-iron alloys directly from blast furnaces under low vacuum conditions have been attempted in many laboratories. However, a combination of good magnetic properties and high glass forming ability is difficult to be attain simultaneously due to adverse effects of oxygen and other impurities [3]. From the point of view of the use of bulk metallic glasses in various applications, it is highly desirable to have the knowledge of structural and magnetic properties and the influence of thermal treatments on the mentioned properties. Present work, reports the influence of thermal annealing as a function of temperature, to obtain better magnetic properties in $Ci_{92}B_4Al_4$ ($Fe_{73.5}Si_{3.8}C_{14}Mn_{0.7}$= Ci – Cast iron) alloy.

2. Experiment details

Ribbons (about 30 μm thick and 1.35 mm wide) of nominal composition $Fe_{73.5}Si_{3.8}C_{14}Mn_{0.7}B_4Al_4$ ($Fe_{73.5}Si_{3.8}C_{14}Mn_{0.7}$= Ci – Cast iron) were prepared using a planar flow casting technique. Thermal annealing induced change in magnetic, structural properties were monitored using, magnetic measurements, differential scanning calorimetry (DSC) and, X-ray diffraction (XRD) to get information on correlation between formed nano-crystalline phases and magnetic properties. DSC measurements were done at four heating rates: 5 °C, 10 °C, 20 °C and 40 °C. Following DSC results, thermal annealing was done at 350, 370, 400, 425 and 450 °C for 60 minutes Bruker D8 Advance X-ray diffractometer was used to perform XRD measurements on as-cast and annealed samples using Cu-K_α radiation (λ = 0.154 nm). XRD patterns were analyzed by fitting a crystalline and, amorphous component using pseudo-Voigt profiles. The lattice parameter a is calculated using Nelson-Taylor-Sinclair correction in order to take into account the peak shift due to sample offset. For the amorphous phase, the first nearest neighbour distance X_m is calculated using the expression $X_m = 1.227\ \lambda/\ 2\mathrm{Sin}\theta$ [5]. Hysteresis loops of as cast and annealed samples were measured at 50 Hz (H_{max} = 2000 A/m) to obtain coercive field (H_c) and, saturation induction (referred as 'B_{2000}'), measured at an applied magnetic field of 2000 A/m. For each specimen two hysteresis loops - one each at low field (*was used to measure* H_c) and the high field (*was used to measure Induction* B_{2000} values) loops were measured.

3. Results and discussions

Figure 1 depicts the DSC curve recorded with four different heating rates (left panel) and, Kissinger plots for calculating activation energy (right panel). Perusal of figure 1 (left panel) shows that the crystallization occurs in two steps. The activation energy of crystallization, calculated using the Kissinger's method, was estimated to be 3.79± 0.6 eV (for main peak) and 3.02± 0.3 eV (for pre peak) and both the peaks can be seen in figure 1 (left panel).

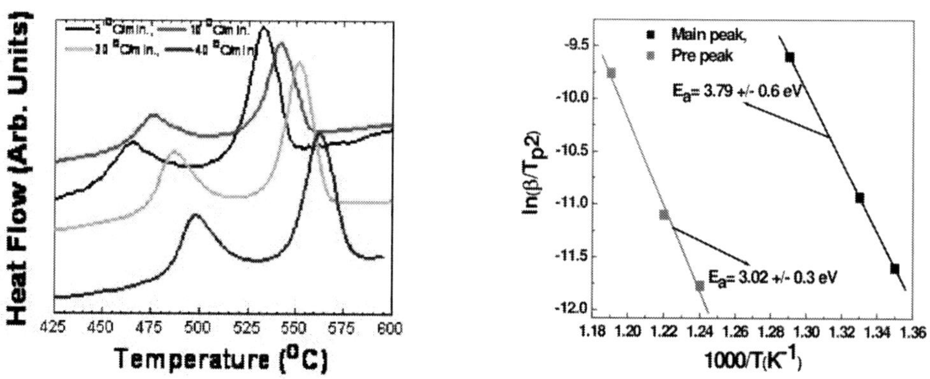

Figure 1: DSC curve of as-cast specimen, measured at different a heating rates (left panel) Kissinger plots of $Fe_{73.5}Si_{3.8}C_{14}Mn_{0.7}B_4Al_4$ for calculating activation energy (right panel).

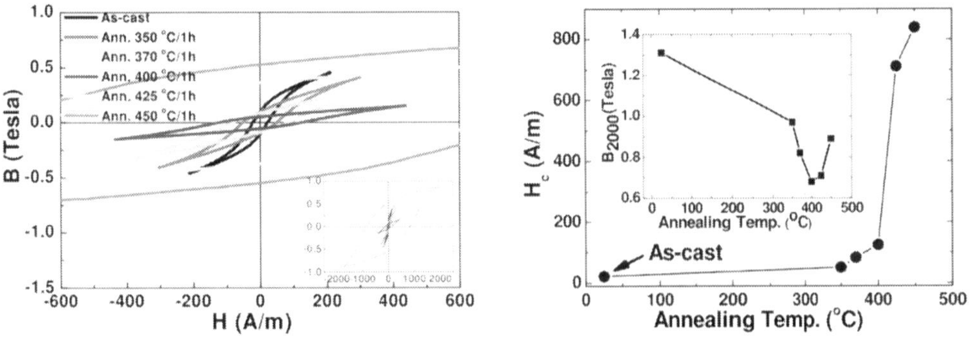

Figure 2: Thermal annealing induced changes in the hysteresis loops (*left panel*) and, Inset : Full loops; coercivity (*right panel*) as a function of annealing temperature and Inset: B_{2000} variation.

Figure2 (*Left panel*) depicts the annealing temperature dependence of hysteresis loops of the studied samples, (INSET: Full loops). Figure 2 (*Right panel*) depicts the annealing temperature dependence of the coercivity (Inset: variation of B_{2000}). Obtained magnetic parameters are summarized in table 1.

Perusal of fig. 2 (*Right panel*) depicts the variation of annealing temperature dependence of coercivity shows the increase of H_c with increase in annealing temperature. Observed sharp increase of H_c after annealing at temperatures higher than 350°C can be ascribed to grain-coarsening with annealing hindering the domain wall motion [see e. g. – 4], precipitation of Fe-Al phase, resulting in the decrease of B_{2000} with increasing annealing temperature, as can be seen in the inset of fig. 2 (Right Panel). For the studied sample as a function of temperature, H_c values range between 22.06-838.67 A/m, whereas the best H_c value of 22.06 A/m is obtain at as-cast sample. B_{2000} (saturation induction) Bs values range between 0.5-1.31 Tesla; best B_{2000} value of 1.31 Tesla is obtained in the studied as-cast sample. Obtained magnetic parameters are summarized in Table 1. Perusal of table 1 shows that, increase of H_c leads to increase of losses. Obtained low lossees after annealing at 400 °C/1h can be ascribed to the measured loop at low field, where full saturation is not obtained (as can be seen in the fig. 2 left panel).

Figure 3 shows the XRD pattern of specimens annealed for various temperatures for 1 hour.Perusal of figure 3 reveals that as-cast sample also shows an amorphous component, a small fraction of crystalline α-Fe phase, and another crystalline phase ascribable to Fe-Al phase. We have shown two dotted vertical lines (see fig. 3) on the peaks corresponding to Fe-Al phase and α-Fe phase. Perusal of figure 3 clearly shows that as the annealing temperature increases, the volume fraction of Fe-Al phase decreases and the volume fraction of α-Fe phase increases. Analysis of XRD data was done using a

Table1. Variation of magnetic properties as a function of temperature Where - $\boldsymbol{H_c}$ – coercive field, $\boldsymbol{B_s}$ – saturation induction measured at field strength of 2000 A/m, core losses (at 50 Hz).

Sample	H_c (A/m)	B_s (Tesla)	Losses (W/Kg)
As cast	22.06	1.31	0.63
Ann. at 350^0 C / 1h	51.14	0.97	0.93
Ann. at 370^0 C / 1h	83.38	0.82	1.6
Ann. at 400^0 C / 1h	124.52	0.68	0.5
Ann. at 425^0 C / 1h	711.50	0.71	12.81
Ann. at 450^0 C / 1h	838.67	0.89	17.03

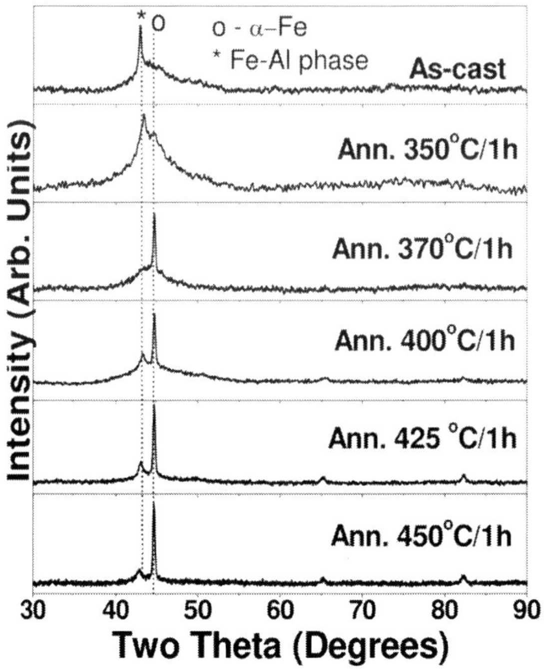

Figure 3: Annealing temperature dependence of XRD patterns.

Matlab® based program, described elsewhere [6], in order to obtain average Scherrer's grain diameter $\boldsymbol{D}$, volume fraction of the nano grains $\boldsymbol{V_x}$,lattice parameter $\boldsymbol{a}$, and for the amorphous phase, the first nearest neighbour distance $\boldsymbol{X_m}$.

Table 2 Grain size < $\boldsymbol{D}$ >, lattice parameter $\boldsymbol{a}$, volume fraction of nano grains $\boldsymbol{V_x}$, for nano crystalline specimens showing temperature transformation, as obtained after analyzing XRD measurements. For comparison coercive field $\boldsymbol{H_c}$ are also given, as obtained from magnetic measurements.

Sample details	$\langle D \rangle$ (nm) (± 1)	a (nm) (± 0.0001)	V_x (%) (± 2)	X_m (nm) (± 0.001)	H_c (A/m) (± 1.0)
As-cast	12	0.2892	17	0.248	22.06
	19	0.2973	12	0.248	
Ann. at	11	0.286	7	0.249	51.14
350 ^{0}C /1h	12	0.2941	6	0.249	
Ann. at	23	0.2855	20	0.252	83.38
370 ^{0}C /1h	13	0.2941	12	0.252	
Ann. at	18.5	0.2859	22.5	0.248	124.52
400 ^{0}C/1h	16.5	0.2945	13	0.248	
Ann. at	23	0.2860	36	0.248	711.50
425 ^{0}C /1h	22	0.2965	22	0.248	
Ann. at	22	0.2864	30	0.246	838.67
450 ^{0}C /1h	19	0.29731	12	0.248	

Table 2 depicts the parameters obtained by analyzing XRD data. For comparison, ***Hc*** values (obtained from magnetic measurements) are also shown in the table 2. Within experimental errors there is modest variation in the first nearest-neighbour distance in the residual amorphous matrix for the studied specimens, showing that the studied samples have rather similar mass densities. Obtained lattice parameter (see Table 2) corresponding to Fe-Al phase range between 0.2945 – 0.2973 nm, which would correspond to the Al content in the alloy ranges between 53 to 64 % (*obtained using linear extrapolation of the variation between lattice parameter and the Al-content in the Fe-Al alloy, as given in [6]*). The grain diameter for Fe-Al phase varies between 12 - 22 nm. It should be noted that the larger atomic size of the Al (Atomic radius: 1.820×10^{-10} m) atom in comparison with the Fe atom (Atomic radius: 1.7×10^{-10} m) is responsible for the increase of the lattice parameter as a function of Al concentration in the precipitated Fe-Al phase. Presence of the Fe- Al alloys (which may contain a non-magnetic component), as observed in [6], would result in the observed reduction of saturation induction and also increase of the coercivity, ascribable to grain coarsening with increase of annealing temperature.

4. Summary

Studied specimen shows two-step crystallization, providing a clue on deciding the annealing temperatures, at which the specimens can be annealed. The activation energy of crystallization, calculated using the Kissinger's method, was estimated to be 3.79± 0.6 eV (for main peak) and 3.02± 0.3 eV (for pre peak). Coercive field Hc of the studied samples varies between 22.06 – 838.67 A/m. Best Hc: 22.06 A/m was obtained for the as cast sample. The measured saturation induction B2000 values range between 0.89 and 1.31 Tesla. Best B2000: 1.31 Tesla was obtained for the as cast sample. XRD data shows that apart from α-Fe, phase (with lattice parameter ~ 0.2859 nm) an additional Fe-Al phase is also formed, with Al ranges between 53 to 64 %, responsible for the reduction of magnetic induction as well as the increase of coercivity which is also owing to grain coarsening.

Acknowledgments

This work is supported by Indo-Portuguese joint project No. Ind/Portugal/P-07/2013. MS (Research fellowship recipient) is supported by UGC-DAE CSR, Indore project CSR-I/CRS51/2011-12.

References

[1] Ohta M and Yoshizawa Y 2007 *Jpn. J. Appl. Phys.* **46** L477
[2] Ohta M and Yoshizawa Y 2008 *J. Appl. Phys.* **103** 07E722
[3] Li H X., Wang S L, Yi S, Jiao Z B, Wu Y and Lu Z P 2009 *J. Magn. Magn. Mater.* **321** 2833
[4] Chakrabarty P K, Mazaleyrat F, L.K. Varga L K 2003 *J. Magn. Magn. Mater.* **447** 254
[5] Kane S N, Coisson M, Tiberto P, Vinai F and Mazaleyrat F 2011 *Current Appl. Phys.* **11** 981
[6] Jartychyk E, Zurawiczy J K, Oleszakz D and M. Pekala M1998 *J. Phys.: Condens. Matter* **10** 4929

Journal of Physics: Conference Series **534** (2014) 012039 doi:10.1088/1742-6596/534/1/012039

Transport and calorimetric studies on $CeNi_2Al_3$

Sankararao Y., S. Shanmukharao Samatham, Venkateshwarlu D., Mohan Gangrade and V. Ganesan
Low Temperature Laboratory, UGC-DAE Consortium for Scientific Research, University Campus, Khandwa Road, Indore-452001, (M.P.) India

E-mail: vganesan@csr.res.in

Abstract. Transport and calorimetric properties of Kondo lattice system $CeNi_2Al_3$ are reported. $CeNi_2Al_3$ shows good agreement with Grand Kadowaki-Woods relation with degeneracy of the quasi-particle N=6. The nonmagnetic nature of $CeNi_2Al_3$ is evident from resistivity measurement even though magnetic elements Ce and Ni are present. Three signatures, deviation from the normal metallic behaviour in resistivity above 140 K, relatively low charge carrier concentration (10^{21} cm^{-3}) and energy gap E_g~11.6 meV obtained from two band model shed light on the most basic notions related to the semimetal nature of $CeNi_2Al_3$.

1. Introduction

Intermetallic compounds are being well studied for exhibiting different ground states because of the hybridization of the 4*f* electrons with the conduction electrons. Ce based systems may fall under the category of heavy fermions (HF) or intermediate valence or delocalised state which causes the enhancement of physical quantities like electronic specific heat coefficient γ, effective mass m^*, thermopower *S* as the strength of the *c-f* hybridization increases [1].

$CeNi_2Al_3$ is a Kondo lattice system with T_K=370 K [2]. Investigation of physical properties like magnetic susceptibility χ, resistivity, specific heat C_p, and X-ray photo emission spectroscopy revealed intermediate valent character of Ce [3] and nonmagnetic, moderately HF metallic like behaviour [4, 5]. Upon Cu doping at Ni site it shows simultaneous optimization of thermoelectric parameters [6, 2].

In this paper, we discuss the structural, transport and calorimetric properties of $CeNi_2Al_3$. Nonmagnetic nature of $CeNi_2Al_3$ is evident from its resistivity, despite of constituting magnetic elements Ce and Ni. The universal behaviour of $CeNi_2Al_3$ to follow Grand Kadowaki-Woods (KW) relation and probable consideration of its semimetal behaviour is discussed.

2. Experimental techniques

Polycrystalline $CeNi_2Al_3$ was prepared by arc melting in high purity Argon atmosphere. X-ray pattern was collected using Bruker D8 Advanced X-ray diffractometer. C_p was carried out by relaxation time method using 14T/2.0K PPMS (QD-USA) from 2 K-300 K. Thermopower measurements were done by differential sandwich method using homemade setup coupled to 4 K Closed Cycle Refrigerator.

3. Results and discussion

3.1. Structural Characterization

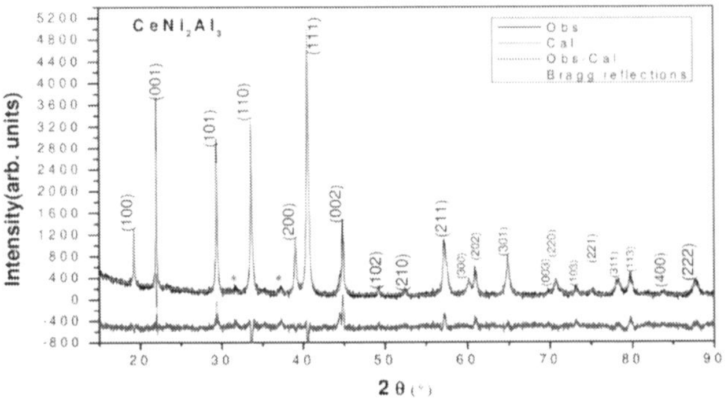

Figure 1. Reitveld refinement of X-ray diffraction pattern for $CeNi_2Al_3$, which crystallizes in hexagonal $PrNi_2Al_3$ type structure with space group P6/mmm [7]. Lattice parameters a =5.3256, c = 4.0413Å. Asterisk represents $CeAl_2$ secondary phase (<1%).

Obtained Chi2 from the Reitveld refinement is 2.72. Earlier investigations have been done on x=0.0–0.6 on $Ce(Ni_{1x-}Cu_x)_2Al_3$ series were prepared by arc melting, also observed impurity peaks [2, 5, 6].

3.2. Transport and Calorimetric properties

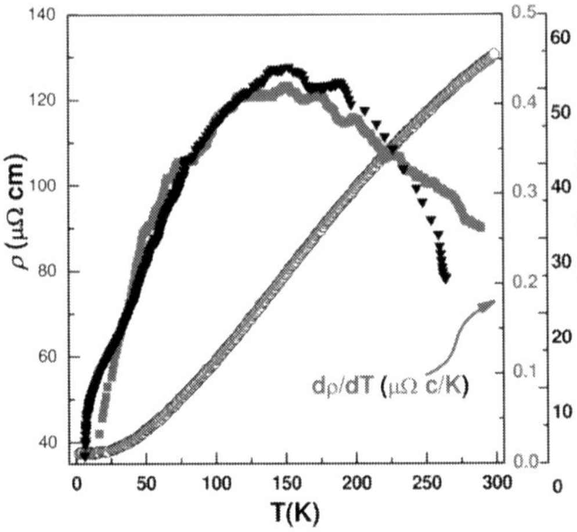

Figure 2. Two band model fit to ρ(T). Obtained E_g is 135.3K. Right Y-axis represent $d\rho/d$T and d ST/dT showing broad hump around 140 K.

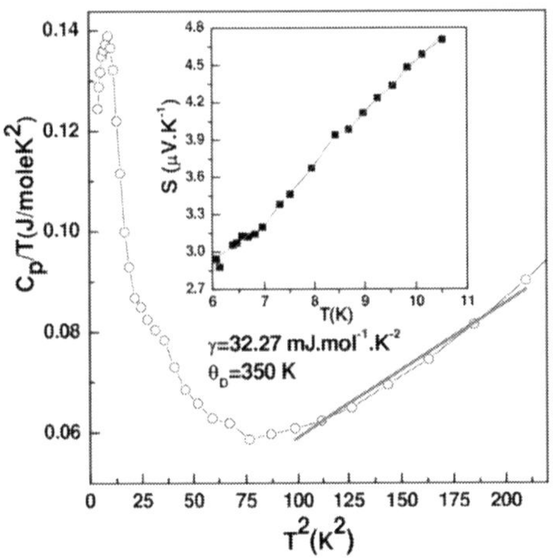

Figure 3. C_p/T vs T^2 of heat capacity is fitted with $C_p/T=\gamma+\beta T^2$. Inset shows low temperature linear fit to thermopower.

figure 2 shows ρ(T) of $CeNi_2Al_3$. It is examined under the light of two band model (TBM) to understand the conduction in $CeNi_2Al_3$. Conductivity of metal and a semiconductor are $(\rho_{01}+AT^2)^{-1}$ and $\sigma_{02}\exp(-E_g/2k_BT)$, respectively. For the ease, $1/\rho = (1/\rho_1 + 1/\rho_2)$

$$\rho = \left[(\sigma_{01}^{-1} + AT^2)^{-1} + \sigma_{02}\exp(-E_g/2k_BT)\right]^{-1} \tag{1}$$

The fit to the data using equation (1) is shown in figure 2. Obtained E_g in Kelvins is 135.3 K. $d\rho/dT$ and $d(ST)/dT$ show a broad hump around 140 K, in figure 3, which agrees with the E_g obtained from TBM. It implies the mixture of metallic as well as narrow gap semiconducting like behaviours. In order to understand better, we have estimated charge carrier concentration 'n' by using the following relation [8].

$$n = \frac{2E_F D(E_F)}{3} \tag{2}$$

where $D(E_F)$ is the density of states and E_F is the Fermi energy [8-10].

$$\gamma = \frac{\pi^2 k_B^2}{3} D(E_F), \quad \frac{S}{T} = B = \frac{\pi^2 k_B^2}{2eE_F} \tag{3}$$

The Sommerfeld coefficient γ is estimated from the fit of $C_p/T=\gamma+\beta T^2$, figure 3. The obtained $\gamma=32.27$ $mJ.mol^{-1}.K^{-2}$ is at par with those reported in literature [2, 6]. Debye temperature $\theta_D=350$ K is estimated using $\theta_D(K) = \sqrt[3]{1944p/\beta}$, where p is the number of atoms in the formula unit. The peak at low temperature is due to antiferromagnetic nature of $CeAl_2$ secondary phase [11, 12] as is shown in XRD. The obtained diffusion thermopower constant $B = 0.418$ $\mu V.K^{-2}$ as $T \to 0$, figure 3. The calculated Fermi energy E_F and $D(E_F)$ from equation (2) are 87.6 meV and 2.43×10^{22} $eV^{-1}.cm^{-3}$, respectively. From equation (1), we get $n=1.4\times10^{21}$ cm^{-3}. Therefore the order of n, hump in resistivity and energy gap E_g, conclusively support the semimetal nature of $CeNi_2Al_3$.

3. 2.1. Grand Kadowaki-Woods relation:

KW relation $A/\gamma^2=1\times10^{-5}$ $\mu\Omega$ cm $(K.mol.mJ^{-1})^2$ is usually applicable to the Fermi liquids (FL). A α m^{*2} and γ α m^* are related to the electrical resistivity as $\rho=AT^2$ and $C_{el} = \gamma T$, respectively [13]. Systematic and significant deviation from the KW relation is observed for Yb based intermediate valent systems and Ce based compounds even though they show FL behaviour at low temperature [14, 15].

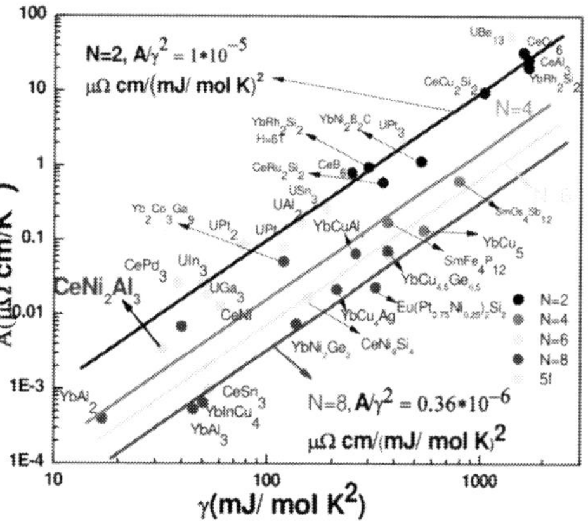

Figure 4. Plot of A vs γ for different HF based on degeneracy of the quasi-particle N=2, 4, 6 and 8. $CeNi_2Al_3$ (N=6) is represented with A= $224\times10^{-5}\mu\Omega$ cm. K^{-2} and $\gamma=32.27$ $mJ.mol^{-1}.K^{-2}$.

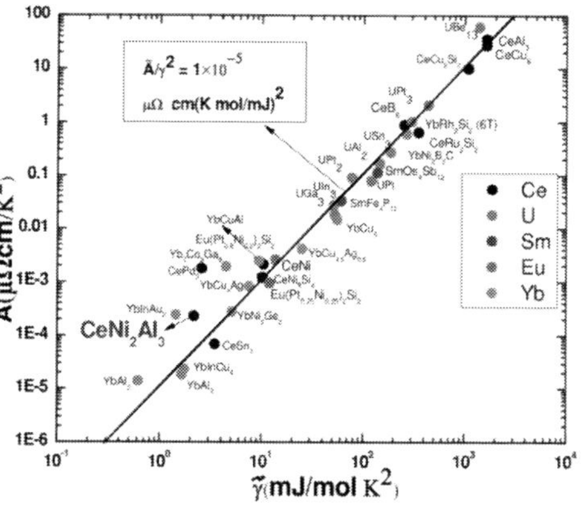

Figure 5. Plot of Ã vs $\tilde{\gamma}$ of HF systems. Universal line represents grand KW relation. $CeNi_2Al_3$ is represented with Ã= 23.5×10^{-5} $\mu\Omega$ cm. K^{-2} and $\tilde{\gamma}$ = 2.15 $mJ.mol^{-1}.K^{-2}$.

A primary report on different HF systems revealed the deviation of A/γ^2 for different degeneracies, N(=2J+1) [14]. Later on, the theoretical work on FL theory also found out the same, mentioning that A/γ^2 ratio would be independent of HF materials but depends on degeneracy N [16]. In $CeNi_2Al_3$, Ce in +3 state having angular momentum J=5/2 with N=6. The obtained values for $CeNi_2Al_3$, $A=354\times10^{-5}\mu\Omega$ $cm.K^{-2}$ and $\gamma=32.27$ $mJ.mol^{-1}.K^{-2}$. According to Ref. [14], $CeNi_2Al_3$, is plotted in figure 4. However, the normalized coefficients of A and γ by absorbing the degeneracy are given as

$$\tilde{A} = \frac{A}{\frac{1}{2}N(N-1)}, \tilde{\gamma} = \frac{\gamma}{\frac{1}{2}N(N-1)}$$

Using the above normalised coefficients $\tilde{A}$ and $\tilde{\gamma}$ the new relation $\tilde{A}/\tilde{\gamma}^2 = 1\times10^{-5}$ μΩ cm $(K.mol.mJ^{-1})^2$ is derived and called as Grand-KW relation. This unique value is independent of materials and degeneracy of quasi-particles [17]. The estimated $\tilde{A} = 23.5\times10^{-5}$ μΩ cm. K^{-2} and $\tilde{\gamma}$ =2.15mJ $mol^{-1}.K^{-2}$ for $CeNi_2Al_3$ with N=6 gives $\tilde{A}/\tilde{\gamma}^2 = 4.9\times10^{-5}$ μΩ cm(K $mol.mJ^{-1})^2$ as is shown in the figure 5, which follows the Grand KW relation along with many other HF systems.

4. Conclusions

In summary, the experimental investigation of moderately HF system $CeNi_2Al_3$ by means of transport, calorimetric and structural characterization is reported. It falls on the universal line of Grand Kadowaki-Woods relation with the estimated normalised coefficients $\tilde{A}$ =23.5×10^{-5} μΩ cm and $\tilde{\gamma}$ =2.15 $mJ.mol^{-1}.K^{-2}$. Nonmagnetic nature evident from resistivity indicates that there is strong hybridization between conduction and Ce-4*f* electrons. It shows hump like behaviour above 140 K. The estimated energy gap by implementing two band model, E_g is about 140 K and the order of charge carrier concentration n is 1.4×10^{21} cm^{-3}. The Fermi energy as estimated from the diffusion thermopower coefficient is E_F=87.6 meV. These three, conclusively, shed light on conduction behaviour of $CeNi_2Al_3$ and support semimetal character.

Acknowledgement
Dr. P. Chaddah, Prof. A. Gupta and Dr. A. K. Sinha, for their support and encouragement. XRD Lab members, especially Dr. Mukul Gupta are thanked for the XRD measurements. The in-mates of Low Temperature Laboratory are thanked for their help and Cryogenics staff, especially, Er. P. Saravanan is thanked for the supply of cryogens.

References:

[1] Liu S. H.: *Handbook on the Physics and Chemistry of Rare Earths*, ed. K. A. Gschneidner, Jr., L. Eyring, G. H. Lander and G. R. Choppin (Elsevier Science, Amsterdam, 1993) **17**, 87
[2] Lu Q., Sun P., Huo D., Mizushima T., Isikawa Y. and Sakurai J., *J. Phys. Soc. Jpn.* **73**, 681
[3] Coldea M., Neumann M., Pop V.and Demeter M., 2001 *J. Alloys Compd.* **323-324** 431
[4] Isikawa Y., Sakai H., Mizushima T., Kuwai T.and Sakurai J. 2002 *Physica B* **312–313** 259
[5] Cava R. J., Ramirez A. P., Takagi H.and Krajewski J. J. 1993 *J. Magn. Magn. Mater.* **128** 124
[6] Peijie Sun, T. Ikeno, T. Mizushima and Y. Isikawa 2009 *Phys. Rev. B.* **80**, 193105
[7] Higashi I., Kitazava H., Kobayashi K., Yoshida M. 1993 *J. Alloys Compd.* **191** L21
[8] Ashcroft N. W. and Mermin N. D. 1976, *Solid State Physics* (Philadelphia: Saunders)
[9] Ziman J. M. 1972, *Principles of the Theory of Solids* (Cambridge University Press)
[10] Abrikosov A. A. 1988, *Fundamentals of the Theory of Metals* (Amsterdam: North-Holland)
[11] Barbara B., Rossignol M. F., Boucherle J. X., and Vettier C. 1980 *Phys. Rev. Lett.* **45** 938
[12] Tomisawa S., Wada S., Ohashi M., and Oomi G. 2006 *J. Phys.: Condens. Matter* **18** 10413
[13] Kadowaki K.and Woods S. B. 1986 *Solid State Commun.* **58** 507
[14] Tsujii N., Yoshimura K., and Kosuge K. 2003 *J. Phys.: Condens. Matter* **15** 1993.
[15] Michor H., Berger St., El-Hagary M., Paul C., Bauer E., Hilscher G., Rogl P. and Giester G. 2003 *Phys. Rev. B* **67** 224428.
[16] Kontani H. 2004 *J. Phys. Soc. Jpn.* **73** 515.
[17] Tsujii N, Kontani H. and Yoshimura K. 2005 *Phys. Rev. Lett.* **94** 057201.

Superconductivity in partially filled skutterudite $Pr_{0.5}Pt_4Ge_{12}$

D Venkateshwarlu, S Shanmukharao Samatham, Mohan Gangrade and V Ganesan
UGC-DAE Consortium for Scientific Research, University Campus, Khandwa Road, Indore (MP), India 452001

E-mail: venkatesh@csr.res.in

Abstract. We have investigated the partially filled Skutterudite $Pr_{0.5}Pt_4Ge_{12}$ for its superconducting properties. Resistivity measurements of $Pr_{0.5}Pt_4Ge_{12}$ down to 2K and in the presence of fields up to 2T are reported. The superconducting transition, T_c of ~7.66K observed from zero fields of Resistivity and Heat Capacity measurements. The interesting observation is that the superconductivity survives even at a 50% filling of Pr in the parent compound and its T_c is equal to that of fully doped $PrPt_4Ge_{12}$. We have estimated relevant parameters of interest like the electron phonon coupling constant, critical field, energy gap ratio, coherence length etc.

1. Introduction

Skutterudites form a class of very interesting materials with exotic thermoelectric properties that also show different ground states such as semiconducting, superconducting, magnetic and heavy fermion etc. $LaPt_4Ge_{12}$, $PrPt_4Ge_{12}$ show the superconducting transitions are respectively 8.27K, 7.91K. It is customary that vibrations of La atoms can be described by a low-energy Einstein term while the polyanionic host $[Pt_4Ge_{12}]$ is treated by the Debye approximation while analyzing the specific heat measurements. From the obtained Heat Capacity jump and energy gap ratio values, $PrPt_4Ge_{12}$ shows strong coupling character than $LaPt_4Ge_{12}$. These systems don't follow a standard BCS formalism [1]. $BaPt_4Ge_{12}$ and $SrPt_4Ge_{12}$ are superconducting below 5.35K and 5.10 K [2]. The normal state of these systems doesn't follow the simple metallic behavior whereas they follow Woodard-Cody (WC) equation. The characteristic temperatures are 123K and 121 K respectively for $BaPt_4Ge_{12}$ and $SrPt_4Ge_{12}$. From DFT calculations, guest atoms strongly stabilize the compound and the calculated DOS around E_F is composed of hybridized of both Ge 4*p*-like and Pt 5*d*-like states in both the systems. Hence the superconducting state is formed from the cage of Pt-Ge structure [2]. A clear T^3 dependence of electronic heat capacity, Gap function has point nodes as evidenced from μSR data and superfluid density data in $PrPt_4Ge_{12}$ [3].

The critical current density and flux pinning forces of $PrPt_4Ge_{12}$ are explained in double exponential model shows the existence of a two gap superconductor [4]. Distinct positive curvatures in H_{c1} (T) & H_{c2} (T), linear temperature dependence of J_C (T, H=0) are the evidence for two superconducting gaps in $PrPt_4Ge_{12}$. The nearly linear temperature dependence of the superfluid density at low temperature is the indication for presence of line nodes in the larger gap. The existence of two gaps with a line node in the $PrPt_4Ge_{12}$ is clear from the above results [5].

In this manuscript we present the results on 50% Pr filled Pt_4Ge_{12} sample that shows the survival of superconductivity even up to a void filling of 50% and the possible presence of point nodes in its energy gap from the evidence of T^3 dependence of electronic heat capacity. We have discussed in detail with different parameters.

2. Experimental Details

Polycrystalline sample $Pr_{0.5}Pt_4Ge_{12}$ is prepared by Argon arc melting using stoichiometric amounts elemental constitutes on water cooled copper hearth. The sample was remelted several times to promote the homogeneity. The sample in quartz tube is evacuated and annealed at 800°C for 12 days [1]. The sample was characterized phase by X-ray diffraction. Resistivity using four probe method, Heat Capacity using relaxation method were carried out using Quantum Design PPMS down to 2K and in the presence of magnetic fields up to 2T.

3. Results and Discussion

The XRD pattern of $Pr_{0.5}Pt_4Ge_{12}$ is as shown in 'figure 1'. Since there is a trend to have threshold value for filling, 50% level of Pr doping segregates impurity phases and are denoted by * symbol. The obtained lattice parameters are 8.615A°. It is in agreement with the reported value [1]. In spite of the impurities; we could observe a strong signal for superconductivity in Resistivity and Heat Capacity. The Superconducting transition of $Pr_{0.5}Pt_4Ge_{12}$ ~7.66K in zero fields from Resistivity and Heat Capacity measurements ('figure 2, 4') and this transition is almost equal of parent compound transition $PrPt_4Ge_{12}$. The Superconducting transition is considered to arise from Pt-Ge cage structure [2] .This zero field transition shifts towards the low temperature by the applying magnetic fields as shown in inset of 'figure 2'.

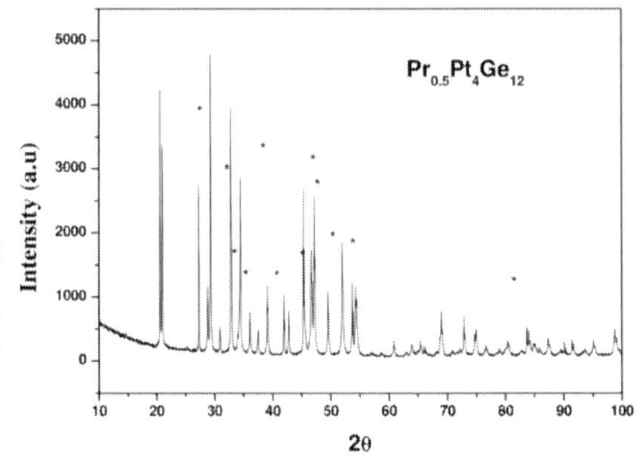

Figure 1. XRD pattern of $Pr_{0.5}Pt_4Ge_{12}$. Due to excess voids, impurities segregates and they are shown as asterisk (*).

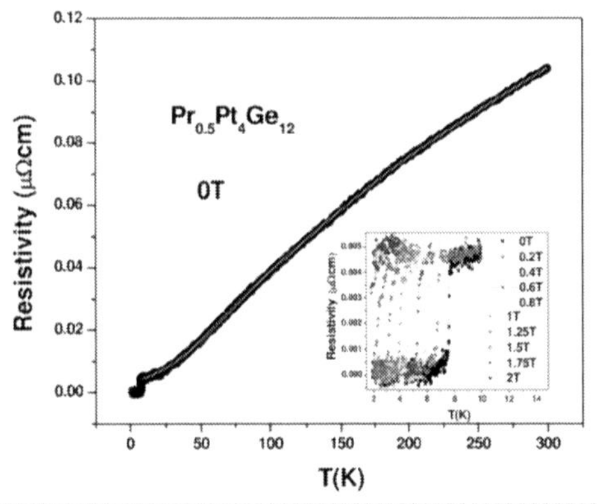

Figure 2. Resistivity as a function of temperature of $Pr_{0.5}Pt_4Ge_{12}$ and red line shows WD-Cody fit. Inset shows in the presence various magnetic fields.

The normal state $T>T_c$, of this compound doesn't follow a simple metallic behaviour. This may be attributed to the electron-phonon interaction strength which is responsible for the formation of cooper pairs. The overall resistivity follows the Woodward and Cody formalism [6] as shown in inset of 'figure 2' and is given by:

$$\rho=\rho_0+\rho_1T+\rho_2(\exp[-T_0/T])$$

The obtained characteristic temperature (T_0) of the present system is 139.89K and almost matches the reported values of $BaPt_4Ge_{12}$ and $SrPt_4Ge_{12}$ [2]. The residual resistivity ratio of the present system is ρ_{300}/ρ_0=22.6 which is less than the $PrPt_4Ge_{12}$, $LaPt_4Ge_{12}$ and more than the $BaPt_4Ge_{12}$, $SrPt_4Ge_{12}$.

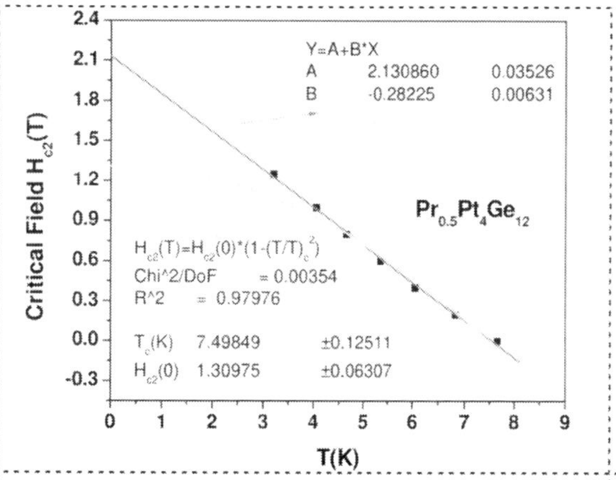

Figure 3. Critical Field as a function of temperature of $Pr_{0.5}Pt_4Ge_{12}$.

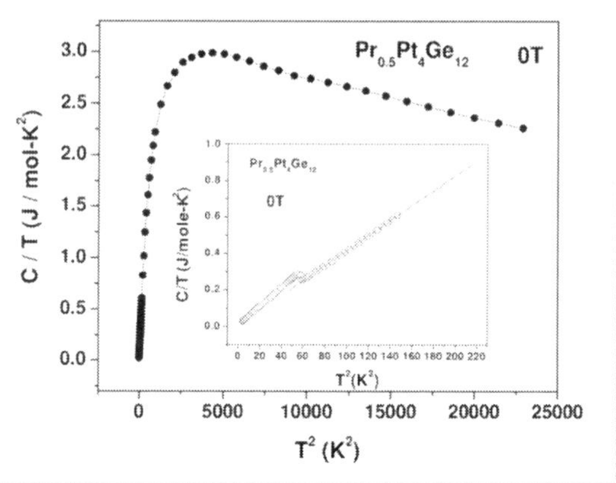

Figure 4. C/T vs T^2 in full region and inset shows up to 12K, red line shows the linear fit.

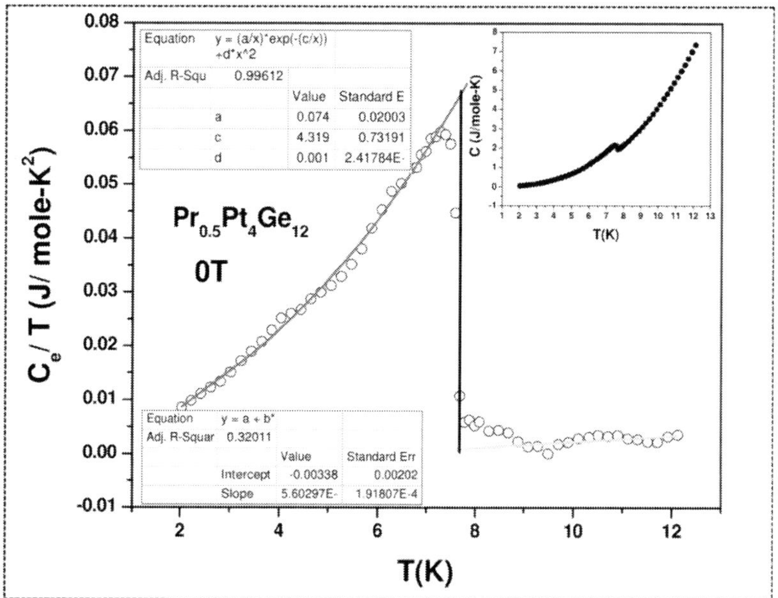

Figure 5. C_e/T vs T of $Pr_{0.5}Pt_4Ge_{12}$ and red line shows BCS fit, Green line shows linear fit for gamma. Inset shows C vs T of $Pr_{0.5}Pt_4Ge_{12}$. Fit parameters are also shown.

We have plotted the critical field with respective temperature as shown in 'figure 3'and fitted with the BCS equation and linear equation. There is a clear deviation from the BCS fit and the data fits well with a linear fit. The obtained critical field from such a linear fit is H_{c2} (0)=2.13T. The estimated coherence length of the present compound is ~12.43nm using the equation $\xi^2 = \varphi_0 / 2\pi H_{c2}(T)$ [7]. This value is similar to ξ value of $BaPt_4Ge_{12}$ and $SrPt_4Ge_{12}$.

The plot of C/T vs T^2 of the present system is as shown in 'figure 4'and its inset. The normal state Heat Capacity of C/T ($C= \gamma T+\beta T^3$, γ is the Sommerfeld coefficient and β is proportional to the Debye temperature θ_D) was fitted by linear equation as shown in inset of 'figure 4'. The obtained γ value is 3.1mJ/mol-K^2 and θ_D is 198K for the present system and the Debye temperature matches with the reported values of MPt_4Ge_{12} (M=Pr, La, Sr, Pr). Debye temperatures of $LaPt_4Ge_{12}$, $PrPt_4Ge_{12}$, $BaPt_4Ge_{12}$ and $SrPt_4Ge_{12}$ are respectively 209 K, 198 K, 247K and 220K.

The C(T) vs T is as shown in inset of 'figure 5'. We have subtracted the phonon part from total Heat Capacity to get the electronic Heat Capacity (C_e) and plotted as C_e/T vs T as shown in 'figure 5'. The Superconducting state of C_e/T (T) doesn't follow the BCS equation. A good fit is obtained by adding a T^2 term in BCS equation, viz $C_e/T = (A/T)\exp[(-\Delta/k_BT)]+BT^2$. The Normal state simply becomes a γ which is almost constant and is shown as linear fit that is quite flat. The line which we have taken at the transition is according to entropy conservation. Here the clear T^3 dependence of $C_e(T)$ is the evidence for point nodes in energy gap of present system. The energy gap ratio is 1.13 which is lower than the BCS value and the corresponding energy gap is 0.37meV only. These values indicate that the present system is in the weak coupling limit which is in contrast to the parent compound that requires further investigation.

4. Conclusions

The superconducting transition of $Pr_{0.5}Pt_4Ge_{12}$ is 7.66K. Superconductivity survives even at 50% Pr filling and shows transition temperature is almost the same as that of its parent counterpart. This supports the fact that Pt-Ge cages are superconducting. The estimated characteristic temperature of Woodard-Cody equation is 140 K, coherence length is 12.4nm and the Debye temperature is 198K. These values are in line with the values reported for the Pr fully filled system. However, the parameters in the superconducting state predicts a weak coupling limit which is in contrast to the fully Pr filled system. This may be due to the fact of pronounced point nodes as revealed by heat capacity data.

Acknowledgement

Authors thank Dr. P. Chaddah, Prof. Ajay Gupta and Dr. A.K. Sinha for their encouragement, members of low temperature lab for their support, Cryogenic team especially Mr.Saravanan for their help in cryogens.

References

[1] Gumeniuk R., Schnelle W., Rosner H., Nicklas M., Leithe Jasper A. and Grin Yu 2008 *Phy Rev Lett* **100** 017002
[2] Bauer E., Grytsiv A., Qiu Chen X, Melnychenko Koblyuk N., Hilscher G., Kaldarar H., Michor H 2007 *Phy Rev Lett* **99** 217001
[3] Maisuradze A., Nicklas M., Gumeniuk R., Baines C., Schnelle W., Rosner H., Leithe Jasper A., Grin Yu. and Khasanov R 2009 *Phy Rev Lett* **103** 147002
[4] Sharath Chandra L S, Chattopadhyay M K and Roy S B 2012 *Supercond. Sci. Technol.* **25** 105009
[5] Sharath Chandra L.S., Chattopadhyay M. K. and Roy S.B. 2012 *Phil. Mag.* **92** 3866
[6] Woodard D.W. and Cody G.D. 1964 *Phys. Rev.* **136** 166.
[7] Ohashi Y., Masubuchi Y., Venkateshwarlu D., Ganesan V., Yakhmi J. V., Yoshida T., Kikkawa S 2014 *J Solid State Chem.* **210** 238.

Synthesis and study of the structural properties of pure $PbTiO_3$ ferroelectric ceramics

Jitendra Singh Rajpoot and Ashutosh Mishra
School of Physics, Devi Ahilya University, Khandwa Road, Indore-452001, India

jitendrarajputsait@gmail.com

Abstract. Lead titanate with perovskite structure ($PbTio_3$) is a Single-phase polycrystalline sample. It has been synthesized using solid-state reaction technique and prepares phase pure, dense, crack-free, and homogeneous samples. Using X-ray diffraction patterns we investigate the sintering behavior of PT-powders. The X-ray powder diffraction data have been analyzed to confirm the phase formation and phase purity, to obtain unit cell parameters and unit cell volume. The porosity of the samples has been obtained through X-ray density and bulk density.

1. Introduction

Crystals of the perovskitefamily ($PbTiO_3$) have been of constant interest because some of these materials show ferroelectric behavior and undergo structural phase transitions [1]. $PbTiO_3$ has been considered to be one of the most important members of this family. It has a high Curie temperature, high pyroelectric coefficient, low dielectric constant, and high spontaneous polarization [2]. Lead titanate ($PbTiO_3$, PT) is a ferroelectric ceramic that has not been proved to be a technologically important material by itself but is a significant component material in electronics such as capacitors, ultrasonic transducers, thermistors, and optoelectronics [3–7]. It is also a promising material for pyroelectric infrared detector applications because of its large pyroelectric coef f icient and relatively low permittivity [8, 9]. In this paper, we have investigated the effect of pro-cessing parameters and optimized them to obtain single-phase, crack-free samples. The sintering behavior of the powders at various temperatures has been investigated using X-ray diffraction (XRD) patterns and scanning electron microscopy.

2. Experimental

PbTiO3 (PT) samples were prepared by solid-state reaction processing using high purity oxide and carbonate as the starting materials, involving several steps such as mixing, wet grinding, calcination, and sintering along with intermediate grinding at each stage. Calcination at temperature ranging from 450 to 550◦C was carried out to synthesize the PT compound before sintering at different temperatures. The polycrystalline samples of $PbTiO_3$ were prepared using the previous high purity ingredient materials in the required stoichiometry, and the powders were wet-milled. The sieved fine powders were then calcined in the temperature range 450–550 °C for 8 hrs/days were again finely ground and sieved through a 75 m. The sieved powders were pressed in to pellets using polyvinyl alcohol (PVA) as a binder. The PT powders were then cold pressed into pellets of diameter 10 mm and thickness 1-2 mm at a pressure of 50 MPa using an isostatic hydraulic press. The pellets were sintered at four different temperatures 700 °C, 750 °C, 800 °C, and 900 °C for 12 hr. The phase structure of sintered specimens was investigated with X-ray diffraction. The morphological studies of all the sintered samples were carried out by using scanning electron microscopy (SEM). For determining the average particle size from full width at half maximum (FWHM) the intensity of Bragg peak (101) was used. The density of the PT ceramic was measured using single pan balance and Archimedes principle.

3. Results and discussion

3.1. Structural and Microstructural Characterization.

X-ray diffraction patterns collected at room temperature for the samples sintered at 700 °C, 750 °C, 800 °C, and 900 °C are shown in Figure 1. All the observed peaks/reflections are indexed in tetragonal crystals systems on the basis of the best agreement between observed and calculated interplaner (d) Values.

Table 1: Unit cell parameters and *c/a* ratio of PT powder samples sintered at different temperatures.

Sintering temp. (°C)	a (A)	C(A)	c/a ratio	Density (X-ray) ρ_x	ρ_m (g/cm^3)
700	3.9008	4.067	1.0426	6.91	6.09
750	3.9013	4.0673	1.0425	7.85	7.18
800	3.9032	4.0679	1.0422	7.91	7.35
900	3.9023	4.0657	1.0419	7.97	7.43

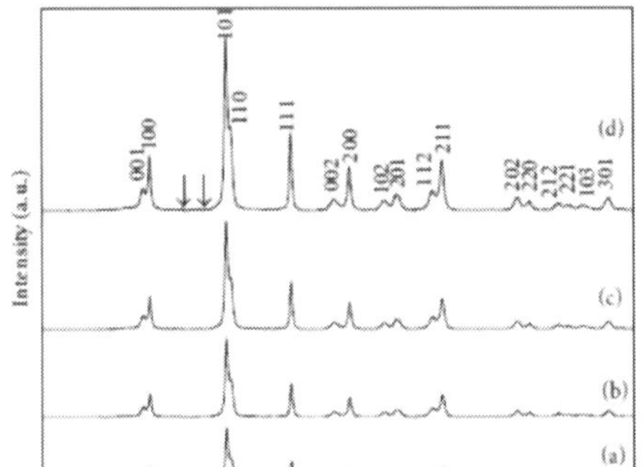

Figure 1: X-ray diffraction patterns of $PbTiO_3$ sample sintered at different temperatures ((a) = 700 °C, b) = 750 °C, (c) = 800 °C, (d) = 900 °C.

The values of unit cell parameters (a, c) and axial ratio (c/a) and density are given in Table 1. The comparative behavior of the tetragonality and unit cell volume with sintering temperature is shown in Figure 2, and the density variation with Sintering temperature is shown in Figure 3.Figure 3 shows the measured and X-ray density as a function of a sintering temperature. A slight increase in density is observed as sintering temperature increases up to 900 °C. However, a density jump occurs at 750 °C, and almost dense ceramics are achieved thereafter. The results are shown in Figure 3.

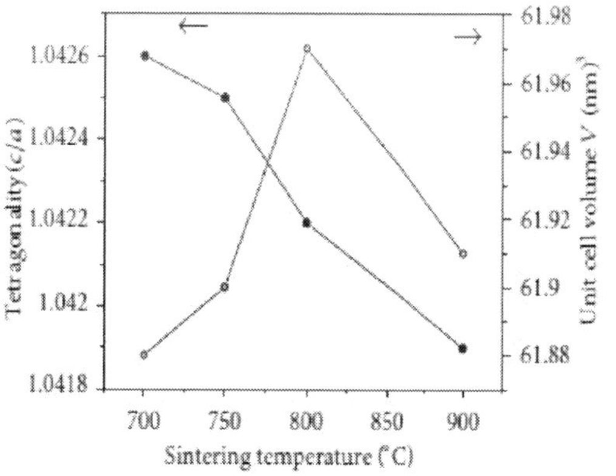

Figure 2: Variation of c/a and unit cell volume (V) with sintering temperature.

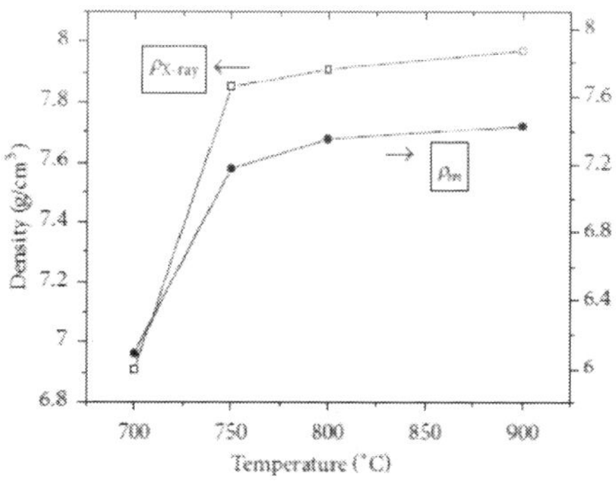

Figure 3: X-ray and measured density of the $PbTiO_3$ Ceramics as a function of sintering temperature.

4. Conclusions

The following interesting trends were revealed in the present study. The unit cell dimensions are not significantly influenced by the sintering temperature. The c/a ratio does not vary significantly with sintering temperature, and its value is nearly 1.04. The SEM micro-graphs do not exhibit the fracture surfaces of sample sintered at 900 °C, and the sample has developed a relative uniform microstructure. $PbTiO_3$ powder can be synthesized directly from the constituent oxide using conventional ceramic technique by optimizing sintering temperature and time and by using starting oxides having fine uniform particle size.

References

[1] Jona F.and Shirane G. 1962 *Ferroelectric Crystals* (New York: MacMillan)
[2] Migoni R., Bilz H., and Bauerle D. 1976 Phys. Rev. Lett. 37 1155.
[3] Cross L. E. 1984 Am. Ceramic Soc. Bull. 63 586.
[4] Sasazawa K., Oshima K. and Yamaoka N. 1987 Jap. J. Appl. Phys. 26 L65.
[5] Shrout T. R. and Halliyal A. 1987 Am. Ceramic Soc. Bull. 66 699 and 704.
[6] Kuwabara M. 1990 J. Am. Cer. Soc. 73 1438.
[7] Chan Y., Chan H. L. W. and Choy C. L. 1998 J. Am. Cer. Soc. 81 1231.
[8] Ting R. Y. 1978 Ferroelectrics 67 143.

Weak localization effect in pulsed laser deposited ZnO film

A Agrawal[1], T A Dar and P Sen
Laser Bhawan, School of Physics, Devi Ahilya University, Khandwa Road, Indore-452001

E-mail: agrawal.arpana01@gmail.com

Abstract. Low temperature longitudinal magnetoresistance measurement in pulsed laser deposited undoped ZnO film has been investigated. The temperature and magnetic field dependence of the resistivity of the film is in good agreement with the theory of weak localization and electron-electron interactions in a disordered system. Negative magnetoresistance is a consequence of the suppression of quantum interference effect in the weak localization theory. The inelastic scattering time $\tau_i \sim T^{-3/2}$ and phase coherence length $L\phi(T) \sim T^{-3/4}$ indicates that the electron-electron interaction is the dominant inelastic scattering process.

1. Introduction

ZnO has engrossed significant scientific and technological concern in the research community due to their interesting physical properties like wide band gap and high exciton binding energy which make them appropriate applicant for optoelectronic devices [1]. When ZnO is doped with appropriate metal atoms (Al, Mg etc), their transparency and conductivity can be changed to much higher values [2, 3]. The attraction is also fuelled by the utility of ZnO for potential spintronic applications. It is known that the spin polarized current or the manipulation of electron spin is a key issue for the utilization of spintronic devices [4]. Therefore the study of the magnetotransport properties of materials is rather important. In view of above, we investigate the magnetotransport properties of ZnO film.

2. Experimental details

ZnO film was grown on glass substrate by pulsed laser deposition technique (PLD) using KrF eximer laser (λ=248 nm) with the repetition rate of 10 Hz, energy density 2J/cm2 and pulse duration 20 ns. After cleaning ultrasonically in acetone and deionised water sequentially, the glass substrates were loaded into the chamber and then heated to 400oC at a base pressure of 5 nbar. The target to substrate distance was maintained to 5 cm. Then ultrapure (99.999%) O_2 gas was introduced in the chamber. Deposition was carried out in oxygen environment maintaining the pressure of 5×10^{-4} mbar. Resistivity measurements were carried out using standard four probe technique. The magnetic field is applied parallel to the direction of applied current giving rise to longitudinal magnetoresistance (MR). Longitudinal MR for both the films was measured at various constant temperatures upto 8 T magnetic field.

3. Results and discussion

It is a well known fact that the application of external magnetic field significantly affects the resistance of the samples and the phenomenon is known as magnetoresistance. For most oxide semiconductors, MR was usually found at low temperature, and several mechanisms of MR were proposed like weak localization, weak antilocalization, Lorentz force, superconducting fluctuations, splitting of spin states by magnetic field (Zeeman Effect) [5]. It is generally considered that the electrons in wide band gap semiconductors are at the localization boundaries and weak localization effects occur in the system. Weak localization is the quantum mechanical phenomenon that occurs in disordered system at low temperature and low fields and primarily occurs due to the self intersecting scattering paths which results in random walk. This random walk manifests itself to the coherent backscattering of electron wavefunctions along two time reversal paths which traverses a closed loop. On applying magnetic field, there will be a phase difference acquire by the two waves and thus the interference condition gets violated so resistivity decreases and hence results in negative MR. We have apply the Kawabata's equation for three dimensional (3D) system which has included the influence of applied magnetic field on the electrical conductivity/resistivity in the weak localization regime at a given temperature. For 3D case, the Kawabata's equation is given by [6],

$$\frac{\Delta\rho(B)}{\rho_0^2} = \frac{-e^2}{2\pi^2\hbar}\sqrt{\frac{eB}{\hbar}}F\left(\frac{B}{B_i}\right) \tag{1}$$

Here $B_i = \hbar(4eD\tau)^{-1}$ refers to inelastic scattering fields and is treated as fitting parameter, D is the diffusion constant, τ_i is the inelastic scattering time and the function F(x) with $x = B(Bi)^{-1}$ can be approximated as,

$$F(x) = 2\left(\left(2+\frac{1}{x}\right)^{\frac{1}{2}} - \left(\frac{1}{x}\right)^{\frac{1}{2}}\right) - \left[\left(\frac{1}{2}+\frac{1}{x}\right)^{\frac{-1}{2}} + \left(\frac{3}{2}+\frac{1}{x}\right)^{\frac{-1}{2}}\right] + \frac{1}{48}\left(2.03+\frac{1}{x}\right)^{\frac{-3}{2}} \tag{2}$$

Fig. 1 depicts the MR of ZnO film in weak field case upto 2 T magnetic field and the solid thick lines are least square fits to our experimental data using the frame work of Kabawata's 3D formula (equation (1)), treating the inelastic scattering fields as the fitting parameter. From this fitting parameter, we have calculated the phase coherence length by the relation $L\phi(T) = (D\tau_i)^{\frac{1}{2}}$ and are given in table 1. From table 1, we find that phase coherence length decreases with increasing temperature. This suggests that temperature can also results in the dephasing of the electron waves.

Table 1. A Calculated values of phase coherence length ($L\phi(T)$) in nm, showing decreasing trend with increasing temperature.

T (K)	$L\phi(T)$ (in nm)
50	14.75
30	20.77
20	28.51
10	54.15
7	61.78

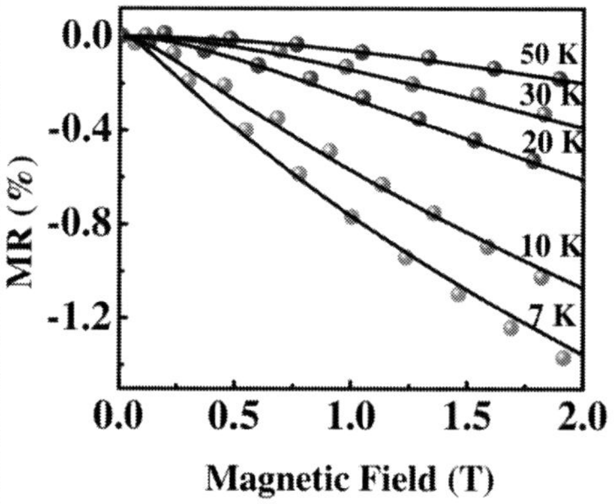

Figure 1: MR of ZnO film upto 2 T magnetic fields, solid thick lines shows least square fit using equation (1).

The variation of inelastic scattering fields as a function of temperature is shown in Fig. 2 and is found to increase with increasing temperature. From the inset of figure 2, we noticed that $B_i \sim T^{3/2}$ and thus the inelastic scattering rate $\tau i \sim T^{-3/2}$ which is expected for electron-electron interaction (e-e interaction).

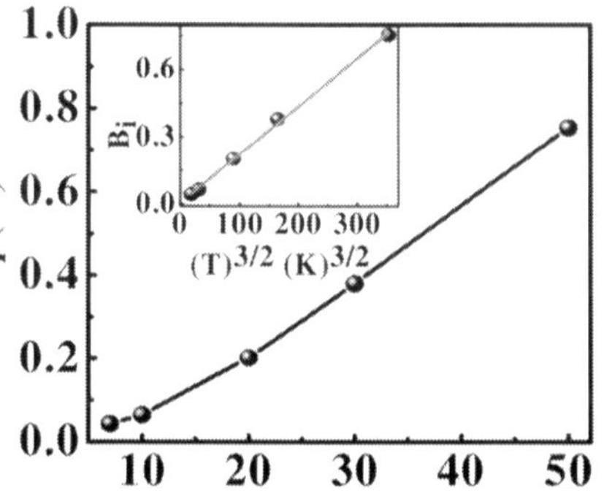

Figure 2. Plot of B_i with T, inset shows $B_i \sim T^{3/2}$

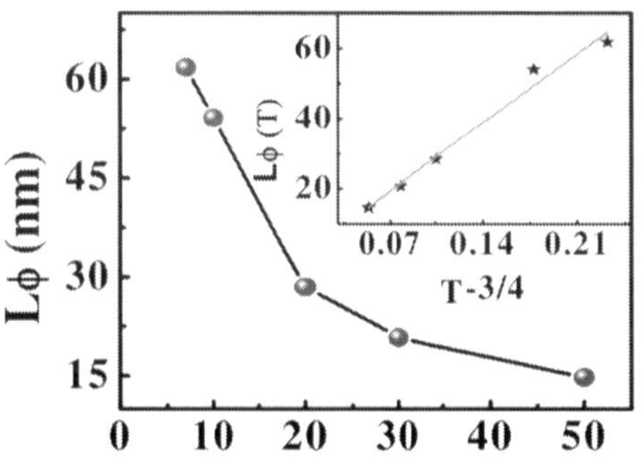

Figure 3. Variation of $L\phi(T)$ with T, inset shows $L\phi(T) \sim T^{-3/4}$

We have also plotted $L\phi(T)$ with temperature in fig. 3 and the inset shows $T^{-3/4}$ dependency of $L\phi(T)$ in a disordered 3D case. Thus $L\phi(T) \approx T^{-3/4}$ suggests that e-e interaction can be the dominated reason for the phase breaking in 3D systems **[7]**. Such a dependency of $L\phi(T)$ on temperature has also been observed by Liu et al. in B doped ZnO films **[8]**.

4. Conclusion

In conclusion, we have presented low temperature magnetoresistance measurements on ZnO film. The temperature and magnetic field dependence of the resisivity of the film is in good agreement with the theory of weak localization and electron-electron interactions in a disordered system. Negative MR is a consequence of of the suppression of quantum interference effect in the weak localization theory. The inelastic scattering time $\approx T^{-3/2}$, indicating that the electron-electron interaction is the dominant inelastic scattering process.

Acknowledgments

Authors are thankful to Dr. R Rawat, UGC-DAE-CSR, Indore for doing MR measurements. We are also thankful to UGC-DAE CSR Indore for financial support.

References

[1] Janotti A and Vande Walle C G 2009 *Rep. Prog. Phys.* **72** 126501
[2] Suchea M S, Christoulakis, Katsarakis N, Kitsopoulos T and Kiriakidia G 2007 *Thin Solid Films* **515** 6562.
[3] Liu X D and Jiang E Y 2007 *Solid State Commun* **141** 394
[4] Gold A 2010 *Appl. Phys. Lett.* **96** 242111.
[5] Behan A J, Mokhtari A, Blythe H J, Ziese M, Fox A M and Gehring G A 2009 *J. Phys.: Condens. Matter* **21** 346001.
[6] Baxter D V, Richter R, Trudeau M L, Cochrane R W and Strom-Olsen J O 1989 *J. Phys. France* **50** 1673.
[7] Andrearczyk T, Jaroszynski J, Grabecki G, Dietl T, Fukumura and T, Kawasaki M 2008 *Phys. Rev. B.* **72** 121309(R)
[8] Liu X D, Jiang E Y and Li Z Q 2007 *J. Appl. Phys.* **102** 073708

Study of hard-soft magnetic ferrite films prepared by pulsed laser deposition

M Satalkar[a], S N Kane[a,1], A Ghosh[a,b], S Raghuvanshi[a], P Tapkir[a], N Ghodke[c], , D M Phase[c], R J Chaudhary[c], A Pasko[d], M LoBue[d], and F Mazaleyrat[d]
[a]School of Physics, D. A. University, Khandwa road Campus, Indore – 452001, India
[b] Dronacharya Group of Institutions, Knowledge Park 111, Greater Noida, India
[c]UGC-DAE CSR, University Campus, Khandwa Road, Indore – 452 001, India
[d]SATIE, ENS Cachan, CNRS, UniverSud Paris, 61 av President Wilson, 94235 Cachan, France

[1]Corresponding author, E-mail: kane_sn@yahoo.com (S. N. Kane)

Abstract. Soft magnetic $Mg_{0.1}Ni_{0.3}Zn_{0.6}Fe_2O_4$ and hard magnetic $BaFe_{12}O_{19}$ bulk nanocrystalline ferrites were synthesized using the sol-gel auto-combustion method, and were used as targets to deposit soft-hard thin films by the pulsed laser deposition (PLD) method. Various soft-hard thin films with different preparation conditions were deposited on Si (100) substrate, which can be effectively utilized to get better magnetic properties. The prepared films were characterized by the X-ray diffraction (XRD), atomic force microscopy (AFM) and magnetic measurements. XRD confirms the presence of soft and hard phases in the thin films. Coercivity of the prepared films ranges from 1.67 to 2.66 kA/m. AFM images show clustering of grains at the film surface with a characteristic columnar growth.

1. Introduction

Exchange spring systems have stimulated intense research activities in the recent years because of the prediction of improvement in the magnetic properties of these systems [1–2]. In the nanocomposite exchange spring media, two magnetically hard layers and a magnetically soft layer are strongly exchange coupled, which can be achieved by their crystallographic coherence [3]. This in turn combines high magnetization of the soft phase with high coercivity of the hard phase, and is determined by the distribution of hard and soft magnetic phases [4]. The objective of the present work is to get the appropriate preparation conditions for depositing soft-hard composite films by the pulsed laser deposition (PLD) and to use them to obtain improved magnetic properties (higher coercivity and magnetization) in the studied composite magnets.

2. Experimental details

The soft magnetic $Mg_{0.1}Ni_{0.3}Zn_{0.6}Fe_2O_4$ and hard magnetic $BaFe_{12}O_{19}$ (BaM) ferrite powders were prepared using the sol-gel auto-combustion method. For synthesis, citrate-nitrate precursors were mixed in stoichiometric proportion with citric acid as fuel. The salt to fuel mass ratio was taken as 1:1. All the precursors were dissolved in de-ionized water, and pH was maintained at 7 by adding ammonia solution (NH_4OH). The solution was heated at 110 °C in air for about 1 hour to obtain the gel and finally to get loose powder known as dry gel) [5]. The composite of hard and soft ferrites was prepared by mixing of the individual ferrite components in a ratio of 2:1. The composite powder was pressed in the form of a pellet by applying a pressure of 3.6 GPa and subsequently annealed at 800 °C for 24 hours. The pellet was used as a target for the deposition of soft-hard ferrite composite thin film on Si (100) substrate which was mounted on a rotating axel. The deposition time was two hours. The PLD setup uses a KrF excimer laser (Lambda Physik model COMPEX-201) with a wavelength of 248 nm and a pulse duration of 20 ns. During deposition the laser energy was 200 mJ/pulse, the substrate temperature and oxygen pressure were maintained respectively at 750 °C and 0.13 Pa (base pressure 0.67 mPa). The focused laser beam was made incident on the target surface at an angle of 45°. The substrate was mounted opposite to the target at a distance of 4 cm on a heater plate. After deposition the sample was cooled slowly to room temperature. During cooling the effect of oxygen pressure on the deposited soft-hard composite thin film was studied.

X-ray diffraction (XRD) measurements were performed in θ-2θ configuration using a Bruker D8 Advance diffractometer with Cu-K_α radiation (λ = 0.154 nm). Atomic force microscopy (AFM) image of the sample was taken by Nanoscope-E using a pyramidal silicon nitride tip in contact mode. Magnetization measurements on the deposited soft-hard composite thin film were done using 7 T MPMS SQUID-VSM setup with a maximum applied field of 2 T. Magnetic properties of the bulk hard ferrite powders were measured using a SQUID magnetometer, Quantum Design MPMS-5S. Magnetic characterization of the synthesized bulk soft magnetic ferrite samples was done using a conventional induction technique at 50 Hz by applying a maximum field of 87.5 kA/m.

3. Results and discussion

In Figure 1, the left panel and its inset show the XRD pattern and magnetization loop of $BaFe_{12}O_{19}$ powder, confirming the formation of hard magnetic phase, whereas the right panel and its inset depict the XRD pattern and magnetization loop of $Mg_{0.1}Ni_{0.3}Zn_{0.6}$ powder, prepared by the sol-gel auto combustion method. The $BaFe_{12}O_{19}$ ferrite serves as a hard phase and the $Mg_{0.1}Ni_{0.3}Zn_{0.6}$ ferrite as a soft phase in the hard-soft composite used for depositing the exchange spring magnet on Si (100) substrate. XRD validates the formation of a single-phase cubic spinel structure, magnetic measurements prove the soft magnetic nature of this powder. Table 1 depicts the structural and magnetic data for the bulk samples (hard and soft ferrites). The obtained *a* and *c* values reveal that the synthesized hard ferrite is a single phase $BaFe_{12}O_{19}$ and confirm the formation of a spinel structure in the soft ferrite powder. The structural and magnetization data match well with the values reported in the literature.

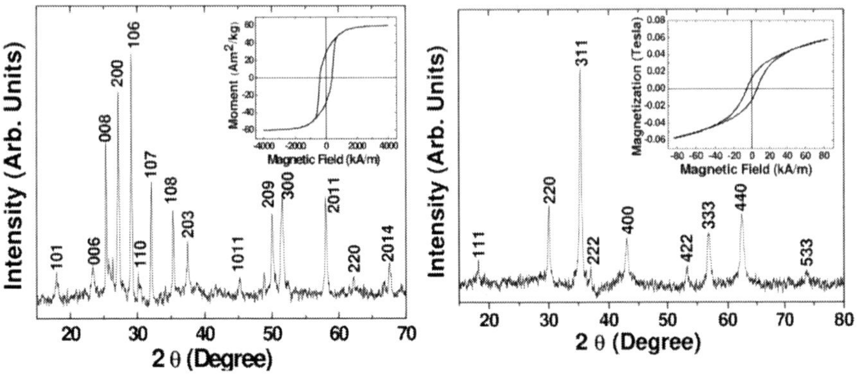

Figure 1. XRD pattern and hysteresis loop for the $BaFe_{12}O_{19}$ powder (left panel and its inset) and the $Mg_{0.1}Ni_{0.3}Zn_{0.6}Fe_2O_4$ powder (right panel and its inset).

Table 1: Structural and magnetic parameters for the $BaFe_{12}O_{19}$ and $Mg_{0.1}Ni_{0.3}Zn_{0.6}Fe_2O_4$ powders.

Phase	Lattice parameters (nm)	Grain Size (nm)	Coercivity (kA/m)	Saturation magnetization
$BaFe_{12}O_{19}$	$a = 0.5589$, $c = 2.3215$	78	394	60.1 (Am^2/kg)
$Mg_{0.1}Ni_{0.3}Zn_{0.6}Fe_2O_4$	$a = 0.8405$	24	5.77	57 (mT)

Figure 2 depicts the XRD pattern (left panel) and magnetization loop (right panel) of the composite $BaFe_{12}O_{19}$ + $Mg_{0.1}Ni_{0.3}Zn_{0.6}Fe_2O_4$ (2:1) thin film deposited on Si (100) substrate, when cooled in the presence of oxygen. The left panel inset shows the surface morphology of the soft-hard composite. AFM image shows clustering of grains on the surface of the film with grain size ~ 130 nm, RMS roughness ~ 0.39 nm, maximum height of the grains ~ 3.46 nm and average height of the grains ~ 1.07 nm. The high vertical resolution of AFM revealed that the film has a characteristic columnar growth with crystallites of height ~ 3.5 nm. Table 2 shows the structural parameters of the phases formed in the prepared thin film. Perusal of a and c values confirm the presence of hard magnetic phase $BaFe_{12}O_{19}$, soft magnetic phase $Mg_{0.1}Ni_{0.3}Zn_{0.6}Fe_2O_4$, and a small amount of secondary Fe_2O_3 phase, which can be ascribed to oxygen deficiency during the sample preparation. The obtained coercivity and magnetic moment values were respectively 26.6 kA/m and 5.89×10^{-7} Am^2.

Figure 3 left panel depicts the XRD pattern of the composite $BaFe_{12}O_{19}$ + $Mg_{0.1}Ni_{0.3}Zn_{0.6}Fe_2O_4$ (2:1) thin film grown by PLD on Si (100) substrate, when cooled in the absence of oxygen. The XRD pattern was compared with available data in the literature, suggesting the formation of $BaFe_{12}O_{19}$ phase.

Table 2: XRD parameters for the soft-hard composite thin film deposited on Si (100) substrate.

Phase	Lattice parameters (nm)	Grain size (nm)
$BaFe_{12}O_{19}$ (hard magnetic phase)	$a = 0.5885$, $c = 2.3462$	1132
$Mg_{0.1}Ni_{0.3}Zn_{0.6}Fe_2O_4$ (soft magnetic phase)	$a = 0.8405$	143
Fe_2O_3 (secondary phase)	$a = 0.5465$, $c = 2.3136$	220

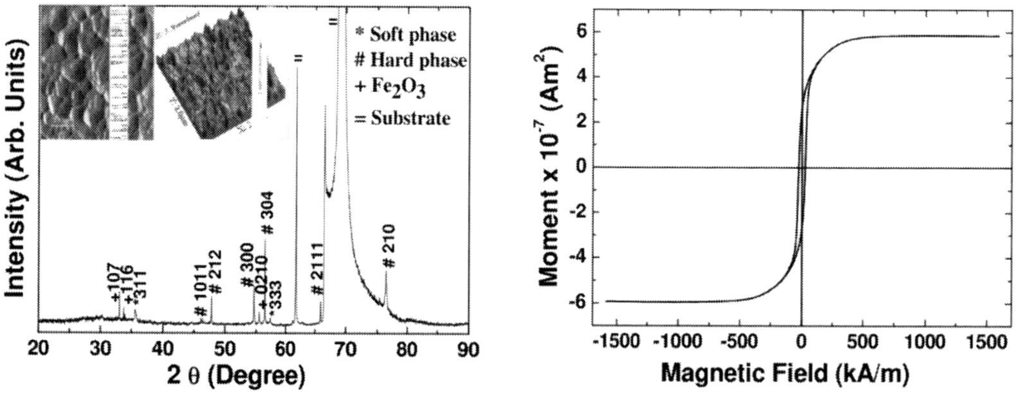

Figure 2. Composite $BaFe_{12}O_{19}$ + $Mg_{0.1}Ni_{0.3}Zn_{0.6}Fe_2O_4$ thin film deposited on the Si (100) substrate, when cooled in the presence of oxygen pressure: XRD pattern (left panel), AFM 2D and 3D images (inset), and hysteresis loop (right panel).

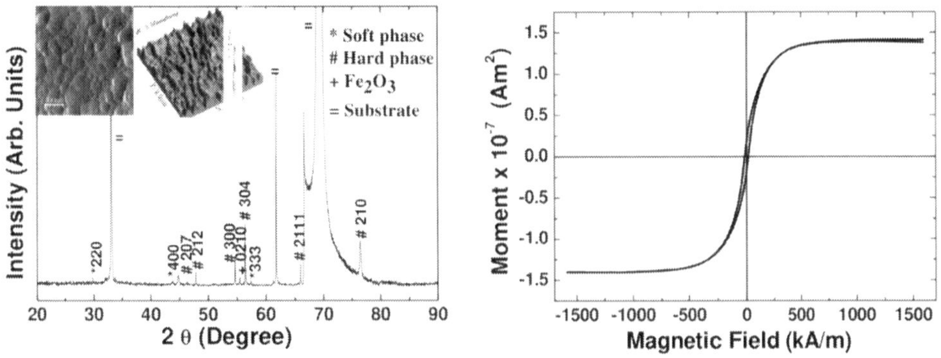

Figure 3. Composite $BaFe_{12}O_{19}$ + $Mg_{0.1}Ni_{0.3}Zn_{0.6}Fe_2O_4$ thin film deposited on the Si (100) substrate, when cooled in the absence of oxygen pressure: XRD pattern (left panel), AFM 2D and 3D images (inset), and hysteresis loop (right panel).

The left panel inset shows the surface morphology of the soft-hard composite, AFM image shows clustering of grains on the surface of the film. Figure 3 right panel depicts the hysteresis loops of the thin film. The measured coercivity is 16.1 kA/m and magnetic moment is 1.40×10^{-7} Am^2. Table 3 depicts the parameters obtained by analyzing the XRD data.

Table 3: XRD parameters of the soft-hard composite deposited on Si (100) substrate, when cooled in the absence of oxygen.

Phase	Lattice parameters (nm)	Grain size (nm)
$BaFe_{12}O_{19}$ (hard magnetic phase)	$a = 0.5885$, $c = 2.3462$	92
$Mg_{0.1}Ni_{0.3}Zn_{0.6}Fe_2O_4$ (soft magnetic phase)	$a = 0.8405$	107
Fe_2O_3 (secondary phase)	$a = 0.5465$, $c = 2.3136$	85

Figure 4 depicts the XRD pattern (left panel) and hysteresis loop (right panel) of the composite $BaFe_{12}O_{19}$ + $Mg_{0.1}Ni_{0.3}Zn_{0.6}Fe_2O_4$ (2:1) deposited on the barium ferrite thin film on Si (100) substrate. Table 5 depicts the parameters obtained by analyzing the structural data. Analysis of the XRD pattern reveals the formation of the following phases: $BaFe_{12}O_{19}$, $Mg_{0.1}Ni_{0.3}Zn_{0.6}Fe_2O_4$, Fe_2O_3 and $BaFe_2O_4$. The measured coercivity is 1.75 kA/m and magnetic moment is 11.5×10^{-7} Am^2.

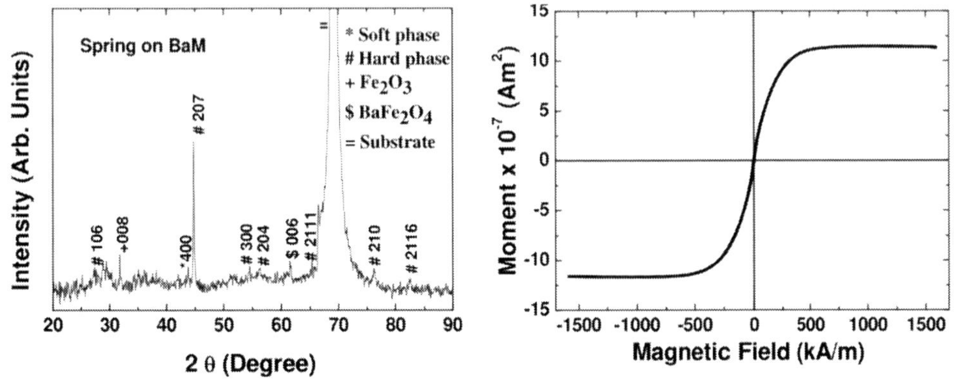

Figure 4. Composite $BaFe_{12}O_{19}$ + $Mg_{0.1}Ni_{0.3}Zn_{0.6}Fe_2O_4$ thin film deposited on the barium ferrite thin film on the Si (100) substrate, when cooled in the absence of oxygen pressure: XRD pattern (left panel), AFM 2D and 3D images (inset), and hysteresis loop (right panel).

Table 4: XRD parameters of the soft-hard composite deposited on $BaFe_{12}O_{19}$ thin film on Si (100) substrate, when cooled in the absence of oxygen.

Phase	Lattice parameters (nm)	Grain size (nm)
$BaFe_{12}O_{19}$ (hard magnetic phase)	$a = 0.5885$, $c = 2.3462$	31
$Mg_{0.1}Ni_{0.3}Zn_{0.6}Fe_2O_4$ (soft magnetic phase)	$a = 0.8405$	66
Fe_2O_3 (secondary phase)	$a = 0.5465$, $c = 2.3136$	55
$BaFe_2O_4$ (secondary phase)	$a = 0.54$, $c = 0.91$	18

Comparison of the results shown in Figures 2 and 3 yields the following important information on the relation between the preparation conditions and the obtained magnetic properties: XRD pattern of the $BaFe_{12}O_{19}$ + $Mg_{0.1}Ni_{0.3}Zn_{0.6}Fe_2O_4$ composite thin film deposited on Si (100) substrate by PLD reveals that the absence of oxygen ensures more phase purity of the soft-hard composite, whereas the presence of oxygen promotes the formation of Fe_2O_3, indicating that the stoichiometry was not transferred effectively in both conditions. Higher amount of Fe_2O_3 helps in the grain growth of soft, hard and secondary phases. For the hard-soft composite thin film deposited on Si (100) substrate, when the film is cooled in the presence of oxygen, the grain coarsening leads to higher coercivity. Therefore, the presence of oxygen during cooling can be utilized effectively in obtaining improved magnetic properties.

4. Conclusion

To summarize, PLD was used to deposit soft-hard composite thin films on Si (100) substrate using different cooling conditions. It is shown that the presence of oxygen during cooling can be effectively utilized to obtain improved magnetic properties. Formation of soft and hard phases in the prepared thin films was confirmed by XRD. The presence of Fe_2O_3 and $BaFe_2O_4$ as secondary phases was also detected in the films. Magnetic measurements show that the coercivity of the prepared thin films ranges from 1.67 to 2.66 kA/m. AFM images of the composite film show clustering of grains at the film surface exhibiting columnar growth.

Acknowledgments

Support in the form of UGC-DAE CSR project No. CSR-I/CSR51/2011-12 is gratefully acknowledged. Authors are grateful to Dr. M. Gupta and Dr. V. Ganesan, UGC-DAE CSR, Indore respectively for doing XRD and AFM measurements. This work is partly supported by MPCST project No. 1783/CST/R & D/Phy and Engg Sc/2013. AG acknowledges financial support from UGC Dr. D. S. Kothari post-doctoral fellowship.

References

[1] Jiang J S, Pearson J E, Liu Z Y, Kabius B, Trasobares S, Miller D J, and Bader S D, 2005 *J. Appl. Phys.* **97** 10K311
[2] Davies J E, Hellwiga O, and Fullerton E E, 2005 *Appl. Phys. Lett.* **86** 262503
[3] Kneller E F, Hawig R 1991 *IEEE Trans. Magn.* **27** 3588
[4] Roy D, Shivakumara C, and Anil Kumar P S 2009 *J. Magn. Magn. Mater.* **321** L11–L14
[5] Ghosh A, Satalkar M, Rathod S, Nag S P, Vyas P, Kane S N, Ghodke N, Prasad R, and Dwivedi R 2013 *Int. J. Mod. Phys.: Conf. Ser.* **22** 28

Synthesis and X-ray diffraction study of some nickel(II) complexes of urea and thiourea

Pradeep Sharma[1,], Jaishree Bhale[2,5], Ashutosh Mishra[3] and Pramod Malviya[4]

[1]Govt. Holkar Science College, Indore(M.P), India
[2]Shree Cloth Market Institute of Professional Studies, Indore (M.P), India
[3]School of Physics, Devi Ahilya University, Indore (M.P), India
[4]Govt.College, Nagda (M.P), India

Email: bhalejaishree@gmail.com

Abstract: XRD of four biologically important nickel(II) complexes having urea and thiourea as primary ligands and sulphate and nitrate as secondary ligands have been studied using Bruker D8 Advance diffractometer at IUC, Indore. The synthesized metal complexes were characterized by XRD measurements in order to elucidate their geometry. The data obtained has been preceded using XRD data analysis program Origin 6.0 Professional. From the experimental measurements, various parameters, e.g., particle size, lattice parameter have been estimated.Particle size for Ni-urea complexes is found in the range of 4.23-9.01 nm while for Ni-thiourea complexes its range is 0.26-9.76 nm.Lattice parameter for all these complexes is found in the range of 7.47-8.06 Å.The XRD analysis revealed the crystalline nature of all the complexes.

1. Introduction:

Urea is physiologically very important. It is the chief nitrogenous product of protein metabolism. Urea is used for preparing formaldehyde-Urea resin (plastics)[1], barbiturates [2], and fertilizers [3-6]. Urea is also extensively used in the paper industry to soften cellulose and has been used to promote healing in infected wounds and many other applications in the field of medicine [7-9].
Recently, urea is used for the manufacture of hydrazine in which urea is treated with alkaline sodium hypochlorite.[2] .Urea is used in active dying [10] which has an effect on the formation and cleavage of covalent bond between the reactive dye and cellulose. Complexes of urea with some metal ions are used as fertilizers [11-14].

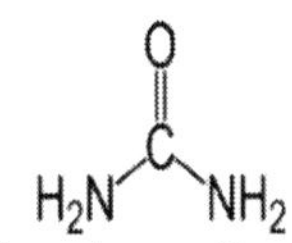

Figure 1. Structure of urea

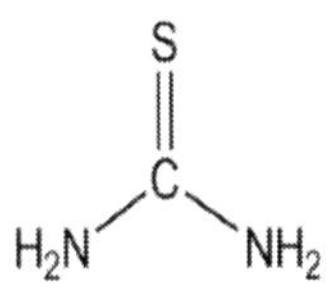

Figure 2. Structure of thiourea

Recently, urea represents not only an important molecule in biology but also an important raw material in chemical industry [15]. The synthesis and characterization of urea based complexes have been studied and reported [2].

Thioureas have a long history of being used as a ligand in coordination chemistry and to coordinate with a metal via sulfur and oxygen atoms [16].These hard and soft donor atoms provide a multitude of bonding possibilities [17]. The complexing capacity of thiourea has been reported in several articles [18,19]. The biological activity of complexes with thiourea has been successfully screened for various biological processes. On the other hand some thiourea and their transition metal complexes are also known to exhibit a wide range of biological activities including antiviral, antibacterial, antifungal, anticancer [20-22] antitubercular, antithyroidal, herbicidal and insecticidal activities [23] and as agrochemicals [24]. Recent advances in organic Nonlinear Optical (NLO) materials have involved a large revival of interest on account of their widespread potential importance such as their high nonlinearity, high flexibility in terms of molecularstructure, high optical damage threshold [25,26]. The origin of nonlinearity in NLO material like thiourea arises due to the presence of delocalized π electrons system, connecting donor and acceptor groups and responsible for enhancing their asymmetric polarizability [27].

We reported here XRD analysis of some selected complexes of nickel(II) carboxylates with urea and thiourea.

2. Experimental

All chemicals used in this study were obtained from Qualikem Company. Urea and thiourea were received from Rankem Company. All chemicals were of analytical grade and they were used without purification The complexes were prepared using the following general method as reported earlier [2].Benzoic acid (5g, 0.0206moles) was dissolved in 50ml of 2M NaOH and stirred with a magnetic stirrer. Ni(II) salt (1.6g, 0.0100moles) dissolved in 10ml of water was added dropwise to the solution and stirred for 30minutes. The resulting solution was filtered through a buchner funnel using a suction pump and washed with ethanol. The precipitate was light green in colour (0.32g). 30ml of water was added to the precipitate formed (0.32g) and it was stirred using a magnetic stirrer. Urea or thiourea (0.0628g, 0.0010moles) was diluted with 10ml of water and added to the solution being stirred dropwisely. The resulting solution was stirred for 30 minutes and filtered to dryness through suction. The precipitate (complex) was weighed and dried over KOH. The colour of the complex is light green.

The sample is irradiated with a beam of monochromatic x-rays over a variable incident angle range.The X-ray were produced using a sealed tube and the wavelength of X-ray as 0.154nm (Cu K-alpa). The X-ray were detected using a fast counting detector based on silicon strip technology (Bruker LynxEye detector). Interaction with atoms in the sample results in diffracted x-rays when the Bragg equation is satisfied. The X-ray were detected using a fast counting detector based on silicon strip technology (Bruker LynxEye detector).X-ray diffraction pattern have been recorded by Bruker D8 advance diffractometer at IUC, Indore. All chemicals are prepared by chemical root method.

3. Results and discussion:

XRD patterns are shown in 'figures 3(a) and 3(b)' for Ni-urea complexes and in 'figures 4(a) and 4(b)' for Ni-thiourea complexes. All the samples are characterized at room temperature by X-ray diffraction using Cu Kα radiation.The diffraction pattern of complexes are recorded between 2θ ranging from 10° to 80°. The particle size of the samples is estimated using the Scherrer's formula. According to Scherrer's equation , the particle size is given by $t = 0.9\ \lambda/B\ \cos\theta$, where t is the crystal thickness (in nm), B is half width (in radians),θ is the Bragg angle and λ is the wavelength. The particle size corresponding to each diffraction maxima are determined from the measurement of the half width of the diffraction peak. Lattice parameter for simple cubic crystal structure is determined by $a^2= \lambda^2(h^2+k^2+l^2)/4\sin^2\theta$. The value of Lattice

Journal of Physics: Conference Series **534** (2014) 012044 doi:10.1088/1742-6596/534/1/012044

parameter and the particle size are shown in 'table 1' for all the four complexes. The particle size was found to be within in the range 4.23-9.01 nm for nickel-urea complexes and 0.26– 9.76 nm for nickel-thiourea complexes.

Table-1 for lattice parameter and particle size

Complexes	Lattice parameter (Å)	Particle size (nm)
$[Ni(CON_2H_4)_6]SO_4.H_2O$	7.58	4.23
$[Ni(CON_2H_4)_6]NO_3.H_2O$	7.47	9.01
$[Ni(CSN_2H_4)_6]SO_4.H_2O$	8.06	9.76
$[Ni(CSN_2H_4)_6]NO_3.H_2O$	7.83	0.26

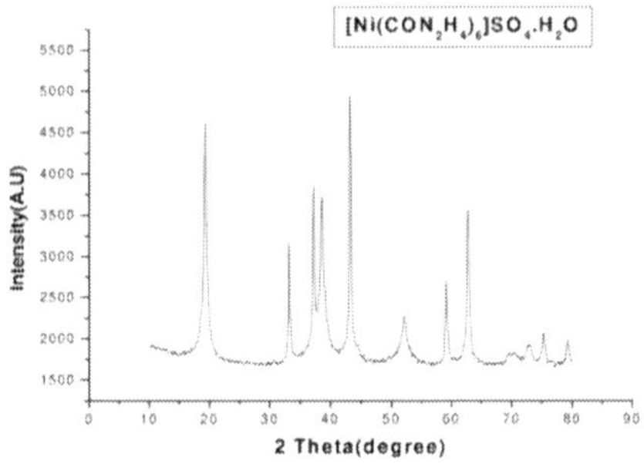

3(a)

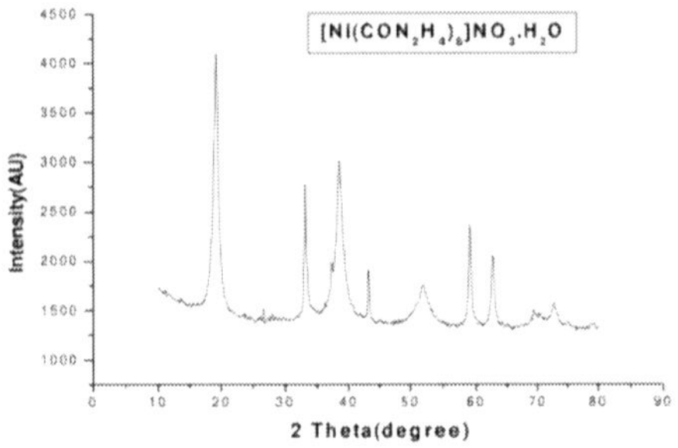

3(b)

Figures 3(a) and 3(b):XRD spectra of Ni-urea complexes

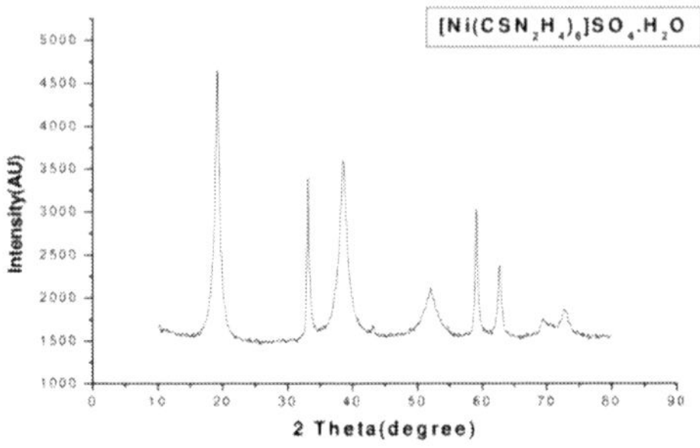

4(c)

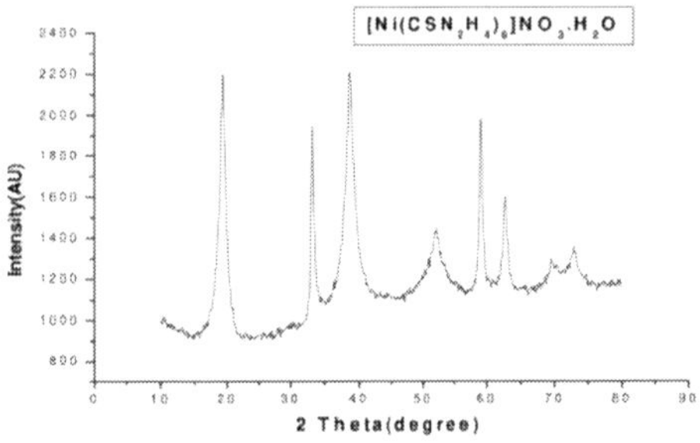

4(d)

Figures 4(c) and 4(d) : XRD spectra of Ni –thiourea complexes

4. Conclusion

The XRD pattern is indicative of their crystalline in nature which is confirmed by 5 main peaks positioned .The X-ray analysis reveals that the sample is cubic in phase as seen from the presence of extra peaks in XRD pattern. All the peaks match with the soft wear JCPDF.

Acknowledgment:
We are grateful to P.Sharma and Dr. Mukul Gupta IUC, Indore for XRD studies of prepared complexes.

References

[1] Feldman D and Barbalata A 1996 *Synthetic Polymers* (London: Chapman & Hall).
[2] Finar I 1973 *Organic chemistry* (London: Longman) p 460.
[3] Rahman M J, Bozadjiev P and Polovski Y 1994 *Fert. Res* **38** 89.
[4] George S , Chellapandian M, Sivasankar B and Jayaraman K 1997 *Bioprocess Eng.* **16** 83.
[5] Wang X J and Douglas L A 1996 *Agrochimica* **40** 209.
[6] Yerokun O A 1997 *S Afr. J. plant soil,* **14** 63.
[7] Heinig R 1996 *SOFW J.* **122** 998.
[8] Gnewuch C T and Sosnovsky G 1997 *Chem. Rev.* **97** 829.
[9] Miyagawa C I 1986 *Drug Intell. & Clin. Pharma.* **20** 527.
[10] Kissa E 1969 *Text. Res. J.* **39** 734.
[11] Srinivasa I, Vishivanathapuram K, Mishra M B and Ghosh G B 1970 *Technolog,* **7** 27
[12] Kaganskii I M A and Babenko M 1970 *Zh. Prikl. Khim* **43** 2390.
[13] Zhang Y ,Bai J ,Wei T and Lu 1996 A *Huaxue Shijie* **37** 178.
[14] Kim Y K , Williard J W and Frazier A W 1998 *J. Chem. Eng. Data* **33** 306.
[15] Meessen J H and Petersen H 2002 *Urea in Ullmann's Encyclopedia of Industrial Chemistry* (Weinheim, Wiley-VCH,) 6th edition.
[16] Moloto M J, MalikMA, O'Brien P,Motevalli M and Kolawole G A 2001 *Polyhedron* **22** 595.
[17] Henderson W, Nicholson B K, Dinger M B and Bennett R.L 2002 *Inorg.Chim. Acta* **338** 210.
[18] Mansuroglu D S ,Arslan H,Flörke U and Külcü N 2008 *J. Coord. Chem.* **61** 3134.
[19] Arslan H, Flörke U, Külcü N and Emen M F 2006 *J .Coord. Chem.* **59** 223.
[20] Saeed S, Rashid N, Hussain R,Ali M and Jones P G 2010 *Eur. J. Med. Chem.* **45** 1323.
[21] Saeed S, Rashid N, Jones P G, Hussain R and Bhatti M H 2010 *Cent. Eur. J. Chem.* **8** 550.
[22] Saeed S, Rashid N, Jones P G and Yunus U J 2010 *Heterocycl. Chem.* **47** 908.
[23] Eweis M, Elkholy S S and Elsabee M Z 2006 *Int. J. Biol. Macromol.* **38** 1.
[24] Lachkova V, Varbanov S,HageleG, Keck, H and Tosheva T 2002 *Phosphorus. Sulfur Silicon Relat. Elem.* **177** 1303.
[25] Bisder-Leib V and Doherty M F 2003 *Cryst.Growth Des.* **3** 221.
[26] Lediux I,Badan J, Zyss J ,migus A,Hulin D,Etchepare J, Grillon G and Antonetti A 1987 *J. Opt. Soc. Am.* **B4** 987.
[27] Selvarajan P, Glorium J,Raj Arual and Perumal S 2009 *J. Crystal Growth* **311** 3835

Theoretical calculations of X-ray absorption spectra of a copper mixed ligand complex using computer code FEFF9

A Gaur* and B. D. Shrivastava
School of Studies in Physics, Vikram University, Ujjain-452001, India

*Email: abhijeetgaur9@gmail.com

Abstract. The terms X-ray absorption near edge structure (XANES) and extended X-ray absorption fine structure (EXAFS) refer, respectively, to the structure in the X-ray absorption spectrum at low and high energies relative to the absorption edge. Routine analysis of EXAFS experiments generally makes use of simplified models and several many-body parameters, e.g. mean free paths, many-body amplitude factors, and Debye–Waller factors, as incorporated in EXAFS analysis software packages like IFEFFIT which includes Artemis. Similar considerations apply to XANES, where the agreement between theory and experiment is often less satisfactory. The recently available computer code FEFF9 uses the real-space Green's function (RSGF) approach to calculate dielectric response over a broad spectrum including the dominant low-energy region. This code includes improved treatments of many-body effects such as inelastic losses, core-hole effects, vibrational amplitudes, and the extension to full spectrum calculations of optical constants including solid state effects. In the present work, using FEFF9, we have calculated the X-ray absorption spectrum at the K-edge of copper in a complex, viz., aqua (diethylenetriamine) (isonicotinato) copper(II), the crystal structure of which is unknown. The theoretical spectrum has been compared with the experimental spectrum, recorded by us at the XAFS beamline 11.1 at ELETTRA synchrotron source, Italy, in both XANES and EXAFS regions.

1. Introduction

X-ray absorption fine structure (XAFS) spectroscopy is a unique tool for studying, at the atomic and molecular scale, the local structure around selected elements that are contained within a material. For this reason, XAFS is a very important probe of materials, since knowledge of local atomic structure, i.e., the species of atoms present and their locations, is essential to progress in many scientific fields: physics, chemistry, biology, biophysics, medicine, engineering, environmental science, materials science, and geology [1]. The XAFS is typically divided into two regimes: X-ray absorption near edge structure (XANES) and extended X-ray absorption fine structure (EXAFS). The terms XANES and EXAFS refer, respectively, to the structure in the X-ray absorption spectrum at low and high energies relative to the absorption edge, with the crossover at about 30 eV above threshold. [2]

As experimental techniques and facilities have improved, including third generation synchrotron sources and the advent of X-ray free-electron lasers, measurements have been taken on wider ranges of systems, exploring the effects of temperature and pressure, and at higher resolutions than before. Hence, theoretical techniques have to be improved to keep abreast with the experimental advancements so that better information may be extracted from the experimental data.

Mostly, the analysis of experimental EXAFS data makes use of EXAFS analysis software packages like IFEFFIT [3]. This package includes Artemis which uses FEFF6L to generate simplified theoretical models in which mean free paths, many-body amplitude factors, and Debye–Waller factors are incorporated. This software does not perform XANES calculations. FEFF9 [2, 4] is a software package which does both XANES as well as EXAFS calculations in an advanced manner. This computer code uses the real-space Green's function (RSGF) approach and includes improved treatments of many-body effects such as inelastic losses, core-hole effects, vibrational amplitudes, and the extension to full spectrum calculations of optical constants including solid state effects. In the present work, we have demonstrated the use of FEFF9 with an example of analysis of XAFS data of a copper complex. The theoretical spectrum generated by FEFF9 has been compared with the experimental spectrum.

2. Experimental

The copper complex aqua (diethylenetriamine) (isonicotinato) copper(II), Cu(dien)(ina) $4H_2O.1/2SO_4$, has been prepared and characterized by following the standard methods [5]. The complex was in powder form (solid state). The absorption screens were in the form of pellets which were prepared by mixing calculated amount of the sample uniformly with a filler/binder material, viz., boron nitride (BN) and then using a press (pelletizer). X-ray absorption experiments were performed at ELETTRA Synchrotron Light Laboratory, Basovizza, Italy at the XAFS beam line 11.1. Other experimental details were similar to those described elsewhere. [6]

3. Description of FEFF9

FEFF uses an ab initio self-consistent real space multiple scattering (RSMS) approach, including polarization dependence, core-hole effects, and local field corrections, based on self consistent, spherical muffin-tin scattering potentials. Calculations are based on an all-electron, real space relativistic Green's function formalism with no symmetry requirements. The code builds in inelastic losses in terms of a self-energy, and includes vibrational effects in terms of Debye-Waller factors. For periodic structures reciprocal space calculations based on periodic boundary conditions are also available. FEFF can use both full multiple scattering based on Lanczos algorithms and a high-order path expansion based on the Rehr-Albers multiple scattering formalism. The name FEFF is derived from f_{eff}, the effective curved wave scattering amplitude in the modern EXAFS equation. [7]

A FEFF calculation starts with two basic tasks: calculation of the potentials, followed by calculation of the scattering phase shifts. Using these phase shifts, scattering paths are found and their scattering amplitudes calculated. There are two ways to do this: either using an explicit enumeration and summation, most appropriate for extended absorption spectra; or using the implicit summation of full multiple scattering, appropriate for near-edge absorption spectra. [7]

FEFF9 can be used for a number of different calculations. In the present work, we have demonstrated its use for XANES and EXAFS calculations. In order to use FEFF9 a program must be used to create the atomic structure that indicates the position of all the atoms in the lattice. ATOMS on the web was used in order to create an input file for the FEFF9 program. The parameters that are required to create the atomic structure in ATOMS is the space group, the edge of interest (Copper K edge in the present case), lattice constants, angles and finally the sites of the elements in the unit cell and the central atom of the crystal structure.

4. Data analysis

The crystallographic data is unavailable for the complex. Hence, for the analysis of the XAFS data of the complex, the crystal structure of an analogous complex aqua(diethylenetriamine) (isonicotinato)-copper(II) hexafluorophosphate $[Cu\text{-}(C_6H_4NO_2)(C_4H_{13}N_3)(H_20)]PF_6$ [5], has been used. The following parameters have been used for generating the feff input file.

Space group: Pnma , Cell constants: a = 22.826 Å, b = 6.971 Å, c = 16.593 Å,
$\alpha = 90.00°$, $\beta = 90.00°$, $\gamma = 90.00°$ and Cluster size = 8 Å.

The remaining input positional parameters are given in ref. [5].

The FEFF program consists of a set of program modules driven by a single input file `feff.inp' [7]. One can either supply this file or can set all input options through the JFEFF GUI, which runs FEFF

Table 1. Some components of feff input file

XANES	EXAFS
TITLE Cu complex	TITLE Cu complex
CONTROL 1 1 1 1 1 1	EDGE K
EXCHANGE 0 0 0 0	S02 1.0
SCF 3.0 0 100 0.2 1	CONTROL 1 1 1 1 1 1
COREHOLE RPA	PRINT 0 0 0 0 0 0
SIG2 0.009	LDOS -30 20 0.1
XANES 8.0 0.07 0.0	COREHOLE FSR
FMS 3.0 0	NLEG 3
POTENTIALS	CORRECTIONS 4.0 1.0
0 29 Cu	SIG2 0.006
1 29 Cu	EXAFS 20.0
2 8 O	RPATH 3.5
3 7 N	POTENTIALS
4 6 C	0 29 Cu
	1 29 Cu
	2 8 O
	3 7 N

without the need for manual text editing. The input file `feff.inp' is a loosely formatted, line-oriented text file. Each type of input read by the program is on a line starting with a CARD: a keyword possibly followed by some numbers or words. The sequence of keyword cards is arbitrary. If any card or optional data is omitted, default values are used. Most calculations do require the POTENTIALS card, and either the ATOMS card or the OVERLAP card. Together, the POTENTIALS and ATOMS/OVERLAP cards specify the physical structure of the molecule or solid one is interested in. FEFF uses Cartesian coordinates and Angstrom units. Hence it is easy to copy atom positions from, e.g., an `xyz' file or other popular file formats. It is also possible to import CIF files (`*.cif'). There are many optional cards to specify the material property to calculate and to control how FEFF calculates it. All distances are in Angstrom (Å) and energies are in electron-volts (eV). Separate input files have to be generated for XANES as well as EXAFS calculations. For example, in the case of present sample, some components of the feff input file for XANES and EXAFS are shown in Table 1.

5. Results and discussions

5.1 XANES calculations

For XANES calculations in FEFF9, the self-consistent potential (SCF) and full multiple scattering (FMS) must be specified in order to run the calculation. The SCF and FMS adjust the amount of atoms used in the calculation. Convergence is reached when increasing the amount of atoms included in the calculation produces no distinct changes in the graph of the simulated spectrum. The XANES calculations performed using FEFF9 include the many-pole self-energy, the treatment of the core-hole within the random phase approximation (RPA), and the many-body spectral function convolution.[8]

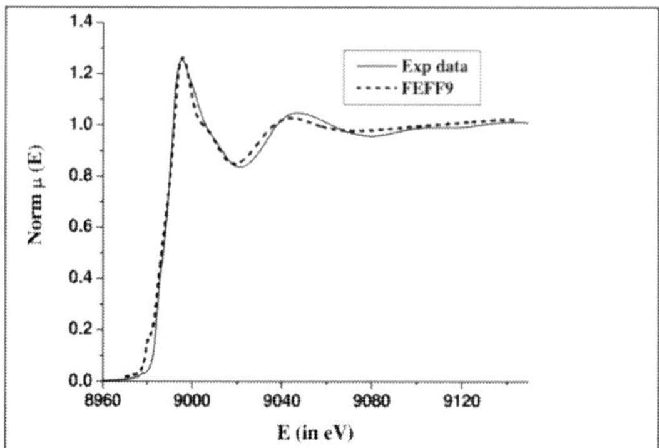

Figure 1. Comparison of experimental Cu K-edge XANES ($\mu(E)$ vs. E) with calculations from FEFF9 for the complex

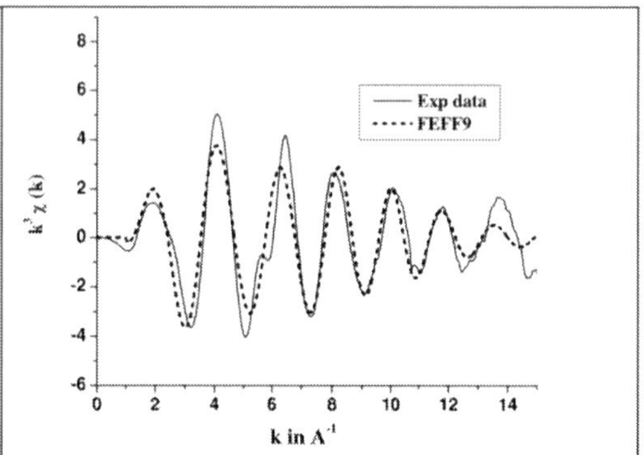

Figure 2. Comparison of experimental Cu K-edge EXAFS ($\chi(k)$ vs. k) with calculations from FEFF9 for the complex

In the present case, for calculating the XANES spectra, self consistent potentials with a cluster radius of 3 Å have been used. The full multiple scattering (FMS) cluster size has also been kept as 3 Å. The calculated spectrum has been shifted to align the white-line peak with experiment. The

XANES spectrum (μ(E) vs. E) thus obtained for the studied copper complex is given in figure 1. In this figure the experimental spectrum has also been given in order to show the comparison between theoretical and experimental spectra.

5.2 EXAFS calculations

In the present case, the calculation of potentials and phase shifts has been done only once and then the results have been used while studying the paths. Because the number of possible paths increases exponentially with total path length, we have started with a short total path length and then examined few paths (representing scattering from the nearest neighbors), and gradually increased the total path length, studying the path importance coefficients and using the filters to limit the number of paths. The paths so generated using FEFF9 for the present complex are listed in Table 2.

Table 2. Paths generated by FEFF9

FEFF 9.6.4, PATH R_{max}= 3.500, S_0^2=1.0

Curved wave amplitude ratio filter 4.000%

File	sig2	cw amp ratio	deg	nlegs	reff
1	0.00600	100.00	1.00	2	2.0049
2	0.00600	99.05	1.00	2	2.0133
3	0.00600	98.80	1.00	2	2.0156
4	0.00600	98.57	1.00	2	2.0176
5	0.00600	70.10	1.00	2	2.3159
6	0.00600	13.26	2.00	3	3.3516
7	0.00600	12.97	2.00	3	3.3633
8	0.00600	10.13	2.00	3	3.4829

The EXAFS spectrum (χ(k) vs. k) has been obtained as the sum of these paths for the studied copper complex and the spectrum so obtained is given in figure 2. In this figure the experimental spectrum has also been given for comparison between theoretical and experimental spectra.

6. Conclusions

In the present work, both XANES spectra as well as EXAFS spectra at the K-edge of copper in a copper complex have been calculated using the computer code FEFF9. The calculations are based on extensions of the quasi-particle theory and the RSGF formalism which make possible general treatment of XAFS, encompassing XANES and EXAFS. New approaches for calculating inelastic losses, self-energies, and Debye–Waller factors yield significant improvements in amplitude and phase for XAFS spectra. Details of generating the feff input file for FEFF9 has been outlined. The theoretical spectra obtained from FEFF9 are comparable with the experimental spectra recorded on a standard XAFS beamline.

The use of recent and advance softwares for XAFS data analysis are important because the EXAFS beamlines at the 2.5 GeV Indus-2 synchrotron at Raja Ramanna Center for Advanced Technology (RRCAT), Indore have become easily available to Indian workers which were not available until now. The beamline BL-8 is a dispersive EXAFS beamline and beamline BL-9 is a DCM EXAFS beamline which have become operational. It is hoped that research workers from Indian laboratories will do experiments on these synchrotron beamlines and the present paper will help them in doing better theoretical calculations.

References

[1] Gaur A, Shrivastava B D and Nigam H L, 2013 *Proc. Ind. Nat Sci Acad* **79** 921
[2] Rehr J J, Kas J J, Vila F D, Prange M P, Jorissen K, 2010 *Phys. Chem. Chem. Phys.*, **12** 5503.
[3] Ravel B and Newville M, 2005 *J. Synchrotron Radiat.* **12** 537. (IFEFFIT is available freely from http://cars9.uchicago.edu/ifeffit/Downloads)
[4] Rehr J J, Kas J J, Prange M P, Sorini A P, Takimoto Y, Vila F D, 2009 *Compt. Rend. Phys.* **10** 548.
[5] Nie F-M, Wang Z-M, Li Y-M, Zhao Y-F and Yan C-H, 1998 *Acta Cryst.* **C54** 1853
[6] Gaur A, Shrivastava B D, Prasad J, Srivastava K and Raghuwanshi V S, 2013 *J Chem Phys* **139** 034303
[7] *User's Guide, FEFF version 9.6.4* (FEFF9 is a licensed software available from http:// www.feffproject.org/feffproject-feff.html)
[8] Rehr J J, Kas J J, Glover J L and Chantler C T 2010 *Nucl Instrum Meth A* **619** 28.

Persistence photo conductivity in Ni doped CdS

M M Patidar[1*], A Ajay[2], A Dewas Wala[3], N Kiran [4], R Panda[5], M Gangrade[1], R Nath[1] and V Ganesan[1]
[1]UGC-DAE Consortium for Scientific Research, Khandawa Road, Indore, M.P 452001, India
[2]Physics Department, University of Delhi, Jawaharlal Nehru Marg, New Delhi, Delhi 110002, India
[3]Holkar Science College, AB Road, Bhanwar Kuan, Indore, M.P. 452001, India
[4]Physics Department, University of Mysore, JLB Rd, Mysore, Karnataka 570005, India
[5]Acropolis Institute of Technology and Research, Ratibad , Bhopal, M.P., 462002,India

[*]E-mail: manjumishra.iuc@gmail.com

Abstract: Persistence Photo Conductivity (PPC) in Ni-doped Cadmium Sulphide [$Cd_{1-x}Ni_xS$, (x=0, 0.03, 0.05 and 0.20)] thin films were investigated. While the morphology of the films prepared by Spray Pyrolysis Technique (SPT) were studied using Atomic Force Microscopy (AFM), the PPC has been studied using a low temperature two probe resistivity set up with a photo-diode excitation. Note worthy changes in the morphology were observed upon Ni substitution and is attributed to the strain induced upon substitution. The simple resistivity analysis shows an evolution to a highly resistive state upon increasing the Ni content and may be attributed to the local oxide formation. Ni doping enhances the photo sensitivity of the films that may be tuned for technological applications.

1. Introduction

Out of the well known photo conducting samples CdS forms an important and integral member of inorganic compounds. This find applications in many areas like photo sensitive applications including in biology, especially in the area of fluorescent dyes and drug delivery. CdS has a direct band gap of 2.42eV, at 300K and is normally a n-type semiconductor. Out of various methods like pulsed laser deposition (PLD), chemical bath deposition (CBD), sputtering and spray pyrolysis technique (SPT), SPT is a very simple and highly reproducible route, is employed in this study. Mn, Fe, Zn and Cu doped CdS has been studied in the literature by various authors [1-2]. Co and Ni doping is sparsely used [3-4]. It is believed that the Ni doping in CdS may improve its electrical properties. Present work is in line with this but with certain improved data collection on the system especially on the persistent photo current scenario.

2. Experimental details

The $Cd_{1-x}Ni_xS$ thin films were prepared for different compositions x=0, 0.03, 0.05 and 0.20 using spray Pyrolysis technique. For pure CdS thin films aqueous solution of 0.1 M monohydrated Cadmium Chloride ($CdCl_2.H_20$) and 0.2 M Thiourea ($CS(NH_2)_2$) were taken as the precursor solution. Deionized water equivalent to 1/3rd part of the main solution is further added. Aqueous solution of 0.1 M Nickel chloride hexahydrate ($NiCl_2.6H_2O$) was added to starting solution for nickel doping. All the solutions were stirred at room temperature for few minutes to yield a clear and homogeneous solution. For all the $Cd_{1-x}Ni_xS$ thin films the quantity of the various constitutes (in grams) has been measured by weight percentile method using the molarity equation i.e. (Molarity x Molecular weight x Volume)/1000. 15ml precursor solution was sprayed with the maintained flow rate of 1ml/min for all the Ni doped and undoped CdS thin films. The atomization of the solution in the fine droplets was done by spray nozzle with the help of compressed dry air as a carrier gas, during the execution of spray. The films have been deposited on chemically and ultrasonically cleaned glass substrate at 360^0C over a time period of 15min. A 2 kW heater was employed for heating the glass substrate and the temperature was measured by using chromel-alumel thermocouple fixed on the hot plate. Distance between substrate and nozzle was fixed at 28 cm. The films were annealed at the same deposition temperature (360^0C) for 30 min to get a good adherent pinhole free and smooth films.

The morphological imaging of the films has been done by using AFM (Digital instruments Nanoscope E with Si_3N_4 100μm cantilever, 0.58N/m force constant) in contact mode at room temperature. Photoconductivity and the electrical properties of the various Ni:CdS films have been studied by a homemade apparatus with Teflon insulated coaxial cables in the temperature range of 150K-300K by using two probe resistivity method in which temperature was read in Lake Shore (DRC-93CA) temperature Controller and resistance in Keithley (617) programmable electrometer. The complete set up was kept in the cryostat to perform the measurement in low temperature region using liquid nitrogen as a coolant. In photoconductivity measurement the sample was illuminated with commercially available 10mm diameter light emitting diode (LED). Its emission has a broad spectrum consisting of prominent emission centered on 470nm and 560nm. Its power is ~ 90mW and has a luminous intensity 0f ~ 4lumen. The sample is kept just an mm away from the LED. Input voltage was set at 10V for all the resistivity and photoconductivity measurements.

3. Results and Discussions

Figure 1(a-d). AFM images of $Cd_{1-x}Ni_xS$ thin films (x=0,0.03,0.05 and 0.20) clearly shows the evolution of hexagonal nano-particles from that of nano-noodle type CdS.

Figures 1(a-d) show changes in morphology of Ni doped CdS for various concentrations being discussed. Pure CdS has a noodle type agglomerated grains which is quite homogenous. This morphology drastically changes to large hexagonal type of particles upon addition of Ni content. A careful analysis of XRD patterns reveal that the crystallite size calculated using Scherrer formula is in fact reduces in a regular fashion. The micro-strain calculated as in literature shows a systematic increase upon Ni doping. This is line as the addition of more strain is expected to induce stresses leading to the crackdown in crystallite size. However, AFM shows a contrasting trend and is due to the agglomeration and is in increasing trend. These results are shown in Figure 3 and 4.

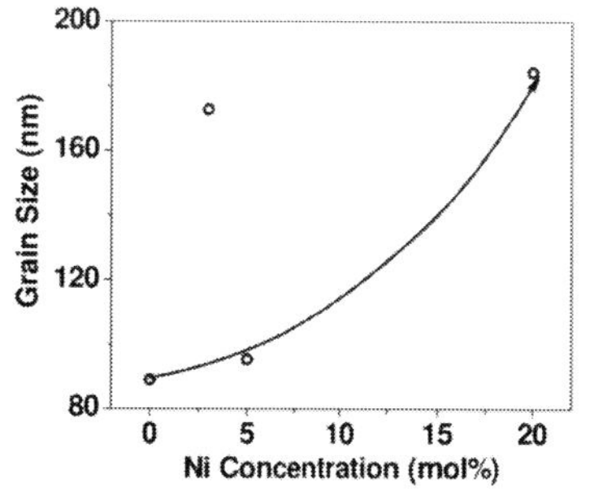

Figure 2. Grain size of $Cd_{1-x}Ni_xS$ thin films (x= 0, 0.03, 0.05 and 0.20) have been calculated from the AFM images.

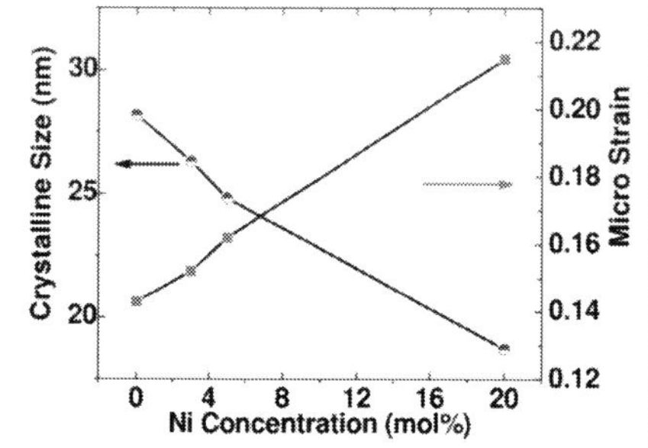

Figure 3. Crystalline Size and Micro Strain has been calculated from the XRD for $Cd_{1-x}Ni_xS$ thin films (x= 0, 0.03, 0.05 and 0.20)

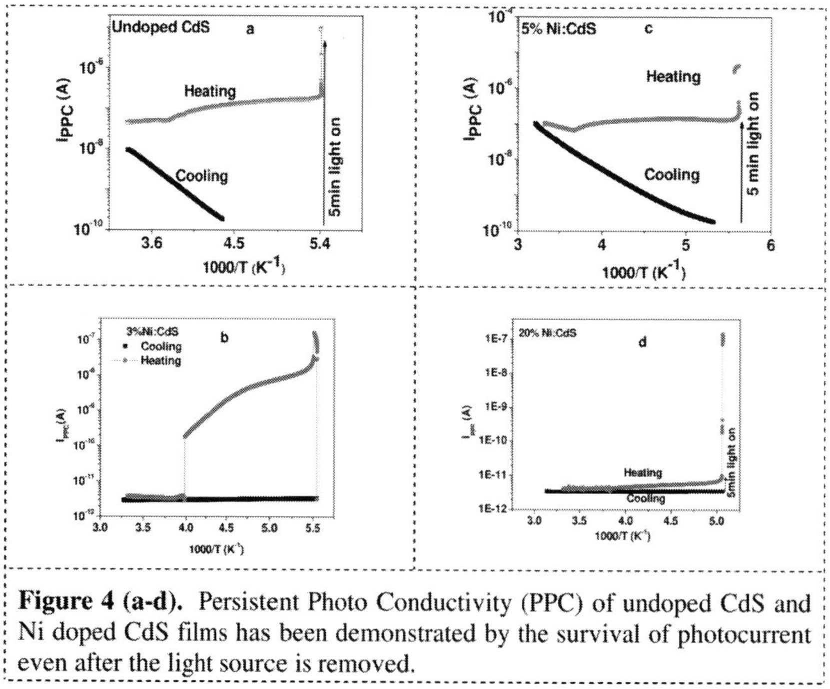

Figure 4 (a-d). Persistent Photo Conductivity (PPC) of undoped CdS and Ni doped CdS films has been demonstrated by the survival of photocurrent even after the light source is removed.

Persistent Photo Current Analysis: In order to understand more about the photoconductive properties of these Ni doped CdS samples, we have analyzed the Persistent Photo Current (PPC) characteristics of the samples. For this, the current verses temperature data has been collected upon cooling in dark and warming in illumination condition. However, the illumination was carried out only for 5 minutes at low temperature. The figures 4(a-d) show that the current persists even in the light OFF condition for a substantial portion of time and one need to warm the system to get back to the initial condition. However, the condition deteriorates upon doping and at 20% Ni doping the drop is quite dramatic. A careful look may reveal quenching of photo currents upon Ni doping at higher levels as far PPC is concerned. However, the sensitivity for light is more upon Ni doping and hence is good for tunable applications like photoconductive switches.

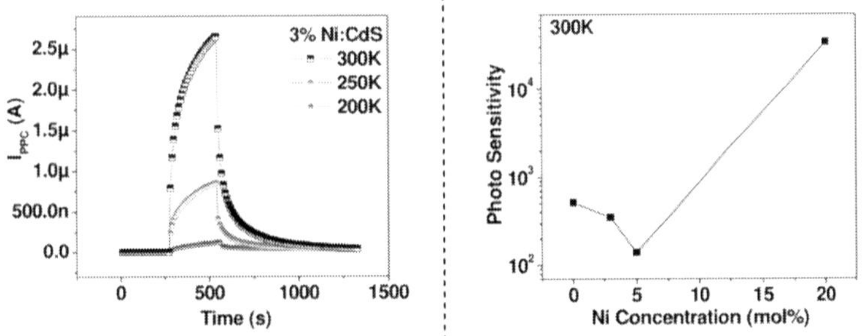

Figure 5 (a-b): The pictures shown above indicate (a) the switching characteristics of 3% Ni doped CdS as a representative case for 300K, 250K & 200K and (b) the photosensitivity as a function of doping of Ni. One could notice the increase of sensitivity at 20% in spite of reduced photoconductivity behaviour as discussed in the text.

4. Conclusions

Good quality thin films of Ni doped CdS prepared by SPT show excellent photoconductivity properties. Doping of Ni gradually induces strain as well as microstructural transformation as seen by AFM. Persistent Photocurrents have been demonstrated.

Acknowledgements

Authors especially from Delhi and Mysore would like to thank Dr. P. Chaddah, Prof. Ajay Gupta and Dr. A.K.Sinha for their encouragement in carrying out science projects at the consortium. MMP is thankful to DST for the fellowship

References

[1] Badera N, Godbole B, Srivastava S B, Vishwakarma P N, Sharath Chandra L S, Jain D, Gangrade M, Shripathi S, Sathe V G and Ganesan V 2008 *Appl. Surf. Sci.* **254** 7042.

[2] Panda R, Rathore V, Rathore M, Shelke V, Badera N, Sharath Chnadra L.S., Jain. D., Gangrade M., Shripathi T., and Ganesan V, 2012 *Appl. Surf. Sci.* **258** 5086.

[3] Rmili A, Ouachtari F, Bouaoud A, Louardi A, Chtouki T, Elidrissi B and Erguig H 2013 *J. Alloys Compd.* **557** 53.

[4] Patidar M M, Ajay A, Dewas Wala A, Kiran N, Panda R, Gangrade M, Nath R and Ganesan V 2013 *AIP Conf. Proc.* [under consideration]

Structural and optical properties of Cd and Mg doped zinc oxide thin films deposited by pulsed laser deposition

Vanita Devi[1], B.C. Joshi[1], Manish Kumar[2], R. J. Choudhary[2]
[1]Department of Physics and Materials Science & Engineering, Jaypee Institute of Information Technology, Noida 201307, India
[2]UGC-DAE Consortium for Scientific Research, Indore-452001, India

E-mail: vanitabishnoi@gmail.com

Abstract. The 7 wt % Cd doped and 15 wt % Mg doped ZnO thin films were deposited on quartz substrate by pulse laser deposition system. The structural and optical properties of the prepared ZnO, Cd:ZnO and Mg:ZnO films were investigated by X-Ray diffraction (XRD), photoluminescence and UV-Vis spectroscopy techniques. XRD results indicate that the doped ZnO films maintain wurtzite crystal symmetry without any defects and are oriented along c-axis. Photoluminescence studies show a sharp band edge emission peak at 384 nm for pure ZnO film. This peak is blue shifted to 381 nm with Cd doping and red shifted to 395 nm with Mg doping. UV visible absorption studies reveals a decrease in band gap with Cd doping and an increase in band gap with Mg doping.

1. Introduction

The increasing demand, of low power optoelectronic devices, such as white LEDs, photo detectors and UV emitters, has been stimulated the research on wide band gap semiconducting materials [1]. Silicon Carbide, Aluminum Nitride, Diamond and Gallium Nitride are very much popular among the researchers. Among these (high brightness white LED and blue laser), GaN has gain lots of interest due to its commercial success in optoelectronic devices [2]. More than 100 billion GaN based white and blue LED's were fabricated in 2013. These LED's were use in backlight of the LCD display, traffic lighting, solid state lighting and other's general lighting applications [3]. Use of this material at commercial level is limited, due to its high cost, maintenance and sophisticated/expensive growth process. ZnO is a low cost wide band gap material, which is easily available, non toxic and environmental friendly [4]. ZnO contains similar hexagonal wurtzite crystal structure as that of GaN materials, and has direct band gap of 3.34 eV. In ZnO the fundamental absorption spectra shows, comparatively higher excitonic binding energy (~ 60 meV) than GaN (~25 meV) at room temperature [5]. These properties of ZnO made it an attractive material for the optoelectronic devices and applications. For fabrication of optoelectronic devices from ZnO, quantum wells structures play an important role. In ZnO quantum well structures can be formed by tailoring the band gap with doping II group element like Cd and Mg [6].

Cd doped ZnO layers in multilayer structures acts as quantum well and undoped ZnO layers acts as barrier [7]. These barrier and quantum well layers can be used for fabrication LEDs and laser diode from ZnO. Mg doped ZnO films can be used as a electron blocking layer in these devices. In this work we have used Cd and Mg dopants successfully for reducing and widening the band gap of ZnO.

2. Experimental

For thin film synthesis, dense bulk targets of ZnO, $Zn_{0.93}Cd_{0.07}O$ and $Zn_{0.85}Mg_{0.15}O$ were prepared by standard solid state reaction route. Thin film from these bulk targets were grown on quartz substrate by Pulsed laser deposition (PLD) technique using KrF (λ=248 nm) excimer laser source. Prior to deposition the quartz substrates were cleaned in ultrasonic bath, first by acetone for 5 minutes, followed by cleaning in methanol for 5 minutes. The background pressure of deposition chamber was ~ 10^{-6} torr. For the deposition, laser pulse rate was kept at 10 Hz with energy density of laser beam at the target surface 2J/cm and the distance between substrate and target was adjusted to 5cm. About 150 nm thick films were grown on 650^0C substrate temperature at Oxygen partial pressure of 1×10^{-4} Torr. For X-ray diffraction Cu-K_α radiations (λ=1.5414A^0) were used. The optical properties of the films were studied by photoluminescence and UV-Vis spectroscopy.

3. Results and discussion

The X-ray diffraction results of ZnO (a), $Zn_{0.93}Cd_{0.07}O$ (b) and $Zn_{0.85}Mg_{0.15}O$ (c) thin film samples grown on quartz substrate are shown in Fig. 1. The X-ray study indicates that the grown films are oriented along c-axis with wurtzite phase. Apart from (002) and (004) peaks, no impurities peaks are observed in these samples. X-ray diffraction results revealed that Cd and Mg atom replaces the Zn atom in wurtzite crystal lattice and there are not apparent changes in crystal lattice structure, of the doped films. We noticed a slight variation in position of reflection peak, corresponding to (002) plane of ZnO, towards lower and higher 2θ values for Cd and Mg doping respectively. The shift occurs due to the change in the lattice parameters of ZnO film with Cd and Mg doping (mismatch in ionic radi of Zn, Cd and Mg ions). The calculated lattice parameters from the XRD data are shown in Table 1and these are in agreement with reported values for ZnO, Cd:ZnO and Mg:ZnO.

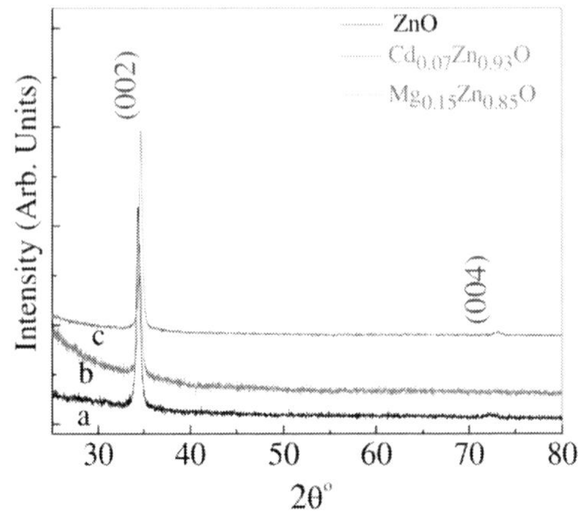

Figure 1. X-ray diffraction pattern of all the studied samples.

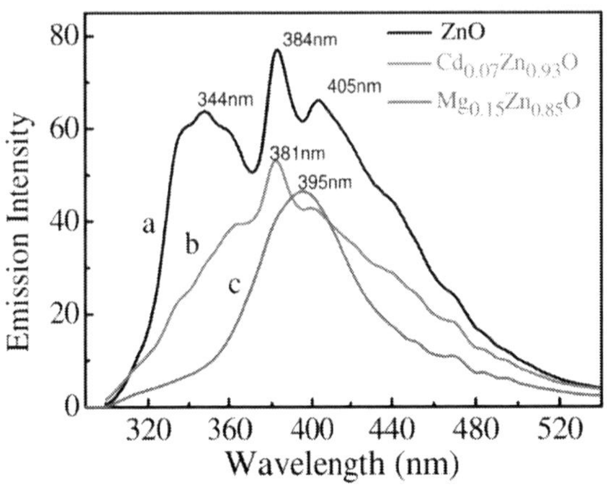

Figure 2. Photoluminescence spectra of doped and undoped ZnO thin film samples.

Room temperature photoluminescence (PL) results for undoped and doped samples are shown in Fig. 2. The strongest peak lines are obtained near band edge UV emission at 384 eV for ZnO with somewhat weaker lines at 344 eV and 405 eV. The main band edge peak is blue shifted towards lower wavelength of 381nm in case of Cd doped ZnO while in Mg doped ZnO; red shifted peak towards longer wavelength (395nm) is seen. For better explanation of red and blue shift in the doped samples requires, the band gap calculation. Therefore, these samples were characterized further by UV-Vis spectroscopy.

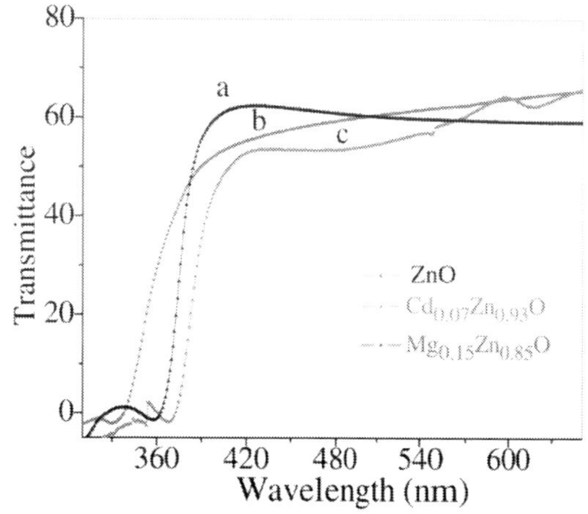

Figure 3. Transmission spectra of doped and undoped ZnO thin film samples.

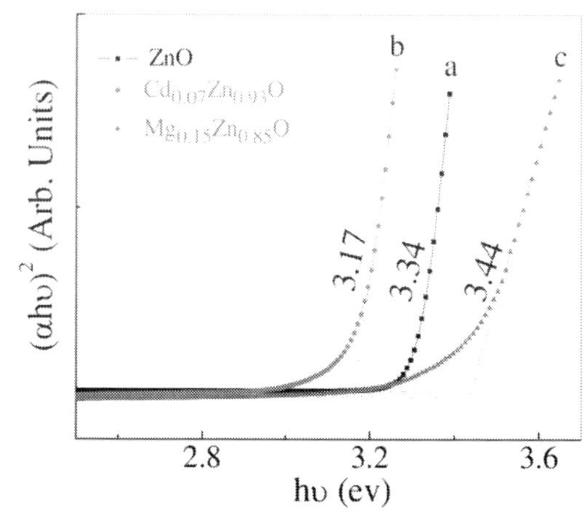

Figure 4. Taut plots calculated from the transmission spectra of studied samples.

Fig. 3 depicts UV-Vis spectra of undoped ZnO along with Cd and Mg doped thin films samples. Fig. 4 shows the taut plot for calculating of band gap of the samples. The band gap of pure ZnO is estimated to 3.34eV. For 7 wt% Cd doped ZnO, the band gap of film is decreases by 0.17eV. This narrow band gap material, can be used in fabricating quantum well devices. However for Mg doped film, the band gap is increase by 0.10eV. This widen band gap material can be used as barrier/ blocking layer in these quantum well structures. The calculated band gaps of $Zn_{093}Cd_{0.07}O$ and $Zn_{0.85}Mg_{0.15}O$ are 3.17eV and 3.44eV respectively. This observed shift in band gap could be responsible for the red and blue shift of band edge emission peaks of photoluminescence (PL) spectra as discussed above. The different important parameters deduced from the present study are summarized in table 1.

Table 1. The significant findings obtained from different thin film samples:

Sample Name	**ZnO**	**Cd:ZnO**	**Mg:ZnO**
Doping(x)	0%	7%	15%
XRD(2θ value for 002 plane)	34.62^0	34.56^0	34.68^0
Lattice Parameter(c)	2.710 nm	2.714 nm	2.706 nm
PL emission Peak	384 nm	381 nm	395 nm
Band Gap	3.34eV	3.17eV	3.44eV

4. Conclusion

Structural and optical properties of ZnO, $Zn_{0.93}Cd_{0.07}O$ and $Zn_{0.85}Mg_{0.15}O$ thin films grown by PLD were studied by XRD, PL and UV-Vis spectroscopy. X-ray diffraction study shows that Cd and Mg doped film are oriented along c-axis. No peak widening is observed in any of these samples. Photoluminescence study shows, blue and red shift of band edge emission peak for Cd and Mg doped ZnO sample, respectively. UV-Vis spectra shows that, by Cd doping the band gap of film is narrowed by 0.17eV and by Mg doping band gap is widen to 3.44eV. This red shift, blue shift of energy gap can be used for fabrication quantum wells and barriers in optoelectronic devices.

Acknowledgement
One of the author V. Devi acknowledge JIIT Noida, for providing research support for the present research work.

References

[1] Mymrin V. F., Bulashevich K. A., Podolskaya N. I. and Yu. S. Karpov 2005 *J. Cryst. Growth* **281**
[2] Zhao H, Liu G and Tansu N 2010 *Appl Phys Lett* **97** 131114
[3] Auf der Maur M., Lorenz K.and Di Carlo A. 2012 *Opt Quant Electron* **44** 83.
[4] Shaolan Li and Lichun Z 2013 *Journal of Semiconductors* **34**
[5] Ali Abbasi M, Hussain Ibupoto Z, Hussain M, Nur O and Willander M 2013 *Nanoscale Research Letters* **8** 320
[5] O'Leary S K, Foutz B E, Shur M S, Bhapkar U V and Eastman L F 1998 *J. Appl. Phys.* **83** 826
[6] Shukla G 2009 *J. Phys. D: Appl. Phys.* **42** 075105
[7] Choi Y S, Kang J W, Hwang D K and Park S J 2010 *IEEE Transcations on Electron Devices* **57**

Journal of Physics: Conference Series **534** (2014) 012048 doi:10.1088/1742-6596/534/1/012048

Extended X-ray absorption fine structure data analysis of copper (II) hydroxamic acid mixed ligand complexes

N Parsai[1], A Mishra[1] and B D Shrivastava[2]

[1]School of Physics, Devi Ahilya University, Indore 452001, India
[2]School of Studies in Physics, Vikram University, Ujjain 456010, India

* E-mail: parsai.neetu@gmail.com

Abstract. The X-ray absorption spectra of copper mixed ligand complexes, having hydroxamic acid as one of the ligands, have been recorded at the K-edge of copper at BL-8 Dispersive EXAFS beamline at the 2.5 GeV INDUS-2 Synchrotron, RRCAT, Indore, India. For the analysis of EXAFS data, crystallographic data of the complex or of its analog is required, which is not available. Hence, for the analysis of EXAFS data, theoretical EXAFS data of the studied complexes have been generated using the EXAFS equation employing computer software program *Mathcad*. Firstly, the experimental data has been processed using the computer program *Athena* to obtain the normalized absorption versus energy data. From the experimental EXAFS data, the phase shift parameter (an energy independent constant δ) has been computed using Lytle, Sayers and Stern's (LSS) method. The backscattering amplitude has been taken from the available theoretical tabulations and other parameters have been taken from crystallographic data of the copper metal. Fourier transforms of both the experimental and theoretical data have been computed, and the two Fourier transforms are found to agree with each other for all the complexes. The position of the first peak in the Fourier transform gives the value of the first shell bond length, which is shorter than the actual bond length as a result of energy dependence of the phase factor ($\delta(k)$) in the sine function of the EXAFS equation. Since, the Fourier transform method and LSS method both are uncorrected for phase and other parameters of the EXAFS equation, the present method gives phase uncorrected bond length of the first coordination shell.

1. Introduction

For the analysis of extended X-ray absorption fine structure (EXAFS) data, the crystal structure of either the complex studied or of its analogous complex is required. The crystallographic information is used to generate theoretical model by using software programs like *Artemis*. In the present work some copper complexes have been studied of which the crystallographic data is not available. Hence, for the analysis of their EXAFS data, the method outlined by us in our earlier communications has been used [1, 2]. The method is simple as it generates theoretical EXAFS data employing computer software program *Mathcad*.. The copper complexes studied have been listed in table 1. The experimental EXAFS data of the copper complexes have been recorded at the K-edge of copper at BL-8 Dispersive EXAFS beamline at the 2.5 GeV INDUS-2 Synchrotron, RRCAT, Indore, India.

2. Analysis of experimental EXAFS data

Firstly, the experimental data has been processed using the computer program *Athena* to obtain the normalized absorption (μx) vs. energy (E) data and then the EXAFS oscillations χ(k) vs. wave vector (k) data is obtained, which are shown in fig. 1. Fourier transforms of the experimental EXAFS data, given fig. 1, have been computed, which are given in fig. 2.

3. Computation of theoretical EXAFS data

The theoretical EXAFS data has been computed for the complexes using the standard form of EXAFS equation [3].

$$\chi(k) = \sum_j \frac{N_j S_0^2 F_j(k)}{kR_j^2} \times \exp(-2\sigma_j^2 k^2) \times \exp(-2R_j/\lambda(k)) \times \sin[2kR_j + \delta_j(k)] \quad \ldots (1)$$

The subscript j denotes the various scattering paths, N_j is the number of atoms in the jth coordination shell, R_j is the average radial distance to the jth atom and σ_j^2 is mean square deviation about the bond length. S_0^2 is the passive electron reduction factor and accounts for the slight relaxation of the remaining electrons in the presence of the core-hole vacated by the photoelectron. S_0^2 usually has a value between 0.7 and 1.0 and is different for different elements. The function $F_j(k)$ is the photoelectron backscattering amplitude and $\delta_j(k)$ is the scattering phase shift for each scattering path. $\lambda(k)$ is the photoelectron mean free path which is dominated by lifetime of the excited state.

In the our earlier works [1, 2], we have developed a simple analysis procedure for extracting bond length for the first coordination shell from the experimental EXAFS data. The method is simple and novel because calculations are done using the computer software *Mathcad*. The theoretical model is generated through the EXAFS equation given in eqn. (1) for the first shell. The values of N_1, σ_1^2 and λ have been taken from the crystallographic data of copper metal [4, 5]. The backscattering amplitude $F_j(k)$ has been taken from the table of Mckale [6]. The phase shift δ_1 has been extracted from the experimental data in the following manner.

The analysis depends on the argument of the sine term in eqn. (1), i.e., $\sin[2kR_j + 2\delta_j(k)]$ where δ_j is assumed as linear in k, i.e., $\delta_j = -\alpha_j k + \beta_j$. Substituting into the argument of sine and rearranging for the first coordination shell (j = 1), it is convenient to define n by [7]:

$$(n + ½)\pi = 2k(R_1 - \alpha_1) + 2\beta_1 \quad \ldots (2)$$

where n = 0, 2, 4... for maxima and n = 1, 3, 5.... for minima. Using the graphical method outlined by Lytle, Sayers and Stern (LSS method) [4, 5], a plot of n versus k for the maxima and minima of the measured EXAFS data determines $(R_1 - \alpha_1)$ from slope and β_1 from intercept.

4. Results and Discussion

From the experimental χ(k) versus k plots, given in fig. 1, the values of k corresponding to maxima and minima of χ(k) have been obtained and are given in table 3. The n versus k curves are given in fig. 3 for the complexes. The values of $(R_1 - \alpha_1)$ and β_1 determined respectively from the slope and the intercept of these plots. Taking α_1 to be 0.31 Å, R_1 and δ_1 have been calculated and used in eqn. (1) and the theoretical EXAFS χ(k) data have been obtained for all the complexes and are given in fig 4. This theoretical EXAFS data has been Fourier transformed which are given in fig. 5. Fourier transforms of both the experimental and theoretical data, given in figs. 2 and 5 respectively, have been found to agree with each other for all the complexes. The position of the first peak in this Fourier transform gives the value of the bond length, i.e., the average distance between the absorbing atom and the neighboring atoms. However, the position of the first peak in these Fourier transforms gives the value of the first shell bond length, which is shorter than the actual bond length as a result of energy dependence of the phase factor (δ(k)) in the sine function of the EXAFS equation [8]. The phase uncorrected bond lengths thus determined from the theoretical EXAFS data are given in table 2.

Thus the present method gives phase uncorrected bond length of the first coordination shell. It can be seen from Figs. 2 to 5 that the theoretical χ(k) curves as well as their Fourier transforms are in good agreement with their experimental counterparts, showing that the simplified method presented in this work is quite satisfactory. The values of bond length for the first shell in all copper complexes obtained from the present analysis of the experimental and theoretical EXAFS data, given in table 2 are in good agreement with each other.

Table 1 Copper (II) hydroxamic mixed ligand complexes with their abbreviations and molecular formulae.

S.No.	Complex	Abbreviation	Molecular formula
1.	Cu(II) 5 chloro-benzohydroxamic acid	Cu 5CBHA	$CuC_7H_6O_2NCl$
2.	Cu(II) malanohydroxamic acid	Cu MHA	$CuC_2H_4O_4$
3.	Cu(II) succenohydroxamic acid	Cu SuHA	$CuC_4H_8O_4N_2$

Table 2 Values of the first shell bond lengths R(Å) obtained from the Fourier transforms of the experimental EXAFS data and from the theoretical data.

S.No.	Complex	R (from FT of experimental data)	R (from FT of theoretical data)
1	Cu 5CBHA	1.81	1.87
2	Cu MHA	1.38	1.44
3	Cu SuHA	2.07	2.11

Table 3 Wave vector k ($Å^{-1}$) for EXAFS maxima and minima at the K-absorption edge of copper in the complexes and their corresponding values of n for LSS plots.

n	Wave vector k ($^{-1}$)		
	Cu 5CBHA	Cu MHA	Cu SuHA
0	1.94	1.90	1.85
1	3.45	3.40	4.05
2	5.00	4.80	4.95
3	5.75	5.60	5.55
4	6.45	6.95	6.45
5	7.80		7.15

5. Conclusions

A simplified technique has been presented for the analysis of extended X-ray absorption fine structure (EXAFS) data and determining the first shell bond length. The method employed here is simple and straight forward, thereby providing a physical picture of the X-ray absorption fine structure. The analysis uses easily available computer program *Mathcad*. The method can be applied for any pure metal and is quite suitable for complex systems also. In the present case, analysis has been carried out only for the first shell, but it can be extended to other shells also.

References

[1] Mishra A, Parsai N & Shrivastava B D 2011 *Ind. J. Pure & Appl. Phys.* **49** 25.
[2] Mishra A, Parsai N, & Shrivastava B D 2012 *X-Ray Spectrom.* **41** 219.
[3] Rehr J J & Albers R C 2000 *Rev. Mod. Phys.* **72** 621.
[4] Lytle F W, Sayers D E & Stern E A 1975 *Phys. Rev. B* **11** 4825.
[5] Stern E A, Sayers D E & Lytle F W 1975 *Phys. Rev. B* **11** 4836.
[6] McKale A G, Knapp G S & Chan S K 1986 *Phys. Rev. B* **33** 841.
[7] Deshpandey A P, Sapre V B & Mande C 1988 *Phys. Stat. Sol. (b)* **145** 77.
[8] Stern, E A & Heald, S M 1983 *Handbook on Synchrotron Radiation* Vol. 1 ed. Koch E E (Elsevier) p. 955.

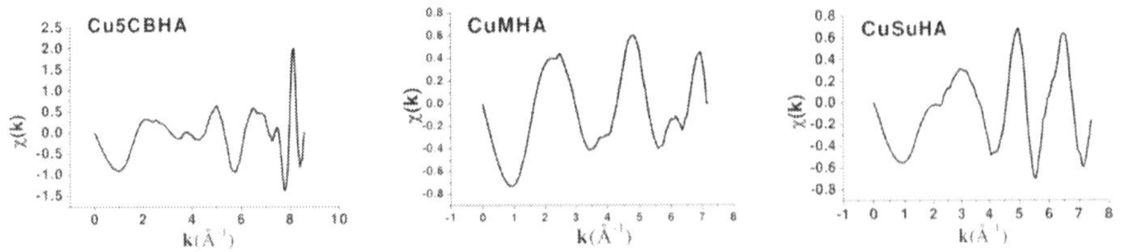

Figure 1 - Experimental χ (k) versus k curves for copper complexes.

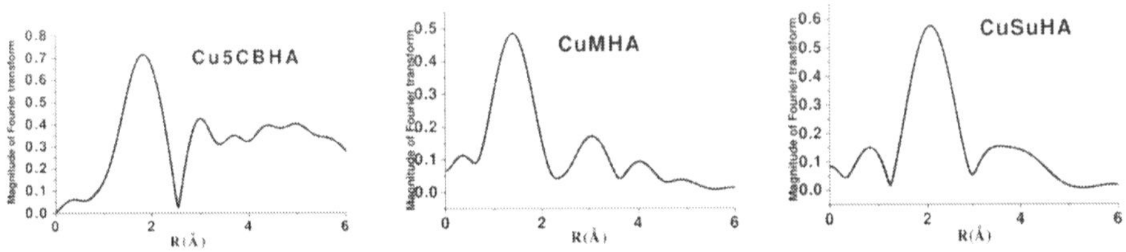

Figure 2 - Magnitude of Fourier transform of χ (k) versus k curves of the experimental EXAFS data, given in figure 2, for copper complexes.

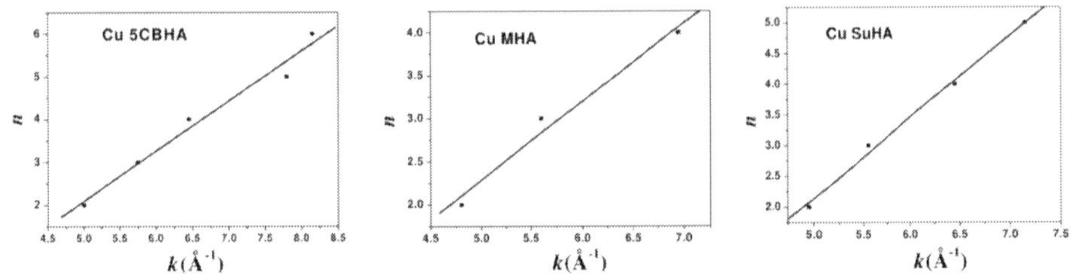

Figure 3 - n versus k curves for the copper complexes.

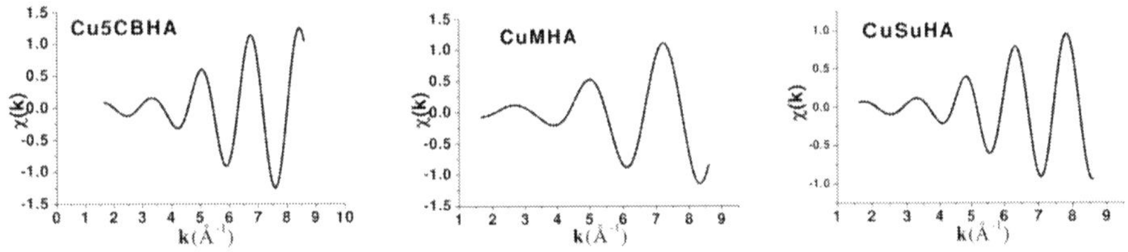

Figure 4 - Theoretical χ (k) versus k curves for copper complexes. For calculations, backscattering amplitude has been taken from Mckale [6] (values for $R = 2.5$ Å)

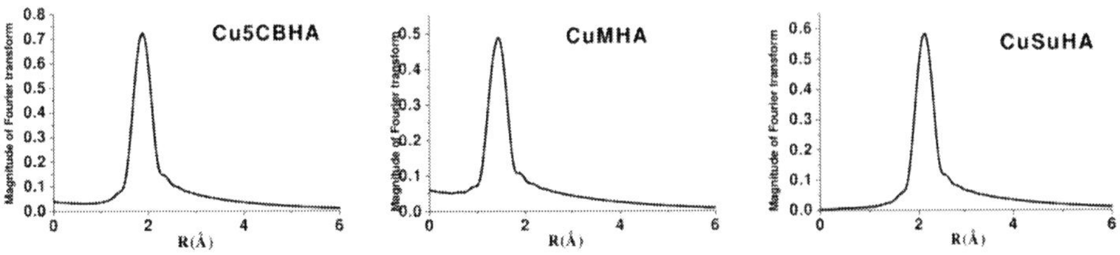

Figure 5 - Magnitude of Fourier transform of χ (k) versus k curves of the theoretical EXAFS data, given in figure 3, for copper complexes.

Propagation and cross-focusing of hollow Gaussian laser beams in plasma

Prerana Sharma[1], A K Bhardwaj[2] and R P Sharma[3]
[1]Department of Physics, Ujjain Engineering College, Ujjain,M.P.465010, India
[2]N.S.C.B. Government Post Graduate College, Biaora, India
[3]Centre for Energy Studies, Indian Institute of Technology, New Delhi, India

Email: preranaiitd@rediffmail.com

Abstract. This paper presents the cross focusing of two high power Hollow Gaussian laser beams (HGLB) by considering ponderomotive nonlinearity in a collisionless plasma. On account of the nonlinearity present in the plasma the two laser beams affect the dynamics of each other and cross-focusing takes place. The expressions for the laser beam intensities by using the eikonal method have been derived and the beam width parameter is calculated in the case when only ponderomotive nonlinearity is operative. The numerical appreciation of the cross focusing has been presented graphically by plotting beam-width parameter vs. dimensionless distance of propagation. Theself focusing of one laser beam is affected due to the optical inhomogeneities introduced by another copropagatingHGLBin the plasma. It is observed thatthe presence of one HGLBalso affects the self focusing of different order's (n) of the other HGLB.

1. Introduction

There has been a great deal ofresearch activity devoted to laser plasma interaction using two laser beams in under dense plasma on account of its relevance to charge particle acceleration [1]. New techniques are being investigated for particle acceleration and one of these techniques is beat wave acceleration [2] ,in which two laser beams simultaneously propagates in the plasma. The Gaussian / Hollow Gaussian laser beam (HGLB) has variation in the irradiance along its wave front. This exerts a ponderomotive force on the electrons arising from the gradient of the non-uniform irradiance [3].This force pushes electrons away from the region where the laser beam is more intense i.e. from a particular distance from the axiswhere the irradiance is maximum.This expulsive force is opposed by the pressure of plasma outside the channel and the balance between these two results in a lowered plasma density in the channel. The result of lowering density increases the refractive index. This nonlinearity is operative on a time scale $\tau = r_t/c_s$, where r_t is the initial beam width and $c_s$is the ion sound speed.

Dynamics of laser beam propagation has significantly affected by the nonlinearity introduced in the plasma and the presence of another laser beam also influence the nonlinearity occur in the plasma.Thus, one can control the focusing/defocusing of laser beams by the choosing the parameters of another beam. This behaviour is known as a cross-focusing of the laser beams in the plasmas.

Recently, for the efficient laser plasma coupling the irradiance distribution along the wave front of the beam is checked and it is observed that it modifies the nonlinear effects in the plasma[4]. Further various intensity distributionshave been taken in the laser plasma interaction studies, theoretically [5] as well as experimentally.The intensity distribution has been taken in the form of zero central intensity called the dark HGLB [6].The physical interpretation for DHBs have been given by various theoretical methods like higher-order Bessel beams, high-order Mathieu beams, TEM_{01}modedoughnut hollow beam.In the present work we have considered a theoretical model for cross focusing of HGLBin the ponderomotive regime using the eikonal method [7].

2. Effective dielectric constant of the plasma

In this communication, we have considered the propagation of two coaxial Hollow Gaussian laser beams of frequencies ω_1 and ω_2. The irradiance distribution (z= 0)of the incident beam with respect to a cylindrical coordinate system is

$$\mathbf{E}_{1,2}.\mathbf{E}^*_{1,2}\big|_{z=0} = E^2_{01,2}\left(r^2/2r^2_{1,2}\right)^{2n}\exp\left(-r^2/r^2_{1,2}\right) \tag{1}$$

where the direction of propagation has been assumed to be along the z axis and r refers to cylindrical coordinate system, $r_{1,2}$ is the spot size of the first and second beam respectively, $E_{1,2}$ is a real constant characterizing the amplitude of the first and second HGLB respectively, $E_{01,2}$ refers to the complex amplitude of the beam which denotes the electric field maximum at $r_1 = r_2 = r_{max} = r_{1,2}\sqrt{2n}$, corresponding to $z = 0$. The positive integer number n decides the order of the HGLB which characterizes the shape of the HGLB and position of its irradiance maximum.Equation(1) represents fundamental Gaussian beams of width $r_{1,2}$ for $n = 0$. These co-propagating beamsexertponderomotive force on plasma electrons given by $F = -\left(e^2/4m\right)\left[\left(1/\omega_1^2\right)\Delta E_1.E_1^* + \left(1/\omega_2^2\right)\Delta E_2.E_2^*\right]^{1/2}$ and modified electron density equation (due to ponderomotive force) can be written as $N_{0e} = N_{00}\exp\left[-3/4\left(m/M\right)\left(\alpha_1 E_1 E_1^* + \alpha_2 E_2 E_2^*\right)\right]$ where $\alpha_{1,2} = e^2 M/18 k_B T_0 m^2 \omega_{1,2}^2$ where N_{00}, k_B, T_0 M and care the electron concentration in the absence of beams, Boltzmann constant, equilibrium temperature, ion massand velocity of light in vacuum respectively.We define a point along the radial distance of the beam and consider that most of the power of HGLB is concentrated in the region around $\eta = 0$. Thus, we can relate the cylindrical coordinate r and variable η as

$$\eta = \left[\left(r/r_{1,2}f\right) - \sqrt{2n}\right] \tag{2}$$

and therefore dielectric constant may be expended around the maximum of HGLB i.e. at $\eta = 0$. Hence for HGLB, one can express the dielectric constant $\varepsilon_{1,2}(\eta, z)$ in terms of $\varepsilon_{1,2}(r, z)$. Expanding dielectric constant around $\eta = 0$ by Taylor expansion, one can write

$$\varepsilon_{1,2}(\eta, z) = \varepsilon_{01,2}(z) - \eta^2\varepsilon_{21,2}(z) \tag{3}$$

$$\varepsilon_{01,2} = 1 - \frac{\omega_{p0}^2}{\omega_{1,2}^2}\exp\left[-\left(\frac{\alpha_1 E_{01}^2}{f_1^2} + \frac{\alpha_2 E_{02}^2}{f_2^2}\right)n^{2n}\exp(-2n)\right], \tag{4}$$

$$\varepsilon_{21,2} = \frac{\omega_{p0}^2}{\omega_{1,2}^2}\left[\left(\frac{\alpha_1 E_{01}^2}{f_1^2} + \frac{\alpha_2 E_{02}^2}{f_2^2}\right)n^{2n}\exp(-2n)\right]\exp\left[-\left(\frac{\alpha_1 E_{01}^2}{f_1^2} + \frac{\alpha_2 E_{02}^2}{f_2^2}\right)n^{2n}\exp(-2n)\right] \tag{5}$$

where $\varepsilon_{01,2}(z)$ and $\varepsilon_{21,2}(z)$ are the coefficients associated with η^0 and η^2 in the expansion of $\varepsilon(\eta, z)$ around $\eta = 0$.

3. Cross-focusing of the HGLB's

For the considered HGLB(in a steady state) theelectric field vector $A_{1,2}$ satisfies the wave equation

$$2ik_{1,2}\frac{\partial A_{1,2}}{\partial z} + iA_{1,2}\frac{\partial k_{1,2}}{\partial z} = \left(\frac{\partial^2 A_{1,2}}{\partial r^2} + \frac{1}{r}\frac{\partial A_{1,2}}{\partial r}\right) + \frac{\omega_{1,2}^2}{k_{1,2}c^2}\left(\varepsilon - \varepsilon_0\right) \tag{6}$$

The complex amplitude $A_{1,2}(r, z)$ may be expressed as,

$$A_{1,2}(r,z) = A_{01,2}(r,z)\exp\left(-ik_{1,2}(z)S_{1,2}(r,z)\right) \tag{7}$$

where $S_{1,2}(r,z)$ is the eikonal associated with the HGLB. Now putting Eq. (7) in Eq. (6) and separating the real and imaginary parts we get twoequations and further using Eq. (2) in these equations we get the following pair of equations as

$$\frac{2S_{1,2}}{k_{1,2}}\frac{\partial K_{1,2}}{\partial z} + 2\frac{\partial S_{1,2}}{\partial z} + \frac{1}{r_{1,2}^2 f_{1,2}^2}\left(\frac{\partial S_{1,2}}{\partial \eta}\right)^2 = \frac{1}{k_{1,2}^2 A_{01,2} r_{1,2}^2 f_{1,2}^2}\left(\frac{\partial^2 A_{01,2}}{\partial \eta^2} + \frac{1}{\left(\sqrt{2n}+\eta\right)}\frac{\partial A_{01,2}}{\partial \eta}\right) + \frac{\omega_{1,2}^2(\varepsilon - \varepsilon_0)}{c^2 k_{1,2}^2} \tag{8}$$

$$\frac{\partial A_{01,2}^2}{\partial z} + \frac{A_{01,2}^2}{r_{1,2}^2 f_{1,2}^2}\left(\frac{\partial^2 S_{1,2}}{\partial \eta^2} + \frac{1}{\left(\sqrt{2n}+\eta\right)}\frac{\partial S_{1,2}}{\partial \eta}\right) + \frac{1}{r_{1,2}^2 f_{1,2}^2}\frac{\partial A_{01,2}^2}{\partial \eta}\frac{\partial S_{1,2}}{\partial \eta} + \frac{A_{01,2}^2}{k_{1,2}}\frac{\partial k_{1,2}}{\partial z} = 0 \tag{9}$$

Substituting Eq. (3) into Eqs. (8) and (9) we get,

$$\frac{2S_{1,2}}{k_{1,2}}\frac{\partial K_{1,2}}{\partial z} + \frac{2\partial S_{1,2}}{\partial z} + \frac{1}{r_{1,2}^2 f_{1,2}^2}\left(\frac{\partial S_{1,2}}{\partial \eta}\right)^2 = \frac{1}{k_{1,2}^2 A_{01,2} r_{1,2}^2 f_{1,2}^2}\left(\frac{\partial^2 A_{01,2}}{\partial \eta^2} + \frac{1}{\left(\sqrt{2n}+\eta\right)}\frac{\partial A_{01,2}}{\partial \eta}\right) - \eta^2\frac{\omega_{1,2}^2(\varepsilon_2)}{c^2 k_{1,2}^2} \tag{10}$$

One canexpresssolution of Eq. (9), with the paraxial approximation $\eta << \sqrt{2n}$ as

$$A_{01,2}^2 = \frac{E_{1,2}^2}{2^{2n} f_{1,2}^2}\left(\sqrt{2n}+\eta\right)^{4n}\exp\left(-\left(\sqrt{2n}+\eta\right)^2\right) \tag{11}$$

$$S_{1,2}(\eta,z) = \frac{\left(\sqrt{2n}+\eta\right)^2}{2}\beta(z) + \varphi(z) \text{ where } \beta(z) = r_{1,2}^2 f_{1,2}\frac{df_{1,2}}{dz} \tag{12}$$

Here $\varphi(z)$ is an arbitrary function of z and $f_{1,2}(z)$ is the beam width parameter for the HGLB for the first and second beam respectively. Further, substituting from Eq. (11) and (12) for $A_{01,2}^2$ and $S_{1,2}$ into Eq. (10) and equating the coefficients of η^0 and η^2 on both sides of the resulting equation, one obtains

$$\varepsilon_{01,2} f_{1,2}\frac{d^2 f_{1,2}}{d\xi^2} = \left(\frac{4}{f_{1,2}^2} - \rho_0^2\varepsilon_{21,2}\right) \tag{13}$$

where, $\xi = \left(cz/r_{1,2}^2\omega_{1,2}\right)$ is the dimensionless distance of propagation, $\varepsilon_{01,2}$ and $\varepsilon_{21,2}$ are given by Eq. (4) and Eq. (5) respectively and $\rho_{01,2} = \left(r_{1,2}\omega_{1,2}/c\right)$ is the dimensionless initial beam width.

4. Discussion and conclusions

In order to have a numerical appreciation of the cross-focusing and the effect of the changing of the parameters of the laser beams, we have performed numerical computation of Eq. (13) with Eqs. (4) and (5). The numerical results have been obtained with typical plasma and laser beam parameters: $r_1 = 15\mu m$, $r_2 = 20\mu m$, $\omega_1 = 1.776\times10^{14}$ rad/S, $\omega_2 = 1.716\times10^{14}$ rad/S and $\omega_{po} = 0.3\omega_1$. For an initial plane wave front of the laser beams, the initial condition used here are $f_{1,2} = 1$, $df_{1,2}/dz = 0$, $\alpha_{01,2} = \alpha_{21,2} = 0$ and $S_{21,2} = 0$ at $z = 0$. We solve Eq. (13) for various combinations of initial fields of the lasers, which are greater than the critical value for self-focusing. In that case f_1 and f_2 are obtained from the numerical solutions. Moreover, the focusing of different orders (n = 0, 1, 2) of firstHGLB is studied when taking different power of second HGLB into consideration.It is clear from the equation (4) and (5) that both $\varepsilon_{01,2}(z)$and $\varepsilon_{21,2}(z)$ depend on the order n of the beam.Figure 1 illustrates the behaviour of the two HGLB's for pure Gaussian mode i.e. n=0 when ponderomotive nonlinearity has been considered. It depicts that both of the beams show oscillatory self-focusing. In this observation, the power of the first beam is kept as constant and power of second is varied. It is clear from the figure that the focusing of the first HGLB is affected due to the change in the power of second HGLB. The beam width parameter of the first beam (which is shown by the solid black line of the figure) changes with the variation in the power of the second beam. In Fig. 2the self-focusing of first laser beam has been shown by considering different orders (n) and the effect of cross focussing on particular order n is observed. Here, the power of the first HGLB kept as fixed and the power of second beam has been varied to observe the beam width parameter for the different orders of the first beam. Figure shows that the different orders of the first beam also affected by the variation in the second beam power. It explicitly illustrates the effect of the coupling term on the focusing of the different order of the first HGLB when the ponderomotive nonlinearity is dominating in the system. Further, this study has been repeated for the fixed higher power of the first HGLB. In conclusion, the cross focusing of various orders of HGLBs is analysed by applying the paraxial like approach when only ponderomotive nonlinearity is operative.; in this approach all the relevant parameters correspond to a narrow range around the irradiance maximum of the HGB. it is observed that the oscillatory divergence of the HGLB becomes narrower with respect to the Gaussian beam. The dependence of the beam width parameter of on distance of propagation has been evaluated. The self-focusing of one HGLB is shown to be affected by the presence of another beam.

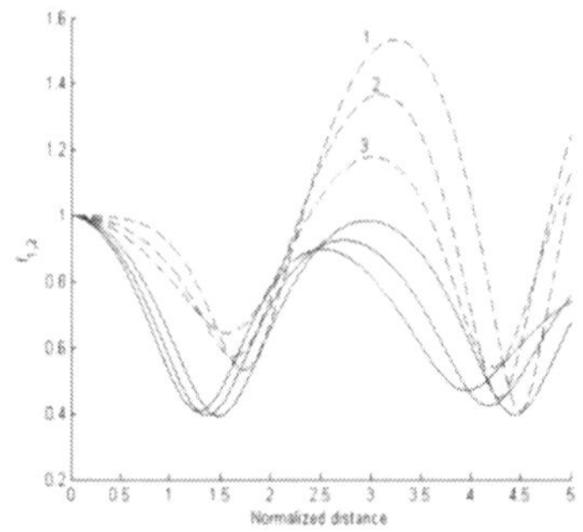

Figure 1: Variation of beam width parameter of first HGLB (for n=0, n=1, n=2) with the distance of propagation for a fixed power of first beam = 0.7 and different power of second beam. Solid line represents = 0.3, dashed line represents= 0.7

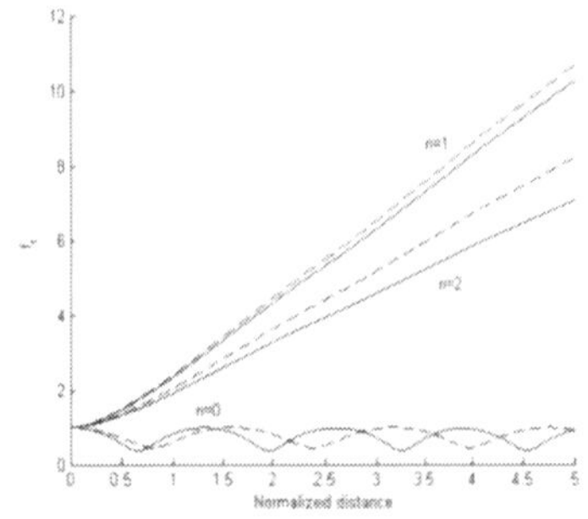

Figure 2: Variation of beam width parameter of first HGLB (for n=0, n=1, n=2) with the distance of propagation for a fixed power of first beam = 0.7 and different power of second beam. Solid line represents = 0.3, dashed line represents= 0.7

Acknowledgment

This work was partially supported by the University Grant commission, NewDelhi, India and Madhya Pradesh Council of Science and Technology, Bhopal, India.

References:

[1] Malka V., Faure J., Glinec Y and Lifschitz 2006 *Phil.Trans. R. Soc. A* **364**, 601
[2] Yousef I. 2003 *Phys. Rev. E* **67** 016501
[3] Gill, T.S., Mahajan, R. and Kaur R. 2010 *Laser & Part. Beams* **28** 521
[4] Sharma R.P. and Singh R. K. 2013 *Laser & Part. Beams* **31** 689
[5] Singh M., Singh R. K. and Sharma R. P. 2013 *Europhys. Lett.* **104** 35002
[6] Gao Z. and Lu B. 2006 *Chin. Phys. Lett.* **23**106
[7] Sodha M. S. Ghatak A. K. and Tripathi V. K. 1976 *Prog. Opt.* **13** 171

Design and fabrication of wire displacement sensors for undulator field measurement applications

G Sharma[1]**, M Gehlot and G Mishra**
School of Physics, Devi Ahilya University, Indore, MP, India.

Email: geetanjali.sharma678@gmail.com

Abstract : In this paper, we describe the design and working principle of two wire displacement measuring sensors. The two sensors are used in a pulsed wire setup for undulator magnetic field measurements. First one is optocoupler sensor consists of a LED - phototransistor pair and the other one is a laser sensor of laser photodiode pair. Both the sensors work on the principle of light interruption and gives comparative results in our experimental setup.

1. Introduction

The magnetic measurement of the undulator is an important issue in design of uniform, precise and quality undulators for synchrotron radiation and free electron laser applications. Such measurements are usually made point to point to by a Hall probe. The pulsed wire method [1-5] was suggested as an alternate method for field integral measurements. In this method, a thin wire is stretched along the undulator axis. When a dc current flows in the wire, a Lorentz force proportional to the local transverse field component is exerted on the wire. The Lorentz force makes the wire to vibrate. These vibrations are picked up by a sensor located at the end of the undulator. The sensor output versus time is the field integral versus position along the wire. In this paper we describe technical design details of two types of sensor used in the pulsed wire setup. One is the optocoupler and the other one is the laser-Photodiode detector pair. We describe the design details, calibration of the sensor. We used these sensors to pulsed wire measurements on our 30 cm long PPM undulator.

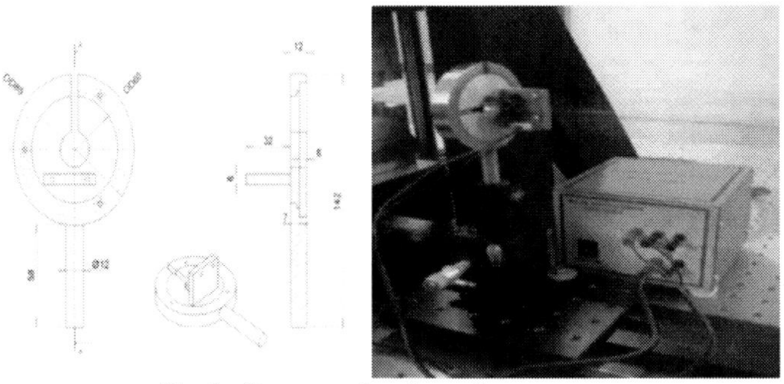

Fig 1. Optocoupler detection system

2. Design and working principle

The optocoupler sensor is a Motorola make Model No is MOC7811 .The optocoupler sensor consists of a infrared LED and a phototransistor which are molded in a plastic housing. The mount assembly consists of two circular rings. The outer diameter of the outer ring is 85 mm and the outer diameter of the inner ring is of 60 mm .The outer ring is marked. The inner ring can be rotated inside the outer ring. The optical switch is fixed on the inner ring .The rotation of the inner ring allows the optical switch to measure both the horizontal and vertical field integral data. There is a slot of 12 mm width for removing and inserting the wire. The schematic and the photograph of complete installation of the system are shown in figure 1. The mount assembly is attached to a post of 12 mm diameter and 58 mm in length and is mounted on a rail through xyz translation stages. The rail allows the switch to be fixed at minimum distance of 148 mm and maximum of 560 mm distance from the undulator end. The xyz stages give vertical, horizontal and height alignments to the switch. The optocoupler works on the principle of current flowing in the circuit by photo illuminated transistor and the principle of operation is represented by [2],

$$V_{CE} = V_{CC} - I_{CE}R \qquad (1)$$

When the wire blocks the LED light path, no light falls on the photo transistor and the current flowing is zero i.e $I_{CE} = 0$ and $V_{CC} = V_{CE}$,V_{CC} is the supply voltage so voltage recorded is maximum when the light path is interrupted. As the wire moves away from the center, the light path is opened up and allows the light to fall on the photo transistor. A current flows and reduces the recorded voltage .The voltage versus distance curve for the optocoupler is shown in figure 3a. The slope of the curve is the sensitivity of the sensor. The sensitivity of the two output curves are calculated in the linear portion of the curve and is 8 mV/μm (250 μm wire) and 6 mV/μm (125 μm wire) respectively.

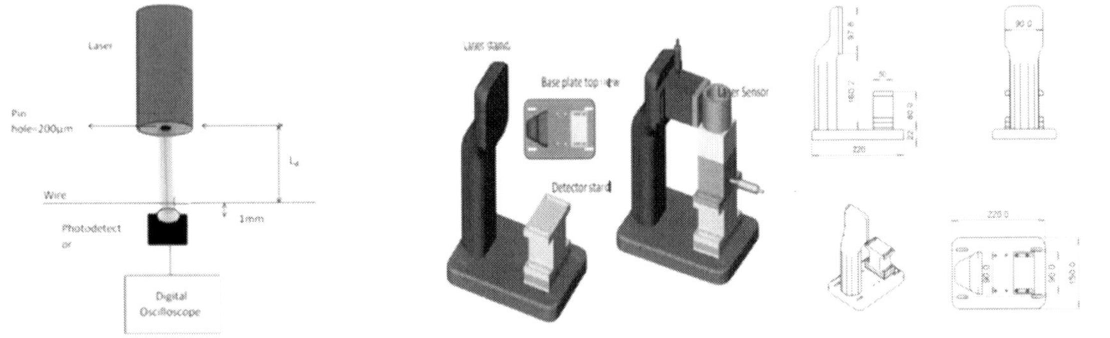

Fig 2. Laser-photodiode detection system

A laser-photodiode sensor is used as a detector at the other end of the undulator. The 5 mW , 635 nm laser is Coherent's Ultra Low Noise series diode laser Model No. 31-0144-000 module. The rms noise typically ≈ 0.06 % . The laser spot size is 1*mm*. The laser is fixed on a laser mount vertical facing downwards through xyz stages. The xyz stages gives a movement of 5 mm each in upward, downward and forward direction .The photodiode for the system is Newport, USA make silicon detector Model No.818-BB-21. The 818-BB-21 High Speed silicon photo detector operates from 300 nm to 1100 nm with a 0.40 mm active diameter and has rise time less than 300 ps. It is connected to Tektronix TDS2024B, 4 channel, 200 MHz digital storage oscilloscope through a 50 ohm BNC connector output. The schematic of principle of laser - photodiode sensor and laser mount assembly is seen in figure 2. The laser – photodiode detector works on the principle of interception of a Gaussian distribution of intensity. The un-interrupted light through a wire is given by [3],

$$P(x) = \frac{2P_0}{\pi w^2}\left[\int_{-\infty}^{\infty} dy \int_{-\infty}^{x-r_w} dx e^{\frac{-2(x^2+y^2)}{w^2}} + \int_{-\infty}^{\infty} dy \int_{x+r_w}^{\infty} dx e^{\frac{-2(x^2+y^2)}{w^2}}\right] \quad (2)$$

where $2\,r_w$ is the wire diameter and w is the laser beam spot size. When the wire is located at the center of the laser beam, the power not blocked by the wire is minimum. It is seen that From figure 3b, that the minimum voltage occurs when the wire completely blocks the laser light. At the center, the intensity of the laser is maximum, the interrupted light is maximum hence this corresponds to minimum voltage .As the wire moves away from the center, more and more light falls on the photodiode and the output voltage rises. The laser used has a spot size of 1000μm. A number of pinhole diameter was used to reduce the beam spot size and the pinhole diameter of 300 μm is placed to reduce the laser spot and gives a sensitivity same as that of the optocoupler. Thus sensitivity of the laser-photodiode with the 300μm pin hole is 8 mV/μm (250 μm wire) and 6 mV/μm (125 μm wire) respectively.

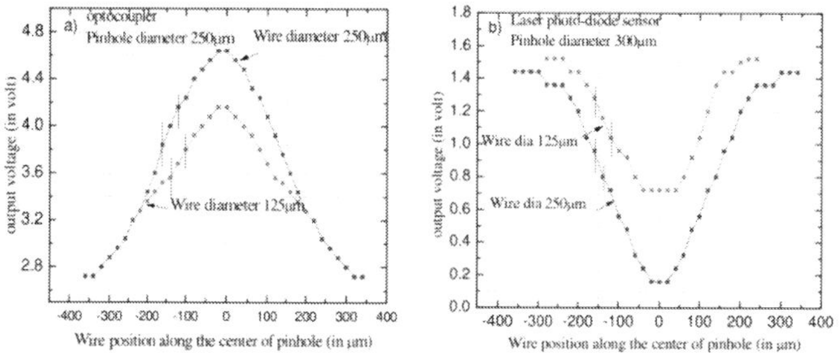

Fig 3a. Calibration curve for a) optocoupler sensor b) laser photodiode sensor

3. Measurement results and discussion

The experimental results of pulsed wire method on 30cm planar undulator. The measurements are taken by both the sensors located at 148 mm away from the end of the undulator and results are compared with the hall probe results. The CuBe wire is fixed at one end and the other end of the wire hangs over a pulley with a weight. The pulsed wire bench is set on the same vibration isolation support [2]. Two CuBe wire having diameter of 250 μm with $\mu = 4.1\times10^{-4}\,kg/m$ and 125 μm with $\mu = 1.01\times10^{-4}\,kg/m$ is installed for magnetic measurements. The wire length is 1.39 meter, which is more than four times the length of the undulator. In our present measurement, with a given SNR, a current of 1.9A and 0.6A is used for wire diameters of 250 μm and 125 μm respectively. The first and second field integral in the pulsed wire method is measured by an appropriate pulse width [4]. Figure 4a and figure 4b shows the comparison of the second field integral and magnetic field obtained by pulsed wire method and hall probe method for wire diameter 250μm. Similarly, figure 4c and 4d shows the same for wire diameter 125μm. Figure 5 shows the ΔB_{rms} versus tension for gap 24mm, 27mm and 30 mm respectively from both the laser sensor and optocoupler sensor, where ΔB_{rms} is the deviation in magnetic field from the hall probe data. It is clear that for wire diameter 250μm as we increase the undulator gap the result discrepancies from both the sensors also increased , while this is not seen for wire diameter 125μm. Possible reason may be the stiffness and thickness of the wire affect the results at higher gap because at higher gap undulator magnetic field decreases.

In conclusion wire displacement sensors are successfully designed and fabricated for undulator field measurement application by pulsed wire method. To ensure the reliability and repeatability of the results, two wires of different diameters are used. The results prove that the laser sensor and the optocoupler works on different principles of light interception, however gives equal results at equal sensitivities.

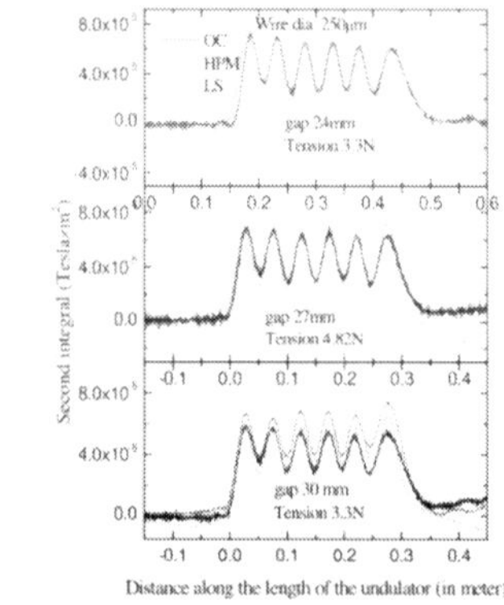

Fig 4a. Second field integral.

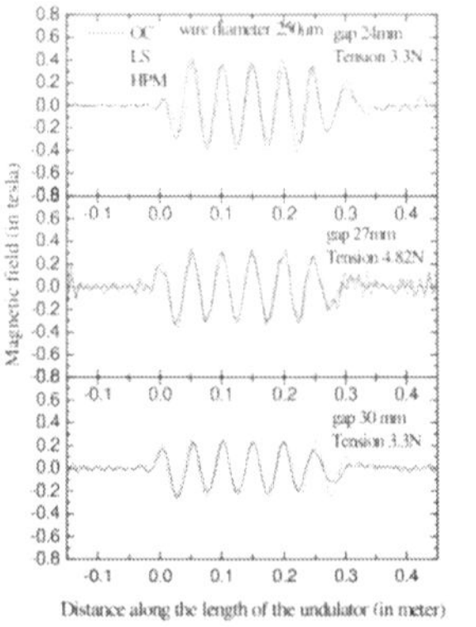

Fig 4b. Magnetic field.

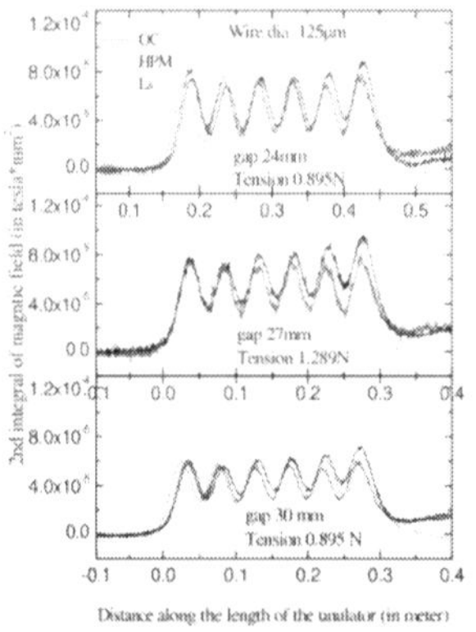

Fig 4c. Second field integral.

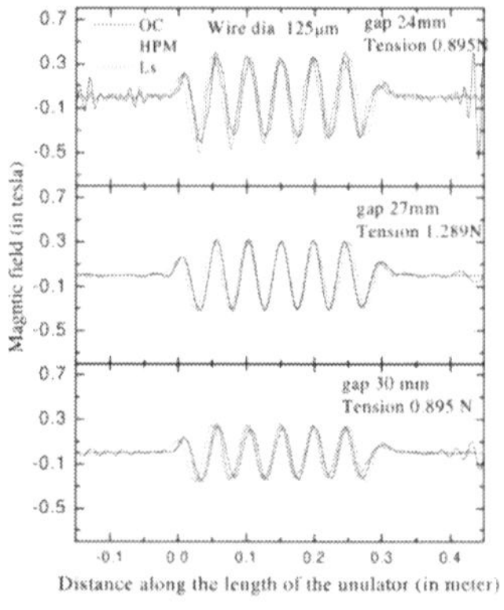

Fig 4d Magnetic field.

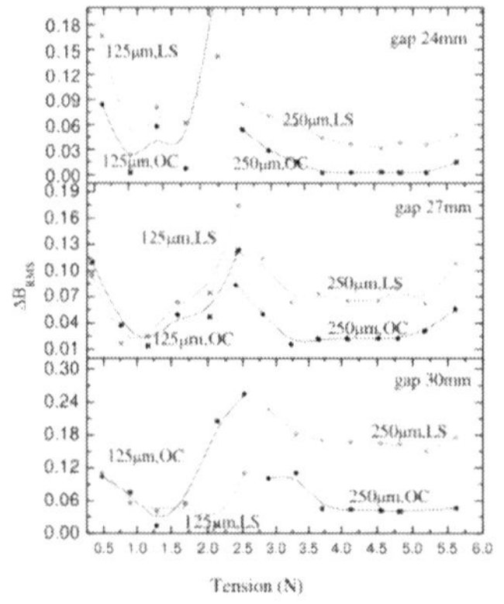

Fig 5 ΔB_{rms} versus tension.

Acknowledgment

We are grateful to Dr. S. Tripathi, Hasy Lab, DESY, Hamburg, Germany for many technical discussions on undulator measurement. This work is financially supported by DST, Delhi, India; DRDO, Delhi, India and CSIR, Delhi, India.

References

[1] Warren R W 1988 *Nucl. Instrum. Meth. Phys. Res. A* **272** 257.

[2] Tripathi S, Gehlot M, Hussain J K, Mishra G, Kumar V and Chouksey S 2011, *Opt. Commun.* **284** 350.

[3] Tripathi S, Gehlot M, Hussain J K, Mishra G, Chouksey S and Kumar V, Kale U and Nerpagar P, 2009 *Nuovo Cimento* **124** 281.

[4] Tripathi S, Mishra G, Kumar V, Chouksey S, Kumar R 2011 *Nucl. Instrum. Meth. Phys. Res. A* **635** 121.

[5] Kumar V and Mishra G 2010 *Pramana J. Phys.* **74** 743.

Application of polygon based laser micrometer for magnet size measurement studies

Roma Khullar*, Geetanjali Sharma, Bramh Prakash, G.Mishra
School of Physics, Devi Ahilya University, Indore, 452001, India

*corresponding author: roma_khullar@yahoo.co.in

Abstract: In this paper, we describe development of a polygon mirror based laser micrometer for magnet size measurement for undulator fabrication. In the laser micrometer, a laser light source is focused on to the facets of a rotating polygon mirror at an angle of 35^0 that reflects the laser beam. The parallel light emerging from the polygon mirror surface falls on the object through an F-theta lens. The F-theta lens is used for collimating the output of a scanning mirror system and to provide a flat field at the image plane of the scan. The object obstructs the light. A collimating lens at the other side of the object focuses the light onto a detector. The shadow caused by the obstruction results in an output voltage for a period of time that is proportional to the size of the object. We have used the laser micrometer for 12.5mm and 6.25mm magnets measurement. This is an efficient noncontact method of magnet sorting for use in undulator fabrication, synchrotron radiation and FEL application.

1.Introduction

There exists interests in undulators for synchrotron radiation and free electron laser applications [1,2]. It is a periodic arrangement of dipole magnets to produce a sinusoidal magnetic field. The electrons execute transverse oscillations in the undulator magnet and emit radiation. The undulator field strength scales with the undulator gap as $B(Tesla) = 1.723B_r \exp(-\pi g / \lambda_u)$, where g is the magnet to magnet gap, λ_u is the undulator period and B_r is the remanent field. A precise measurement of gap is crucial as a variation of the magnet to magnet gap introduces errors in undulator parameter along the length of the undulator.

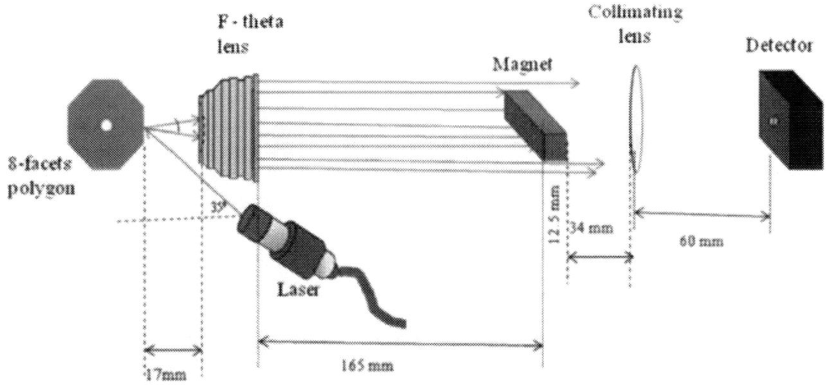

Fig 1.Principle of laser micrometer.

The laser micrometer [3-5] is a non contact magnet size measurement method (Fig1)where a laser light is focused onto the facets of a rotating polygon mirror. The parallel light emerging from the polygon facets falls on the object through an F-theta lens. The object obstructs the light. A lens on the opposite side of the object focuses the interrupted and the un-interrupted light onto a detector. This causes an output voltage to be recorded by a DSO that is connected to a detector through a BNC. The voltage versus time curve shape is proportional to the object size. A complete setup of laser polygon scanner at DAVV is shown in Fig 2.

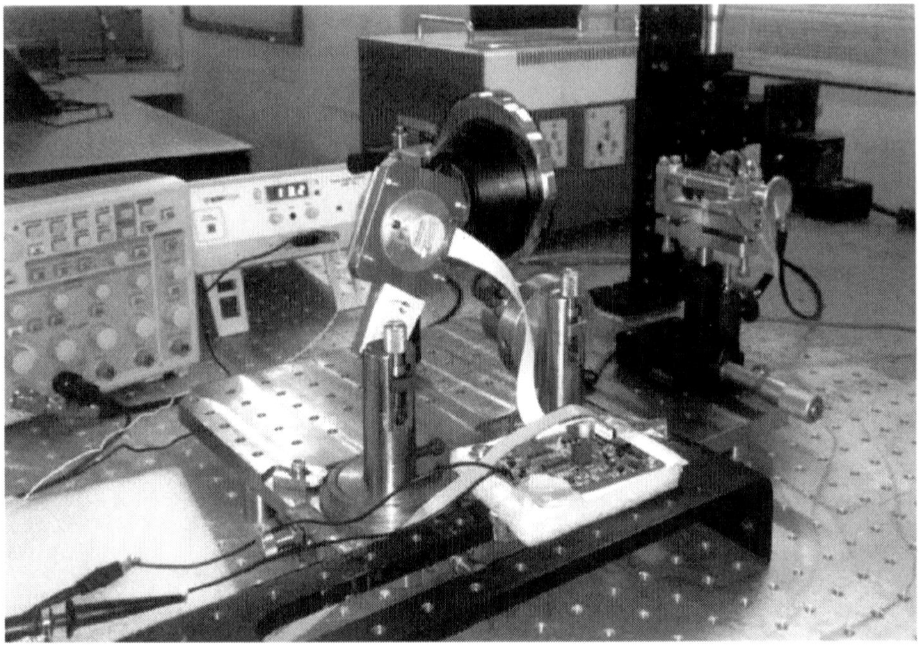

Fig 2. Experimental setup of laser micrometer .

2. Laser scanning of the magnets

The RPM of the polygon mirror is fixed at 6974.The magnet is placed at the focal length of the F-theta lens. In the measurement method it is required to direct the light at a particular location on the magnet. Depending on the magnet size at that location, a voltage versus time is recorded by the DSO. The inner width of the pulse on DSO is a measure of the magnet size measured in μs. To convert the magnet size from μs to mm, we multiply the data by calibration factor. Fig 3 and Fig 4 represents the magnet size measurement for magnets of 12.5mm and 6.25mm size respectively. It is observed that the magnet sizes can be measured at an accuracy of 20μm. It is required to pick up magnets that show close agreement with the data.

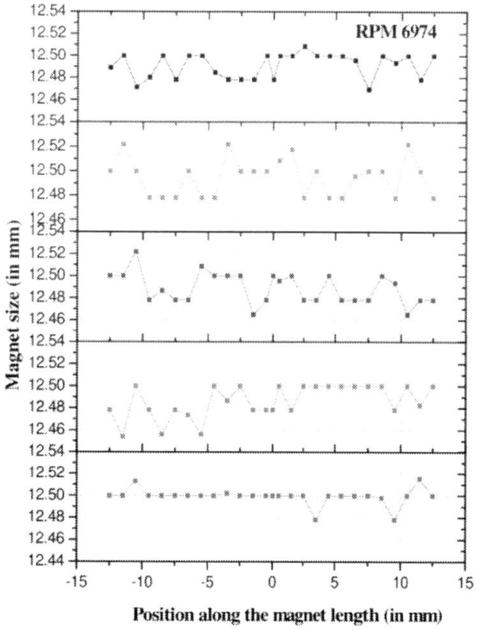

Fig 3. Magnet size measurement of 12.5 mm magnets along the length.

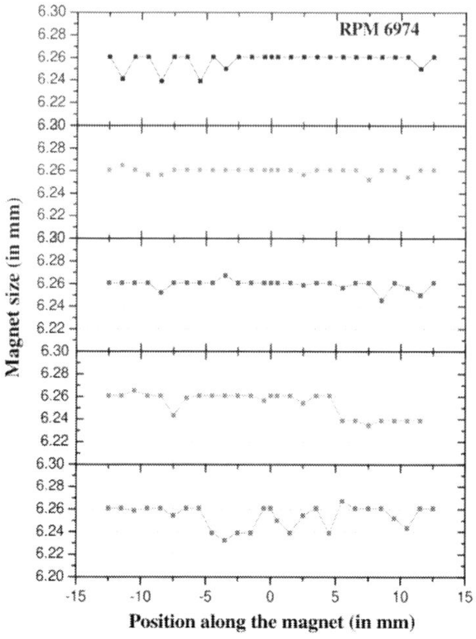

Fig 4. Magnet size measurement of 6.25 mm magnets along the length.

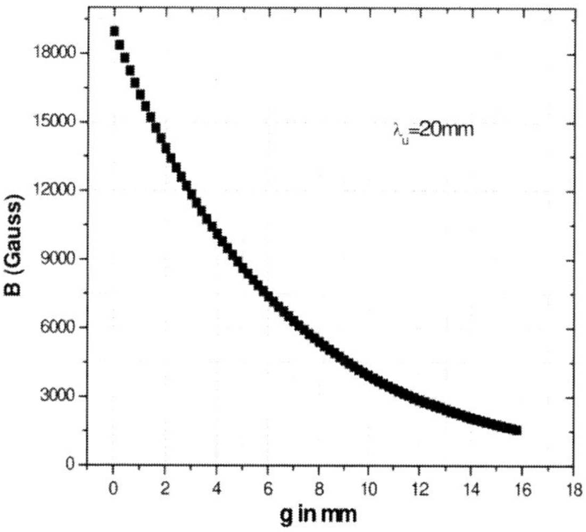

Fig 5. Magnetic field measurement with gap

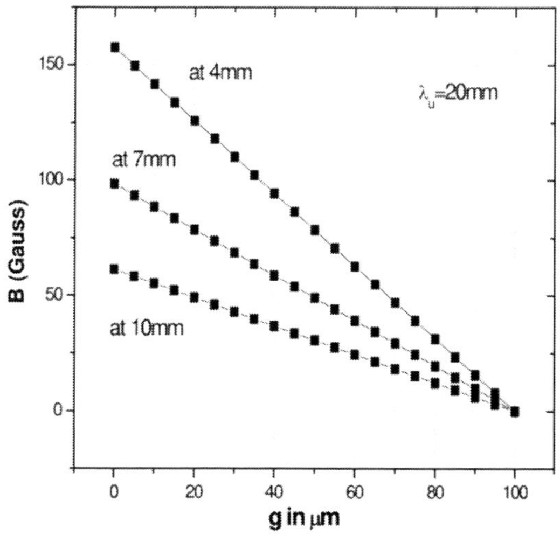

Fig 6. Magnetic field measurement nearly three different undulator gap.

3. Conclusion

The undulator field strength is a function of the undulator gap. For this precise magnet size selection are required to assemble the magnets to the undulator. In the Fig 5 we plot the variation of the undulator field strength verses the mean undulator gap. For a mean gap of 4mm, a deviation of 1mm gap mismatch results a field variation of 1615 Gauss. This is an important issue in the design of narrow gap undulators and will lead to non uniform magnet field strength along the undulator axis. In Fig 6 we plot the magnet field strength versus undulator gap in μm near three different mean undulator gaps. The magnet size of mismatch around 30 μm will be used. In that case the magnet field strength variation will be around 50Gauss.This is the tolerance value of our undulator.

ACKNOLEDGEMENT

This work has been done on financial support from DRDO, Delhi and UGC, Delhi, Government of India.

We are grateful to Dr.Sendhil Raja S, RRCAT, Indore, Dr. Viraj P Bhanage, RRCAT, Indore, Shradha Tiwari, RRCAT, Indore, for various technical discussions on the topic.

REFERENCES

[1] W.A.Barletta etal, 2010, "Free electron lasers: Project status and future challenges", *Nuclear Instruments & Methods in Physics Research A*, **618**, p-69.
[2] M.F.Couprie, 2013, "Undulator Technologies for future free electron laser facilities & storage rings", *MOZB102, Proceedings of Particle Accelerator Conference, Shanghai, China.*
[3] Leo Beiser, 1991, "Design Equations for a polygon laser scanner", *Proc. SPIE*, Vol.**1454**, G.F Marshall and L.Beiser (eds).
[4] Leo Beiser , 2003, "Unified optical scanning technology", *John Wiley & Sons, New Jersy, USA.*
[5] Leo Beiser, 1995, "Fundamental architecture of optical scanning systems", *Applied Optics*, Vol.**34**, No.31 p-7307.

Influence of hot carriers on parametrically interacting polaron mode in semiconductors

Ratna Agrawal, Swati Dubey and S Ghosh
School of Studies in Physics, Vikram University, Ujjain 456010, India

E-mail: ratnagarwal@gmail.com

Abstract. In the present paper effect of hot carriers due to parametrically interacting electron-longitudinal optical phonons in polar semiconductor is analytically investigated. Presence of hot carriers is found to significantly modify the threshold and amplification characteristics in the presence of external magnetic fields. Expressions for threshold pump field required for the onset of polaron induced parametric interaction and amplification characteristics are explicitly derived. It is found that at moderate magnetic field and high carrier concentrations hot carriers affect threshold and amplification characteristics strongly. Resonance between polaron frequency and plasma frequency is found to be favourable for the minimum threshold field. Presence of hot carriers and magnetic field along with mass modulation effects are found to be additive and resulted into increment in the parametric gain. Typical dependence of parametric gain on magnetic field and carrier concentration could be utilized for the construction of optical switches.

1. Introduction

Parametric interactions offer an efficient way to heat semiconductor plasma, since the incoming photons participating in the process are completely converted into plasmons. The parametric amplification of collective-waves in polar semiconductors such as phonons and polarons at the expense of the pump wave has already aroused much attention for modern optoelectronic device applications [1] e.g. for the generation of tunable laser with high conversion efficiency, photovoltaic effect, far-infrared diagnostic systems and optical parametric amplifiers (OPAs).

Large changes at room temperature and below have been shown to be due to impact ionization of the lattice by hot carriers [2, 3] in polar semiconductors. History of inclusion of carrier heating effects in semiconductor plasmas dates back to 1973 [4]. Recently authors analytically investigated polaron induced parametric interactions in magnetized semiconductors [5] (here after referred as paper I). Nonlinear electromagnetic effects connected with free carriers in homogeneous semiconductors result from heating up of the carriers and peculiarities in the dynamics of the carriers in the bands or the dependence of their effective mass m^* on the field. However which of the mechanisms play the dominant role in the processes of Infrared parametric generation is still a question of important. A more fundamental motivation for such studies is that they help in elucidating the interaction of the carriers with the scattering mechanism: the lattice, impurities, and other defects. Therefore in the present paper emphasis will be on the heating of carriers due to Fröhlich interaction and the underlying theory, significant results will be cited. Several factors, including acceptor concentration, carrier density, temperature, magnetic field, mass modulation, all affect hot carrier scattering, with different mechanisms dominating in various regimes. Here we examine how these external parameters can be used to extract analytically the carrier heating dependence of this interaction in polar semiconductors.

2. Theoretical Formulation

In this section, the parametric amplification of polaron mode arising due to three-wave interaction induced by $\chi^{(2)}$ in polar semiconductor duly irradiated by a relatively high-power laser with photon energy much below the forbidden energy gap of the crystal has been studied. Well-known hydrodynamic model of homogeneous semiconductor plasma, satisfying the condition $kl << 1$ (k and l being the wave number and electron mean free path, respectively) has been used. We assume that a spatially uniform $(|k_0| \approx 0)$ pump electric field as $\vec{E_0} = \hat{x}E_0 \exp(-i\omega_0 t)$ irradiates a polar semiconductor medium immersed in a transverse dc-magnetic field $\vec{B_o} = \hat{z}B_0$. Since we are interested in three wave interactions, the phase matching conditions are $\omega_0 \approx \omega_1 + \omega_{pl}$ and $|k_1| \approx |k_{pl}| \approx k_0$ (say). Here the parametric interaction of the pump generated a polaron wave at $(\omega_{pl}, \vec{k}_{pl})$ and scattered side-band wave (SBW) at $(\omega_1, \vec{k_1})$ supported by the lattice and electron plasma in the medium, respectively.

2.1. Nonlinearity due to mass modulation

The effective mass is taken to be energy dependent based on Kane's model [6] as

$$m^* = \frac{2\left\langle \dfrac{\xi}{k_B T_0} \right\rangle + \dfrac{\varepsilon_g}{k_B T_0}}{L} \tag{1}$$

where, $L = \dfrac{4p^2}{3\hbar^2 k_B T_0}$ and $\langle \xi \rangle = \dfrac{3}{2} k_B T_e$ is the average thermal velocity of an electron (the drift velocity being much smaller than the random velocity), ε_g is the energy band gap and $p\left(\approx 8.5 \times 10^{-8} eV\, cm\right)$[7] a matrix element.

2.2. Nonlinerarity due to electron collision frequency

When the mobility is high, heating of the carriers occurs in the fields as low as a few volts per centimeter. Particularly in high mobility semiconductors like InSb, it is an established fact that due to high intensity pump (which is one of the pre-requisite conditions for the onset of parametric instability) heating of carrier becomes inevitable. This heating of electrons modifies the electron collision frequency (ECF) through the relation [8]

$$\Gamma_e = \Gamma_{e0}\left(\frac{T_e}{T_0}\right)^{1/2} \tag{2}$$

where, T_e is the effective temperature of electrons, T_0 is the lattice temperature, and Γ_{e0} is the ECF when $T_e = T_0$. As a result, the momentum-transfer collision frequency (MTCF), mobility of the carriers and conductivity of the medium become function of the pump amplitude and hence produce refinement effects. The momentum transfer of carriers is assumed to be due to acoustical phonon scattering and the energy transfer due to polar optical phonon (POP) scattering mechanisms in an n-type III-V semiconductor. In steady state, the power absorbed per electron from the pump is just equal to the power lost per electron in the POP scattering. Hence for moderate heating, carrier temperature is-

$$\frac{T_e}{T_0} = 1 + \frac{e^2\Gamma_e}{2m_e}\frac{\tau(\omega_c^2+\omega_0^2)}{[(\omega_c^2-\omega_0^2)^2+4\Gamma_e^2\omega_0^2]}E_0.E_0^* \tag{3}$$

where $\tau^{-1} = \left(\frac{2k_B\theta_D}{m_e\pi}\right)^{1/2} eE_{po}x_0.K_0\left(\frac{x_0}{2}\right).\frac{x_0^{1/2}\exp\left(\frac{x_0}{2}\right)}{\exp(x_0)-1}$

2.3. Threshold and amplification characteristics

Using basic equations from paper I [5] second-order nonlinear susceptibility $\chi^{(2)}$ can be derived as

$$\chi^{(2)} = \frac{-ike(NM)^{0.5}}{m_e\varepsilon_0\omega_{pl}}\left(\frac{\omega_0\Omega_p^2}{\omega_0^2-\omega_c^2}\right)\frac{QA_1T}{F_1}G^* \tag{4}$$

Here we have neglected the Doppler shift under the assumption $\omega_0 >> \Gamma_{e0} > \vec{k}_0.\vec{\vartheta}_0$ and

$G = \left[\delta_2^2 - 2i\Gamma_{pl}\omega_{pl} - \frac{k^2A_2^2|\bar{E}|^2}{\delta_1^2 - 2i\Gamma_e\omega_1}\right]^{-1}$, $\delta_1^2 = A_1\overline{\omega}_P^2 - \omega_1^2$, $\delta_2^2 = A_1\overline{\omega}_P^2 - \omega_{pl}^2$, $\bar{E} = \frac{-e}{m_e}E_{eff}$, $\omega_p = \sqrt{\frac{n_0e^2}{m_e\varepsilon}}$,

$\omega_{cpl} = \left(\frac{-e}{m_e}+\frac{q}{M}\right)B_0$, $A_1 = \frac{\omega_1^2}{\omega_1^2-\Omega_p^2-\omega_c^2}$, $A_2 = \frac{\omega_0^2}{\omega_0^2-\Omega_p^2-\omega_c^2}$, $Q = X_1 + Y_1$, $X_1 = \frac{4\Gamma_{pl}^2}{4\Gamma_{pl}^2+\omega_c^2}$

, $Y_1 = \frac{\omega_{cpl}^2}{4\Gamma_{pl}^2-\omega_{cpl}^2}$, , $T = \frac{n_0eq}{M} + \frac{Neq}{m_e} - \frac{n_0e^2}{m_e} - \frac{Nq^2}{M}$, $F_1 = \omega_{0,pl}^2 - \omega_{pl}^2 - 2i\Gamma_{pl}\omega_{pl}$,

$\overline{\omega}_P^2 = \left[\omega_p^2 + k_1^2\left(\frac{k_BT_e}{m^*}\right)\right]\left(\frac{4\Gamma_{pl}^2}{4\Gamma_{pl}^2+\omega_c^2}\right)$.

It is clear that $\chi^{(2)}$is influenced by the wave vector k, carrier concentration n_o and by the transverse dc magnetic field through ω_c.

Threshold pump amplitude for the onset of parametric amplification may be obtained as

$$[E_{0th}]_{para} = \frac{m_e}{ekA_2}\left(1-\frac{\omega_c^2}{\omega_o^2}\right)\delta_1\delta_2 \tag{5}$$

The parametric amplification can be achieved at excitation intensity above this threshold value under favourable conditions and nonlinear absorption coefficient is given by

$$\alpha_{para} = \frac{k}{2\varepsilon_1}\left[\chi^{(2)}\right]_i E_0 \tag{6}$$

The nonlinear parametric gain of the signal as well as the idler waves can be possible only if α_{para} obtained from equation (6) is negative.

3. Results and discussions

Numerical estimations were carried out by using the relevant parameters chosen for InSb medium are listed in [9]. Figure 1 depicts the variation of threshold pump electric field $[E_{oth}]_{para}$ (with or without CH effects) with carrier density. The threshold pump electric field achieves its minimum value carrier density when dependent modified plasma frequency becomes smaller and smaller and finally resulting into $A_1\overline{\omega}_P^2 \approx \omega_{pl}^2$ in δ_1^2 term. This figure clearly illustrates that the polaron frequency plays an important role in the appreciable reduction of threshold pump electric field. CH effects tend to shift minimum threshold pump electric field towards the smaller concentration of the carriers.

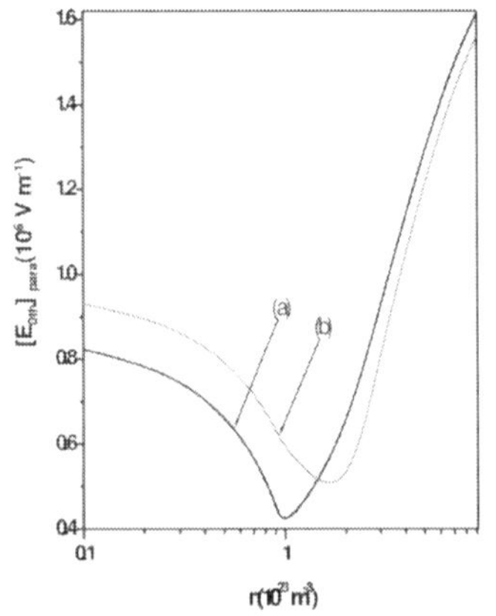

Figure 1. Variation of threshold pump field as a function of carrier density. [Curve (a) with CH effects and Curve (b) without CH effects].

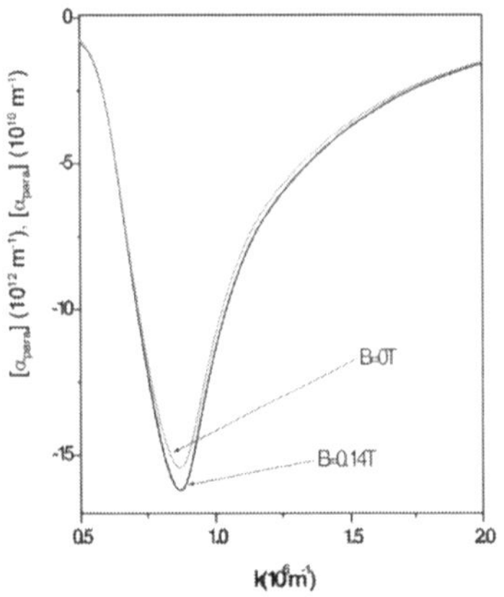

Figure 2. Variation of absorption coefficient as a function of wave vector with and without magnetic field.

Figure 2 represents the variation of absorption coefficient with wave vector. It is inferred from figure 2 that at $\left[k \approx \frac{\left[\left(\delta_2^2 - 2i\Gamma_{pl}\omega_{pl}\right)\left(\delta_1^2 - 2i\Gamma_e\omega_1\right) \right]^{0.5}}{A_2 \,|\, \bar{E} \,|} \right]$ the absorption coefficient α_{para} decreases suddenly and achieves its minimum. Above or below this resonance condition α_{para} is nearly independent of wave vector. Presence of magnetic field significantly increases the parametric gain by a factor of $\approx 10^2$ in comparison with that in the absence of magnetic field.

It is found that presence of magnetic field is favourable for polaron induced parametric gain in heavily doped semiconductor. It is also noticed that the nonlinearity due to mass modulation and electron collision frequency modifies threshold and amplification characteristics strongly. CH effects are found to lower the threshold pump field and effectively raise the parametric gain.

Acknowledgement
One of the authors (S. Ghosh) is thankful to MPCST, Bhopal for financial assistance.

References

[1] Tian L, Allman M S and Simmonds R W 2008 *New J. Phys.* **10** 115001
[2] Glicksman M and Steele M C 1958 *Phys. Rev.* **110** 1204; 1959 *Phys. Rev. Lett.* **2** 461
[3] Prior A C 1958 *J. Electron Control* **4** 165
[4] Sodha M S, Kumar A, Tripathi V K and Kaw P K 1973 *Opto-electronics* **5** 509
[5] Agrawal R, Dubey S and Ghosh S 2012 *J. Phys.: Conf. Ser.* **365** 012045
[6] Kane E O 1957 *J. Phys. Chem. Solids* **1** 249
[7] Ehrenreich H 1957 *J. Phys. Chem. Solids* **2** 131
[8] Conwell E M 1969 *High field transport in semiconductors*, Suppl. 9 (New York: Academic press) 159
[9] Dubey S and Ghosh S 2009 *New J. of Phys.* **11** 093030

Design and fabrication of PPM and hybrid undulator for free electron laser applications

G. Mishra, Geetanjali Sharma, Roma Khullar[#], Bramh Prakash, Mona Gehlot
Insertion Device Development and Measurement Lab, School of Physics, DAVV, Indore.
Corresponding author: roma_khullar@yahoo.co.in

Abstract: Two undulators for free electron laser applications are under fabrication at Insertion Device Development & Measurement Laboratory, DAVV, Indore. The first one is of PPM type with twenty periods, 50 mm each period. The second one is a hybrid undulator with twenty five periods, 20 mm each period. NdFeB magnets of rectangular cross section 12.5mm, 12.5mm & 6.25 mm, 6.25 mm will be used for the undulators respectively. Each magnet length is of 50 mm. Each magnet is scanned by a laser micrometer to ensure gap uniformity along the length of the undulator. In this paper we describe the technical design details of both the undulators.

I.Introduction

The undulator [1,2] is the key component of a free electron laser device. A precise fabricated undulator gives improved free electron laser performance. Two undulators are under fabrication at Insertion Device Development and Measurement Laboratory, Devi Ahilya University, Indore, India. The first one is of one meter in length with twenty periods, 50mm each period PPM type. It will be of four magnets per period structure. The rectangular magnets of size 12.5mm × 12.5mm × 50mm will be used in the undulator. The second one will be a hybrid undulator with two magnets and two iron poles. Each period of the undulator will be of 20mm and there will be twenty five periods in total. The magnet size is 6.25mm × 6.25mm × 50mm. It is proposed to design the hybrid undulator with the following optimized parameters. The salience will be kept at zero mm [3]. This will enable to have maximum values of the magnetic field. Another parameter of design interests is $\xi = 2b/\lambda_u$, where b is the magnet width, λ_u is the undulator period. The parameters are describe in fig 8a. It is proposed to design the hybrid undulator with

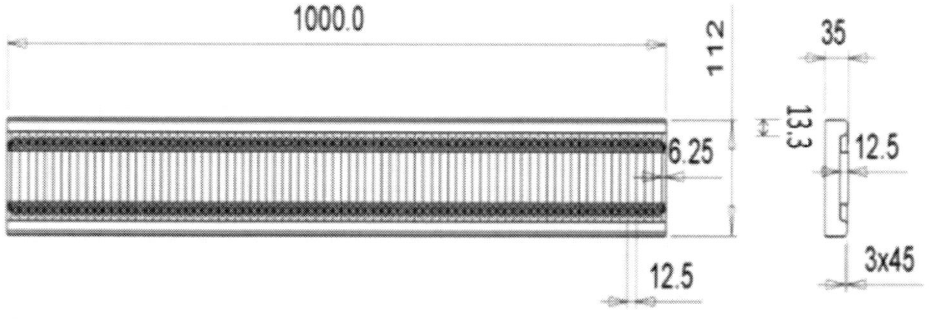

Fig 1.Design of 1m PPM undulator.

$\xi = 0.625$. For $\lambda_u = 20mm$, this calculates the magnet width as 6.25mm. The iron pole width is 3.75mm. The pole height to the magnet height ratio is another important parameter in the design of the hybrid undulator. The magnet height is 6.25mm. In order that this ratio is kept around 0.75, the pole height comes around 4.68mm. The pole width, height and length are 3.75mm, 4.68mm and 50mm respectively. Fig1 and Fig2 represent the design details of these two undulators. The NdFeB magnets are procured from Shanghai Strong Magnet Industrial and Commercial Co. Ltd, Shanghai, China.

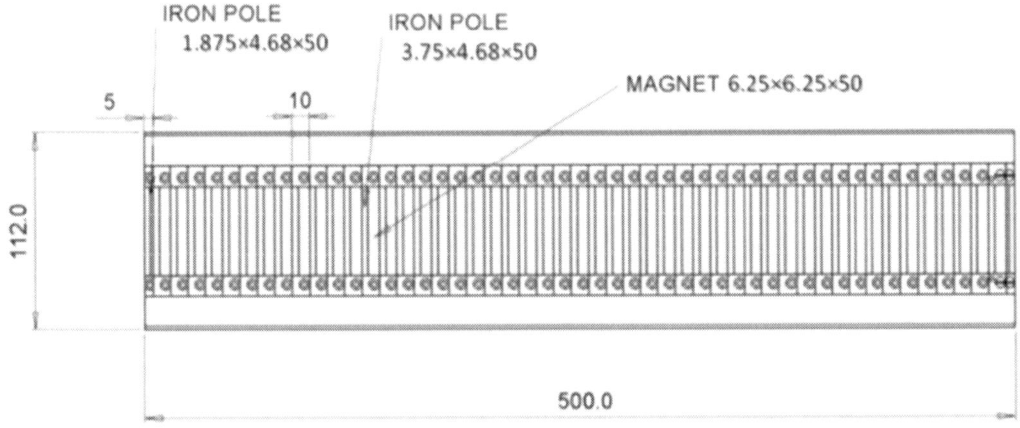

Fig 2.Design of hybrid undulator.

II. Initial Measurements and fabrication

Two magnet measurements are crucial for uniform field and uniform gap undulator. The magnet measurements of the single magnets are shown in Fig 3 and Fig 4. The measurements are done at 1mm away from the magnet surface. The uniform field magnets will be chosen for the undulator. A precise uniform gap undulator requires uniform size magnets. The magnets are scanned by a laser micrometer setup. The results are shown in Fig 5 and Fig 6 for the two magnets. It is observed that a size variation of 30μm will provides a magnet field variation of around 50Gauss and will be the tolerance value in our design. A magnet fixture is designed to clamp the magnets over the jaws shown in Fig 7. A magnet fixture has been designed and fabricated to clamp the magnets in the undulator jaw. The magnets have to be fixed to the undulator jaws. Part (1) is made of brass with two magnetic points. Magnet is attached to part (1) with two magnetic points. On both side of part (1) there are two slots represented by point (2) which can be moved back and forth. These slots are set in front first as shown in the Figure. Now the whole assembly of jaw is placed across the fixer who is designed according to the size of the jaw at point (3). The first magnet is placed manually. By rotating the handle at point (4) the magnet is placed in the jaw, now the slots on either side of the magnet i.e.at point (2) are moved at the back and the magnet is fixed by the clamp on either side of the magnet. Similarly other magnets are placed in the whole assembly.

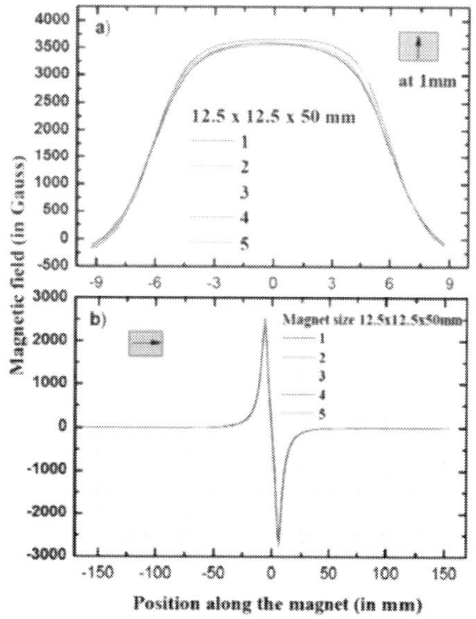

Fig3.Magnetic field measurements of 12.5mm magnets

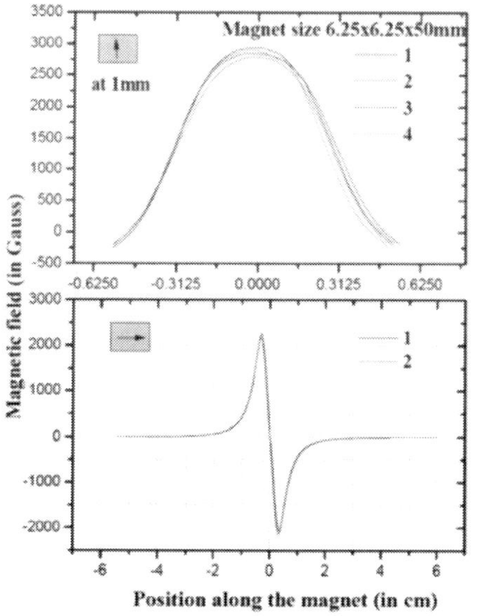

Fig4. Magnetic field measurements of 6.25mm magnets

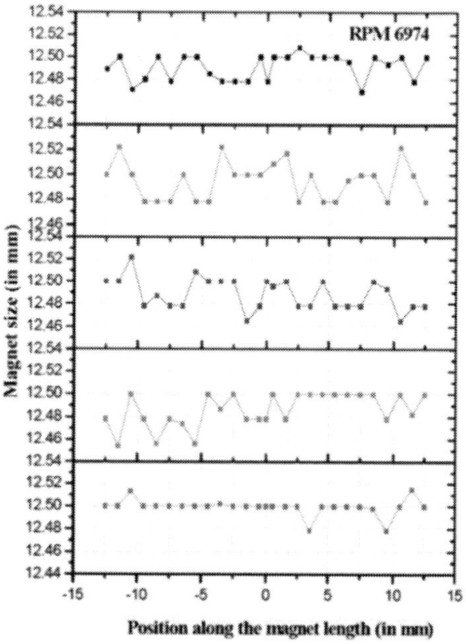

Fig 5. Magnet size measurement of 12.5mm magnets along the length.

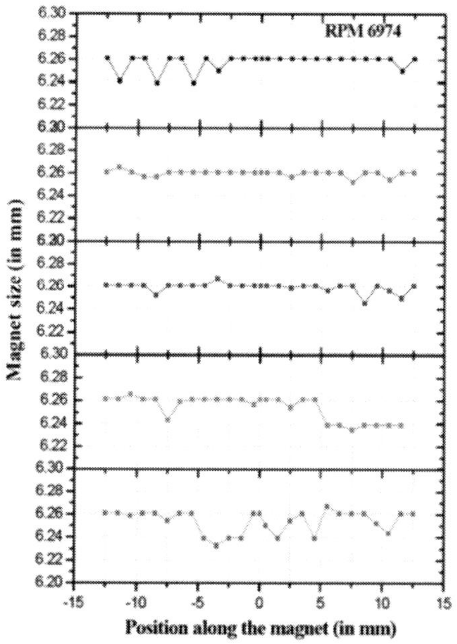

Fig 6. Magnet size measurement of 6.25mm magnets along the length.

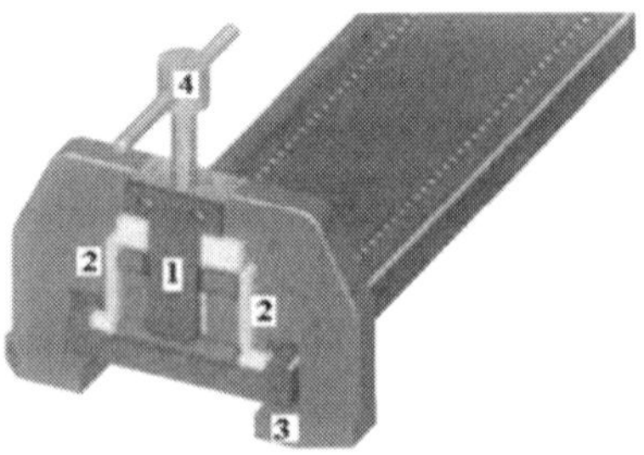

Fig 7.Magnetic fixer with jaw

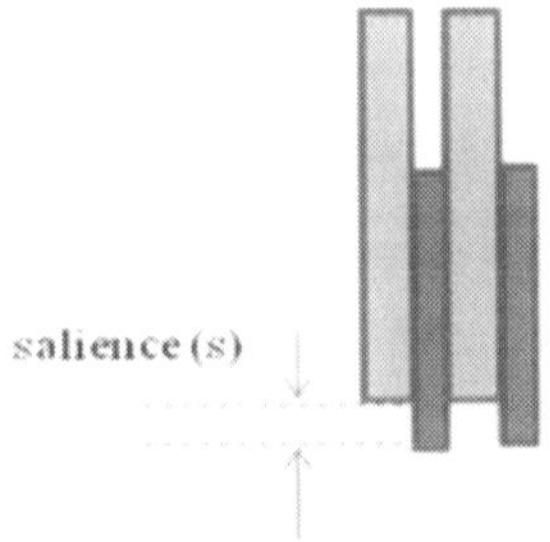

Fig 8a.Salience

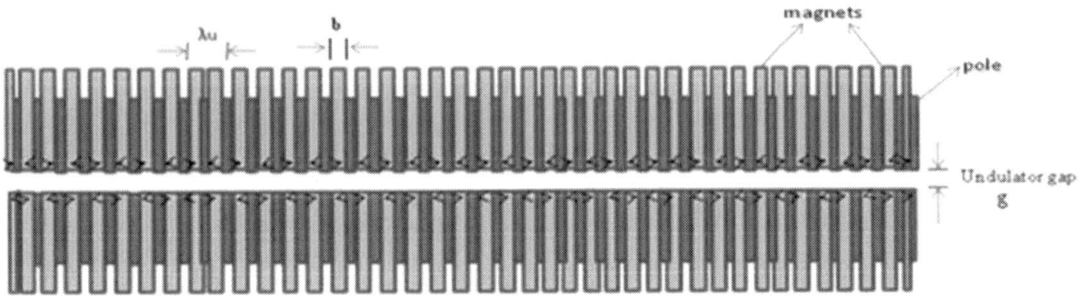

Fig 8b.Hybrid undulator side view.

III. Conclusion

Two undulators with PPM and hybrid configurations are underway at Insertion Device Development Lab, DAVV, Indore. The PPM undulator consists of 156 magnets made up of NdFeB, (12.5×12.5×50)mm each and 8 magnets of (6.25×12.5×50)mm. The hybrid undulator will be made up of 98 magnets of size (6.25×6.25×50)mm, 4 magnets of (3.125×6.25×50)mm, 98 poles of (3.75×4.68×50)mm and 4 pole of (1.875×4.68×50)mm. The side view of the hybrid design is shown in Fig8b.

ACKNOLEDGEMENT

This work is supported by UGC, Delhi and CSIR, Delhi, India.

REFERENCES

[1] George Brown, Klaus Halback, John Harris & Herman Winick, 1983,"Wigglers and Undulator magnets", *Nuclear instruments and Methods*, Vol **208**, pp.65-67.

[2] James A. Clarke, "The Science and Technology of Undulators and Wigglers", *Oxford Science Publications*.

[3] F. Sircilli, A.C. J. Paes, V.A. Serrao and Y.C. De Polli, 2001, "Design of a Hybrid Undulator for Free-Electron Lasers", *IEEE Transaction on Magnetics*, Vol **37**, No.5.

[4] Ma Youwu, Zhang Zhooming, Liu Bo, 1990, "Hybrid Undulator Design and Field error analysis", *Inspire Conf. Proc.* **C900612**, 1377-1379.

[5] G.Sharma, Roma Khullar, Mona Gehlot, G.Mishra, 2013 "Design consideration of one meter long undulator", *Indian Particle Accelerator Conference, VECC, Kolkata, India.*

Rayleigh Taylor instability of two superposed compressible fluids in un-magnetized plasma

P K Sharma[1], A Tiwari[1], S Argal[1] and R K Chhajlani[2]
[1] Barkatullah University Institute of Technology, Barkatullah University, Bhopal- 462026 India
[2] School of Studies in Physics, Vikram University, Ujjain – 456010 India

E-mail: pks_buit30@yahoo.com, anitatiwari7987@gmail.com

Abstract. The linear Rayleigh Taylor instability of two superposed compressible Newtonian fluids is discussed with the effect of surface tension which can play important roles in space plasma. As in both the superposed Newtonian fluids, the system is stable for potentially stable case and unstable for potentially unstable case in the present problem also. The equations of the problem are solved by normal mode method and a dispersion relation is obtained for such a system. The behaviour of growth rate is examined in the presence of surface tension and it is found that the surface tension has stabilizing influence on the Rayleigh Taylor instability of two superposed compressible fluids. Numerical analysis is performed to show the effect of sound velocity and surface tension on the growth rate of Rayleigh Taylor instability. It is found that both parameters have stabilizing influence on the growth rate of Rayleigh Taylor instability.

1. Introduction

Rayleigh Taylor instability is a phenomenon that is arised when heavier fluid is supported by lighter fluid in the presence of gravitational force [1]. Its occurrence in space and astrophysical phenomenon's, e.g. supernova explosion, implosion, crab nebula interplanetary medium, intergalactic medium, hydrodynamic phenomenon's e.g. pouring of oil into water, inverting glass of water, geophysics and ICF (inertial confinement fusion) intends to discuss it in various system [2, 3]. Chandrasekhar [4] studied Rayleigh Taylor instability of incompressible fluids comprehensively. Sharma et al. [5] carried out the effect of surface tension and rotation on Rayleigh Taylor instability of two superposed fluids with suspended particles. Ghasemizad [6] et al.have studied the growth rate of Rayleigh Taylor instability in inertial confinement fusion. Hoshoudy [7] studied the Rayleigh Taylor instability in quantum magnetized viscous plasma. Sharma et al. [8] have discussed the Rayleigh-Taylor (R-T) and Kelvin-Helmholtz (K-H) instabilities of a plasma in the incompressible porous medium with horizontal magnetic field and suspended dust particles.

Compressibility is an important aspects of fluid dynamics. In this context Vandervroot [9] investigated the character of the equilibrium of compressible fluid of varying density. Verma and Verma [10] have discussed the Rayleigh Taylor instability for an interface separating two compressible media the presence of magnetic field. Furthermore, Ariel [11] studied the effect of vertical magnetic field on the Rayleigh Taylor instability of compressible fluid of stratified variable density by using variational principle. Bhatia [12] has discussed the Rayleigh Taylor instability of viscous compressible fluid. More recently Diaz et al. [13] have reported the Rayleigh Taylor instability in partially ionized compressible plasmas.

Thus, for the better insight we have considered the effect of surface tension on the Rayleigh Taylor instability of two superposed compressible fluids in un-magnetized plasma where magnetic field is negligible. Several authors have discussed Rayleigh Taylor instability of incompressible fluids. However the study of Rayleigh Taylor instability for compressible fluids is more realistic.

2. Formulation of problem

We assume that two homogenous compressible nonviscous superposed fluids are separated by a plane interface z = 0 in presence of gravitational force **g** (0, 0, g). The region z < 0 and z > 0 are denoted for fluid 1 and 2 respectively. The effect of surface tension is also considered on the interface of superposed compressible fluids. The physical quantities of system are taken as density of fluid ρ, pressure p, and velocity of fluid U.

In order to investigate the stability or instability of the system, we have used the normal mode method. First we expect the basic fluid equation, the adiabatic equation and the equation of state which concern the flow of compressible fluids in the equilibrium configuration. A small perturbation is applied to the system, therefore the physical quantities of system becomes as

$$\boldsymbol{U}=\boldsymbol{U}_0+\boldsymbol{u},\ \rho=\rho_0+\delta\rho,\ p=p_0+\delta p,\ z_s=z_s+\delta z_s \tag{1}$$

Here $\boldsymbol{U_0}$, ρ_0, and p_0 indicates the fluid velocity, fluid density and pressure respectively in the equilibrium and $\boldsymbol{u}$ (u_x, u_y, u_z), $\delta\rho$, and δp indicates the perturbed fluid velocity, fluid density and pressure respectively. The governing equations of the problem are linearized. To seek solutions into normal modes whose dependence on x, y, and t is given by

$$exp(ik_x x + ik_y y + int) \tag{2}$$

Where in is growth rate and k_x, k_y $(k^2 = k_x^2 + k_y^2)$ are the wave numbers.
The governing linearized equations are in scalar form as

$$\rho in u_x = -ik_x \delta p \tag{3}$$

$$\rho in u_y = -ik_y \delta p \tag{4}$$

$$\rho in u_z = -D\delta p - g\delta\rho - \sum T_s (k_x^2 + k_y^2)\delta z_s \delta(z - z_s) n_s \tag{5}$$

$$in\delta\rho + u_z D\rho + \rho(ik_x u_x + ik_y u_y + Du_z) = 0 \tag{6}$$

$$in\delta p - \rho g u_z = V^2 [in\delta\rho + u_z D\rho] \tag{7}$$

On using Eq.(2) in Eqs (3-7) and eliminating some variables we finally obtain an equation in the z component of velocity (u_z) as

$$(in)^2 D\left(\frac{\rho D u_z}{Q'_j}\right) - (in)^2 gD\left(\frac{\rho u_z}{Q'_j V^2}\right) + g u_z D\rho + (in)^2 g \frac{\rho D u_z}{Q'_j V^2} - (in)^2 \rho u_z + \frac{g^2 k^2 \rho u_z}{Q'_j V^2} - k^2 \sum T_s u_z \delta(z - z_s) = 0 \tag{8}$$

$$Here \quad Q'_j = \left(\frac{(in)^2}{V_j^2} + k^2\right)$$

3. Dispersion relation

We consider that two superposed compressible fluids are separated by plane interface z = 0, and bounded by z = ±∞. We shall now derive the dispersion relation of compressible fluid from Eq.(8) for the constant density and sound velocity, we get

$$(D^2 - q_j^2) u_z = 0 \tag{9}$$

$$q_j^2 = \left[\frac{(in)^2}{V_j^2} + k^2 - \frac{g^2 k^2}{(in)^2 V_j^2}\right] \quad \text{Where } j = 1, 2 \tag{10}$$

The general solution of Eq. (9) written as

$$u_z = Ae^{q_1 z} + Ae^{-q_2 z} \tag{11}$$

Here A is a arbitrary constant. The appropriate boundary conditions to solve the problem are

[1] Velocity is continuous at the interface and satisfied as $u_{z_1} = u_{z_2} = u_{z_0}$.

[2] The total pressure is continuous at the interface z = 0.

By using boundary conditions and assuming ω = (*in*), the dispersion relation for the two superposed compressible fluids can be written as,

$$\omega^2\rho_2\left[\frac{\omega^2}{V_2^2}+k^2-\frac{g^2k^2}{\omega^2V_2^2}\right]^{1/2}\left(\frac{\omega^2}{V_2^2}+k^2\right)^{-1}+\omega^2\rho_1\left[\frac{\omega^2}{V_1^2}+k^2-\frac{g^2k^2}{\omega^2V_1^2}\right]^{1/2}\left(\frac{\omega^2}{V_1^2}+k^2\right)^{-1}$$
$$+k^2T = g\rho_2k^2\left(\frac{\omega^2}{V_2^2}+k^2\right)^{-1}-g\rho_1k^2\left(\frac{\omega^2}{V_1^2}+k^2\right)^{-1} \tag{12}$$

Eq. (12) is dispersion relation for the Rayleigh Taylor instability of two superposed compressible fluids in presence of surface tension. The dispersion relation (12) is similar with Vandervroot [7] and Verma and Verma [8] in absence of surface tension. In case $V_1 \to \infty$ and $V_2 \to \infty$, the dispersion relation (12) is reduced to

$$\omega^2 = \frac{gk(\rho_2-\rho_1)}{\rho_2+\rho_1} - \frac{k^3T}{\rho_2+\rho_1} \tag{13}$$

The dispersion relation (13) represents the effect of surface tension on the Rayleigh Taylor instability of two incompressible fluids and it is identical to Sharma et al. [5].

4. Discussion

The dispersion relation (12) is complicated to discuss, therefore, we simplify the dispersion relation by making the assumption that the superposed fluids are highly compressible. For this case the Eq. (10) is solved by using binomial expansion as

$$q_j = \left[k + \frac{\omega^2}{2kV_j^2} - \frac{g^2k}{2\omega^2V_j^2}\right] \qquad j=1,2$$

The dispersion relation becomes as

$$\omega^6+\omega^4\left[k^2V'^2(\beta_1+\beta_2+4)+\frac{2k^3T}{\rho_1+\rho_2}\right]+\omega^2\left[\begin{matrix}2k^4V'^4(\beta_1+\beta_2+2)-g^2k^2+\\ \frac{2k^5TV'^2}{\rho_1+\rho_2}(\beta_1+\beta_2+2)\end{matrix}\right]$$
$$+\left[\frac{2k^7TV'^4}{\rho_1+\rho_2}(\beta_1+\beta_2+2)+2gk^5V'^4(\beta_1-\beta_2)-g^2k^4V'^2(\beta_1+\beta_2)\right]=0 \tag{14}$$

Where $\beta_1 = \frac{\rho_1}{\rho_2}$, $\beta_2 = \frac{\rho_2}{\rho_1}$, $V'^2 = \frac{p}{\rho_2+\rho_1}$

4.1 In absence of surface tension

The dispersion relation of the Rayleigh Taylor instability for compressible fluids in absence of surface tension is written as

$$\omega^6+\omega^4\left[k^2V'^2(\beta_1+\beta_2+4)\right]+\omega^2\left[2k^4V'^4(\beta_1+\beta_2+2)-g^2k^2\right]$$
$$+\left[2gk^5V'^4(\beta_1-\beta_2)-g^2k^4V'^2(\beta_1+\beta_2)\right]=0 \tag{15}$$

We can able to define the criterion for stability of system through the constant term of Eq. (15) as

$$k_c = \frac{g}{2V'^2}\frac{(\rho_2^2+\rho_1^2)}{(\rho_2^2-\rho_1^2)} \tag{16}$$

Here k_c is critical wave number. The system will be stable if $k > k_c$ and unstable if $k < k_c$.

4.2 In presence of surface tension

In order to perform numerical calculations on the growth rate of the Rayleigh Taylor instability of two compressible fluids in presence of surface tension, we write the dispersion relation (14) in the dimensionless form as

$$\omega^* = \frac{\omega}{\sqrt{gk}},\ V^* = \frac{V'}{gk},\ T^* = \frac{k^2 T}{g(\rho_l + \rho_2)},\ k^* = k\sqrt{gk}$$ [ω*,v*,T* and k* are dimensionless parameters]

$$\omega^{*6} + \omega^{*4}\left[k^{*2} V^{*2}(\beta_2 + \beta_l + 4) + 2T^*\right] + \omega^{*2}\begin{bmatrix} 2k^{*4} V^{*4}(\beta_2 + \beta_l + 2) + \\ 2T^* k^{*2} V^{*2}(\beta_l + \beta_2 + 2) - l \end{bmatrix} \tag{17}$$
$$+ \left[2T^* k^{*4} V^{*4}((\beta_l + \beta_2 + 2)) + 2k^{*4} V^{*4}(\beta_l - \beta_2) - k^{*2} V^{*2}(\beta_l + \beta_2)\right] = 0$$

To examine the effect of surface tension, sound velocity, on the growth rate of Rayleigh Taylor instability of two superposed compressible fluids following figures have been drawn. In figure 1, we have potted the growth rate against wave number with increasing sound velocity V^* = 4.5, 5.5, 6.3 and we find that the growth rate is decreasing on increasing the value of sound velocity of fluid which indicates the stabilizing effect of sound velocity. In this case we have taking constant values of $\beta_2 = 2$, $\beta_l = 0.5$, and $T^* = 0.8$.

The figure 2 illustrates that the growth rate is decreasing on increasing the value of surface tension while the parameters have been taken as $\beta_2 = 2$, $\beta_l = 0.5$, $V^* = 4.5$, T^* =0.4, 0.8, 1.2. This figure reveals the stabilizing effect of surface tension on the unstable mode of RT instability.

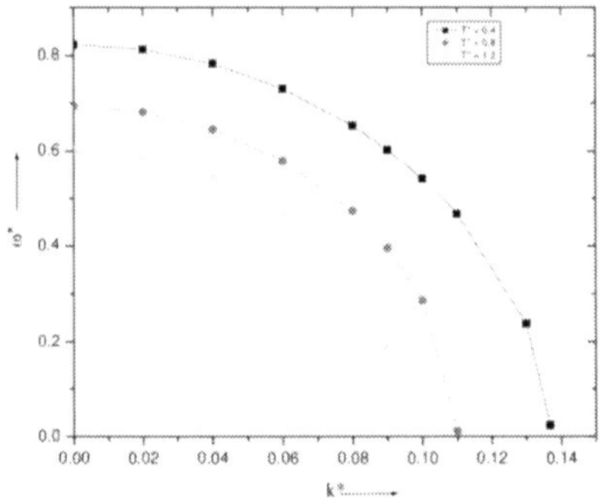

Figure:- 2 The growth rate (ω^*) verses wave number (k^*) in variation of surface tension (T^*) .

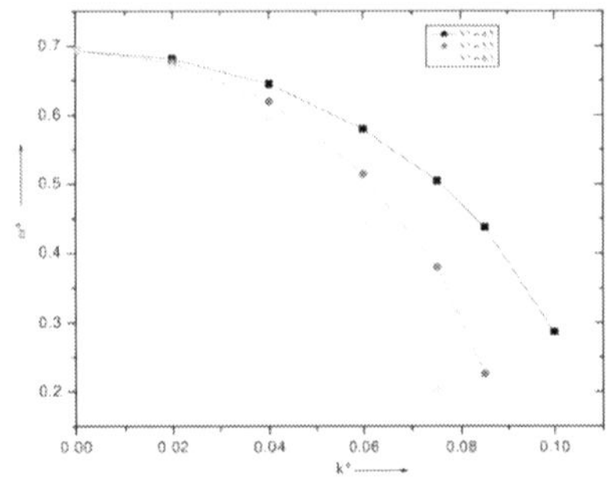

Figure 1. The growth rate (ω^*) verses wave number (k^*) in variation of sound velocity (V^*).

References

[1] Rayleigh L1882 *Proc. of the London Mathematical Society* **14** 170
[2] Ghasemizad A, Zarringhalam H, and Gholamzadeh L 2009 *J. Plasma Fusion Series* **8** 1234
[3] Hester J J et al 1996 *Astrophys. J.* **456** 225
[4] Chandrasekhar S 1961 *Hydrodynamic and Hydromagnetic Stability* (Oxford: Clarendon press)
[5] Sharma P K, Prajapati R P and Chhajlani R K 2010 *Acta Phys. Polon. A* **118** 576]
[6] Ghasemizad A, Zarringhalam H and Gholamzadeh L 2009 *J. Plasma Fusion Res. SERIES* **8** 1234
[7] Hoshoudy G A 2011 *Plasma Physics Reports* **9** 775
[8] Sharma P K, Tiwari Anita and Chhajlani R K 2013 *J. Porous Media* **16(12)** 1063
[9] Vandervroot P O 1961 *Astrophys. J.* **134** 699
[10] Verma Y K and Verma P 1963 *Proc. Nat. Inst. Sci. India A* **29** 309
[11] Ariel P D 1971 *Appl. Sci, Res.* **24** 294
[12] Bhatia P K 1974 *Astrophys. Space Sci.* **26** 319
[13] Diaz A, Soler R and Ballester L J 2012 *Astrophys. J.* **754** 1

Journal of Physics: Conference Series **534** (2014) 012055 doi:10.1088/1742-6596/534/1/012055

Effect of magnetic field on Jeans instability of self gravitating strongly coupled dusty plasma

R. K. Chhajlani[1] and Prerana Sharma[2]
[1]School of Studies in Physics, Vikram University Ujjain, M. P. - 456010, India
[2]Department of Physics, Ujjain Engineering College, Ujjain, M.P.465010, India

Email: preranaiitd@rediffmail.com

Abstract. The effect of the magnetic field on the Jeans instability of self-gravitating strongly coupled dusty plasma is investigated. The medium consists of extremely massive, negatively charged strongly correlated dust grains and weakly correlated Boltzmann distributed electrons and ions. The basic equations of the problem are constructed using generalized hydrodynamic model. In deriving dispersion relation the plane wave solutions are used on the linearized perturbation equations. The analysis is done by normal mode analysis theory. The dispersion relation is found to be affected due to the presence of viscoelastic effects and magnetic field. The dispersion relation is further reduced for hydrodynamic and kinetic regime of propagation. The condition of Jeans instability is also obtained for both the regimes of propagation. Numerical results are presented to show the effect of magnetic field and time relaxation parameter which is found to have stabilizing influence on growth rate of Jeans instability in strongly coupled dusty plasma. The condition of Jeans instability are compared in kinetic and hydrodynamic regime.

1. Introduction.

Dust particulates in the plasma acquire electric charges due to collecting ions and electrons and are generally strongly coupled. A wide-ranging system in diverse astrophysical environments such as neutron stars, white dwarfs and interior of heavy planets are observed to find in the strongly coupled state. The presence of strongly coupled species adds new effects in the study of instability of the system. The response of strongly coupled plasma to the wave propagations made the subject interesting for both theoretical and experimental research, In recent years much attention has been devoted to study the waves and instabilities in strongly coupled dusty plasmas. Banerjee et al [1] have examined the influence of the shear flow induced viscosity gradient in a strongly coupled dusty plasma. The effect of polarization force with effective dust temperature on dust acoustic solitary waves in strongly coupled dusty plasma has been discussed by Mamun et al [2]. Thus, from the above mentioned works it is found that collective effects has been very well studied in the strongly coupled dusty plasma. Furthermore, the strong correlation effects of particles affect the structure formation due to gravitational collapse. Therefore, Jeans instability in strongly coupled dusty plasma is needed to be discussed in context of formation of astrophysical objects. A limited number of theoretical investigations have also been made on Jeans instability in plasma which is in strongly coupled state. Moreover, the importance of magnetic field in strongly coupled dusty plasma dynamics has been recognized [3-4]. The Jeans instability of viscoelastic electron ion fluid is investigated by Janaki et al. [5] using the magnetohydrodynamic approach when ions are considered as strongly coupled. The self-gravitational instability in magnetized finitely conducting strongly coupled magnetized plasma is discussed using the magnetohydrodynamic model by Prajapati and Chhajlani [6] considering the strong correlation of ions. None of the works deal with Jeans instability in presence of magnetic field when dust is considered as strongly coupled. Thus looking to the importance of the problem, we investigate the influence of magnetic field on the Jeans instability of strongly coupled dusty plasma.

2. Perturbation equations and dispersion relations

Let us consider a strongly coupled dusty plasma whose constituents are negatively charged dust, electrons, and ions embedded in uniform magnetic field $\boldsymbol{B}_0(0, 0, B)$. In the present work the electrons and ions are assumed to be weakly coupled due to their higher temperatures and smaller electric charges, and that dust is strongly coupled because of its lower temperature and larger electric charge. The self-gravitational contribution of electrons and ions are not considered as they are assumed to be inertia-less. The basic equations of the problems are linearized using the standard linearization process. We suppose all the physical quantities are sum of their equilibrium and perturbed parts as $n_{d,e,i} = n_{d,e,i0} + n_{d,e,i1}$, $\phi = \phi_0 + \phi_1$, $\psi = \psi_0 + \psi_1$, $\boldsymbol{v}_d = \boldsymbol{v}_{d1}$. The electron and ion number densities obey the Maxwellian distribution, we obtain the linearized perturbation equations for electron and ion fluids are $n_{e1} = n_{eo}\ (e\phi_1/T_e)$ (1) and $n_{i1} = n_{io}\ (-e\phi_1/T_i)$ (2)

where T_i and T_e are the ion and electron temperature respectively. However, employing plane wave analysis considering all the perturbed quantities vary as $\exp(i\boldsymbol{k.r} + i\omega t)$, (where ω is the frequency of the harmonic disturbances and $\boldsymbol{k} = k_x x + k_z z$ is the wave number).

The perturbed momentum transfer equation for dust grain

$$\left(1+\tau_m \frac{\partial}{\partial t}\right)\left[m_d n_{d0}\frac{\partial v_{d1}}{\partial t} + m_d n_{d0}\nabla\psi_1 - Z_d e n_{d0}\nabla\phi_1 + Z_d e n_{d0}\left(v_{d1}\times B\right)\right] = \eta_l \frac{\partial^2 v_{d1}}{\partial x^2} \quad (3)$$

The perturbed continuity equation is $-\omega\ \left(n_{d1}/n_{d0}\right) = \boldsymbol{k}\ .\boldsymbol{v}_{d1}$ (4)

The Poisson equations for electrostatic potentials $k^2\phi_1 = 4\pi e(n_{i1} - n_{e1}) + 4\pi Z_d e n_{d1}$ (5)

The Poisson equations for gravitational potential $-k^2\psi_1 = \left(\omega_J^2/n_{d0}\right)n_{d1}$ (6)

where τ_m is the relaxation time parameter, $\omega_J^2 = 4\pi G m_d n_{d0}$ is the square of the Jeans dust frequency. Further, from Eq. (3) with the help of Eqs. (4) and (6), we obtain a relation for perturbed dust density as

$$n_{d1} = -\frac{(1+\tau_m i\omega)Z_d e n_{d0} k^2/m_d}{(1+\tau_m i\omega)\left(\omega^2+\omega_J^2 - i\omega\omega_{cd}\right) - \left(\eta_l k^2 i\omega/m_d n_{d0}\right)}\phi_1 \quad (7)$$

Substituting the values of n_{i1}, n_{e1} and n_{d1} from equations (1), (2) and (7) respectively into Poisson's equation given by Eq. (5), then we obtain the following dispersion relation

$$(1+\tau_m i\omega)\left(\omega^2+\omega_J^2 - i\omega\omega_{cd}\right) - \frac{\eta_l i\omega k^2}{m_d n_{d0}} - \frac{\omega_{pd}^2 k^2\lambda_D^2(1+\tau_m i\omega)}{1+k^2\lambda_D^2} = 0 \quad (8)$$

where $\omega_{cd} = Z_d e B_0/m_d$ is the dust cyclotron frequency, $\omega_{pd}^2 = 4\pi Z_d^2 e^2 n_{d0}/m_d$, is the dust-plasma frequency, $\lambda_D = \lambda_{Di}\Big/\sqrt{1+\left(\lambda_{Di}/\lambda_{De}\right)^2}$ with $\lambda_{Di(e)} = T_{i(e)}\big/4\pi e^2 n_{i(e)}$. Equation (8) represents the required dispersion relation for self-gravitating strongly coupled dusty plasma in presence of magnetic field. Further, we discuss the dispersion relation given by Eq. (8) for the low frequency hydrodynamic regime and relatively high frequency kinetic regime of propagation.

2.1 Hydrodynamic regime

The dispersion relation for low frequency wave modes in strongly coupled dusty plasma under the limit $\omega\tau_m << 1$ and substituting $i\omega = \sigma$ the dispersion relation given by Eq. (8) can be discussed within the hydrodynamic regime as

$$\sigma^2 + \omega_{cd}\sigma + V_{cd}^2\tau_m k^2\sigma + \frac{\omega_{pd}^2 k^2\lambda_D^2}{1+k^2\lambda_D^2} - \omega_J^2 = 0 \quad (9)$$

where $V_{cd}^2 = \eta_l / m_d n_{d0}\tau_m$. The dispersion relation given by Eq. (9) represents self-gravitating magnetized mode strong coupling effects of the dust in the hydrodynamic regime. It is clear from the above dispersion relation that this low-frequency mode suffers viscous damping in the hydrodynamic regime with strongly correlated dust and Maxwell distributed electrons and ions. Moreover, in absence of strongly coupled effects of the dust, the dispersion relation resembles the result obtained by Prajapati [7]. However, if the constant term is less than zero, the condition of Jeans instability for the above considered system can be written using Eq. (9) as

$$\frac{\omega_{pd}^2 k^2\lambda_D^2}{1+k^2\lambda_D^2} < \omega_J^2 \quad (10)$$

The relation shows that the condition of Jeans instability in the SCDP does not affect due to magnetic field and strong coupling effects of dust. For investigating the effects of magnetic field, we normalize Eq. (9) in terms of dust Alfven speed and Jeans dust frequency as

$$\gamma^2 + \omega_{cd}^*\gamma + V_{cd}^{*2}\tau_m^* k^{*2}\gamma + \frac{\omega_{pd}^{*2} k^{*2}\lambda_D^{*2}}{1+k^{*2}\lambda_D^{*2}} - 1 = 0 \quad (11)$$

where the following dimensionless parameters have been used

$$\gamma = \sigma/\omega_J, k^* = kV_{da}/\omega_J, V_{cd}^* = V_{cd}/V_{da}, \tau_m^* = \tau_m\omega_J, \omega_{pd}^* = \omega_{pd}/\omega_J, \omega_{cd}^* = \omega_{cd}/\omega_J, \lambda_D^* = \lambda_D\omega_J/V_{da} \quad (12)$$

The effect of magnetic field on growth rate of Jeans instability is observed in the form of Fig. 1 when the fixed value of τ_m^*, ω_{pd}^*, λ_D^* and V_{cd}^* are taken into account. It is obvious that as ω_{cd}^* increases the growth rate decreases hence magnetic field has stabilizing influence on the growth rate of the perturbation. In addition to that the effect of relaxation parameter on the growth rate of Jeans instability is shown in Fig. 2. It is clear from curves that the growth rate decreases as the value of τ_m^* increases. Therefore, the dimensionless time relaxation parameter also plays stabilizing role on the growth rate of Jeans instability.

2.2 Kinetic regime

In the kinetic regime $(\omega\tau_m >> 1)$ the wave motion will become relatively fast and the features of the wave mode become different from the case of hydrodynamic regime and substituting $i\omega = \sigma$ in Eq. (8) we get the dispersion relation as

$$\sigma^2 + \omega_{cd}\sigma + V_{cd}^2 k^2 + \frac{\omega_{pd}^2 k^2\lambda_D^2}{1+k^2\lambda_D^2} - \omega_J^2 = 0 \quad (13)$$

This dispersion relation represents a gravitating mode in the presence of strong coupling effects and magnetic field with Maxwellian distributed electrons and ion in the kinetic regime. Above equation clearly shows that the magnetic field and strong coupling effects of dust significantly affect the mode of propagation in the kinetic regime. If we compare Eq. (9) to the Eq. (13) it is clear that in the kinetic mode the term $V_{cd}^2 k^2$ acts as source for perturbation while in the hydrodynamic regime it acts as a damper for the perturbation of propagation in the gravitating mode. From the constant term of Eq. (13) the condition of instability can easily be obtained as

$$V_{cd}^2 k^2 + \frac{\omega_{pd}^2 k^2 \lambda_D^2}{1 + k^2 \lambda_D^2} < \omega_J^2 \tag{14}$$

This equation shows the condition of Jeans instability in the strongly coupled dusty plasma in the kinetic regime. The above condition clearly shows that the strong coupling effects of dust surely modify the Jeans condition of instability but not affected due to magnetic field in the kinetic regime of propagation.

Further, it is interesting to compare the Jeans condition for the kinetic regime (14) to the condition for the hydrodynamic regime (10). It is to be noted that the additional term comes into the condition (14) is the velocity due to compressional effect of dust which is not coming into the Jeans condition for hydrodynamic regime. So it is to be mentioned that the dust thermal compressional velocity plays role in the kinetic regime only.

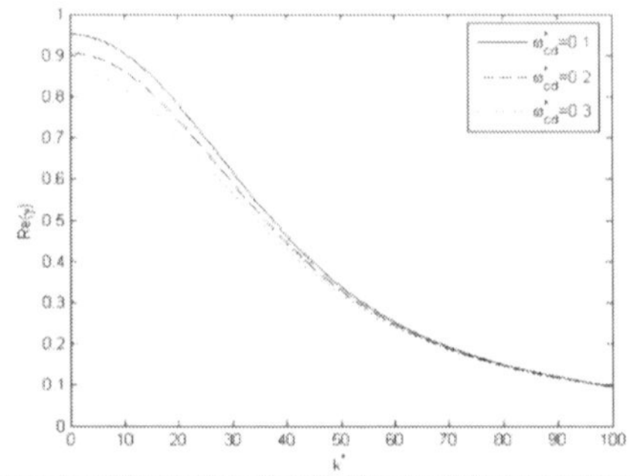

Fig.1. The growth rate of Jeans instability Re (γ) versus wavenumber (k^*) for various values of magnetic field $\omega_{cd}^* = 0.1, 0.2, 0.3$ in the hydrodynamic regime. The values of constant parameters are taken to be $\tau_m^* = 0.1, V_{cd}^* = 0.1, \omega_{pd}^* = 0.1, \lambda_d^* = .001$

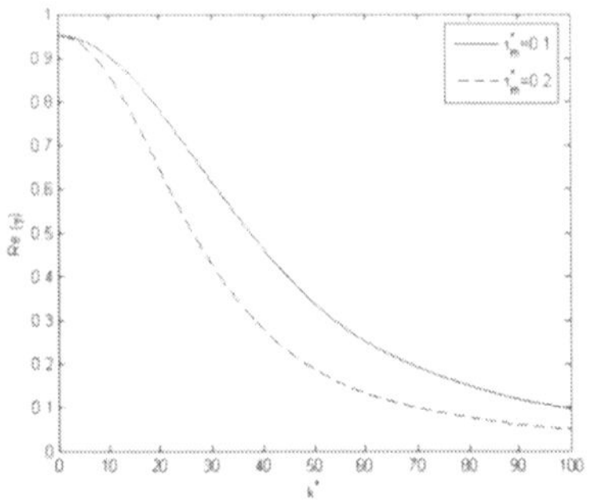

Fig.2. The growth rate of Jeans instability Re (γ) versus wavenumber (k^*) for various values of dimensionless time relaxation parameter $\tau_m^* = 0.1, 0.2$ in the hydrodynamic regime. The values of constant parameters are taken to be $\omega_{cd}^* = 0.1$, $V_{cd}^* = 0.1, \omega_{pd}^* = 0.1, \lambda_d^* = .001$

3. Conclusions

We have studied the effect of magnetic field on the Jeans instability of strongly coupled dusty plasma. The modified GH model is used to describe the dynamics of the system. The analysis of the results is done for the hydrodynamic and kinetic limits. The Jeans condition of instability is determined for both the regimes. The effect of magnetic field on growth rate of Jeans instability is numerically observed and found to have stabilizing influence on the strongly coupled system.

Acknowledgement

This work was partially supported by the Madhya Pradesh council for science and Technology Bhopal (M.P.), India. One of the authors R. K. Chhajlani is thankful to Vikram University Ujjain.

References

[1] Banerjee D, Janaki M S, Chakrabarti N and Chaudhuri M 2010 *New J Phys* **12** 123031
[2] Mamun A. A., Ashrafi K. S. and Shukla P. K. 2010 *Phys. Rev. E* **82** 026405
[3] Bonitz M, Kahlert H, Ott T and Lowen H 2013 *Plasma Sources Sci. Technol.* **22** 015007
[4] Xie Bai-Song 2002 *Chinese Phys. Lett.* **19** 1463
[5] Janaki M. S., Chakrabarti N. and Banerjee D. 2011 *Phys. Plasmas* **18** 012901
[6] Prajapati R. P. and Chhajlani R.K. 2013 *Astrophys. & Space Sci.* **344** 371
[7] Prajapati R. P. 2013 *Phys Letts* A **377** 291.

Influence of rotation and FLR corrections on self-gravitational Jeans instability in quantum plasma

Shweta Jain[1], Prerana Sharma[1] and R.K. Chhajlani[2]
[1] Department of Physics, Ujjain Engineering College, Ujjain, M.P.465010, India
[2] School of Studies in Physics, Vikram University Ujjain, M. P. - 456010, India

Email: jshweta09@gmail.com

Abstract. In the present work, the self-gravitational instability of quantum plasma is investigated including the effects of finite Larmor radius corrections (FLR) and rotation. The formulation is done employing quantum magnetohydrodynamic (QMHD) model. The plane wave solutions are employed on the linearized perturbed QMHD set of equations to obtain the general dispersion relation. The rotation is assumed only along the z- direction. The general dispersion relation is further reduced for transverse and longitudinal directions of propagation. It is found that in transverse direction of propagation the Jeans criterion is modified due to the rotation, FLR and quantum corrections while in longitudinal direction of propagation it is observed that the Jeans criterion is modified by quantum corrections only. The growth rate of perturbation is discussed numerically including the considered parameters FLR and quantum corrections. The growth rate is observed to be modified significantly due to the quantum correction and FLR effects.

1. Introduction.

The field of quantum plasmas is a subject of great interest in both the space and astrophysical plasmas. When a plasma is cooled to an extremely low temperature the de Broglie wavelength of the charge carriers could be comparable to the spatial scale of the plasma system. In such a situation, ultra cold plasma behaves like a Fermi gas and quantum mechanical effects are expected to play a central role to explain the dynamics of charged particles. It is a well-established fact that gravitation plays a key role in both classical as well as in quantum plasmas producing Jeans instability. Moreover, Jeans instability has been studied with the quantum mechanical effects by many researchers. The quantum hydrodynamic (QHD) model is employed by Masood et al. [1] to study the effect of Jeans term in a multicomponent self-gravitating quantum magnetoplasma. The Jeans instability of quantum magnetoplasma in the presence of Hall current and resistivity has been discussed by Wu et al. [2]. Sharma [3] has discussed the Jeans instability in presence of FLR corrections and rotation. Bhatia and Chhonkar [4] have investigated the effect of FLR corrections on the Jeans instability in rotating plasma. Recently Sharma and Chhajalani [5] have studied the self-gravitational instability in the presence of FLR corrections in quantum plasmas but they neglect the effect of rotation. Thus in the present work, we have examined the effect of rotation, quantum and FLR corrections on the Jeans instability.

2. Model equations and dispersion relation

We start with a magnetized quantum plasma system described by quantum magneto-hydrodynamic (QMHD) model with FLR corrections and rotation. The plasma is assumed to be embedded with a magnetic field $\mathbf{B}(0,0,\mathrm{B})$ along the z-direction and rotating with frequency $\boldsymbol{\Omega}(0,0,\Omega)$. The linearized perturbed QMHD equations of the problem with these effects are written as:

$$\frac{\partial\delta\rho}{\partial t}=-\rho\nabla.\delta\boldsymbol{v}\ \ (1),\ \ \rho\frac{\partial\delta\boldsymbol{v}}{\partial t}=-\nabla\delta p+\rho\nabla\delta\phi-\nabla.\delta\boldsymbol{P}+\frac{1}{4\pi}(\nabla\times\delta\boldsymbol{B})\times\boldsymbol{B}+\frac{\hbar^2}{4m_e m_i}\nabla\nabla^2\delta\rho+2\rho(\boldsymbol{v}\times\Omega)\ \ (2)$$

$$\frac{\partial\delta\boldsymbol{B}}{\partial t}=\nabla\times(\delta\boldsymbol{v}\times\boldsymbol{B})\ \ (3),\ \ \nabla^2\delta\phi=-4\pi G\delta\rho\ \ (4),\ \ \delta p=c_s^2\delta\rho\ \ (5)$$

where $c_s=(dp/d\rho)^{1/2}$ is the modified ion-acoustic velocity due to the quantum effects given as $c_s^2=v_{ti}^2+\frac{m_i}{m_e}\left(v_{te}^2+\frac{3}{5}v_{Fe}^2\right)$ having v_{ti} and v_{te} as the ion and electron thermal velocities and $v_{Fe}=\hbar\left(3\pi^2 n_e\right)/m_e$ is the Fermi velocity. The symbols $\delta\rho$, $\delta\boldsymbol{v}$, δp, $\delta\boldsymbol{B}$, $\delta\phi$ and $\delta\boldsymbol{P}$ represents the perturbations in density, fluid velocity, pressure, magnetic field, gravitational potential and pressure tensor.

G is the gravitational constant, $m_{e,i}$ is the electron and ion mass respectively, $\hbar=h/2\pi$, where h is the Planck constant. In momentum equation the perturbed components of pressure tensor are given as

$$\delta P_{xy}=\delta P_{yx}=\rho\upsilon_0\left(\frac{\partial\delta v_x}{\partial x}-\frac{\partial\delta v_y}{\partial y}\right),\delta P_{xz}=\delta P_{zx}=-2\rho\upsilon_0\left(\frac{\partial\delta v_y}{\partial z}+\frac{\partial\delta v_z}{\partial y}\right),\ \delta P_{zz}=0,$$
$$\delta P_{yz}=\delta P_{zy}=2\rho\upsilon_0\left(\frac{\partial\delta v_z}{\partial x}+\frac{\partial\delta v_x}{\partial z}\right),\delta P_{xx}=-\rho\upsilon_0\left(\frac{\partial\delta v_y}{\partial x}+\frac{\partial\delta v_x}{\partial y}\right),\ \delta P_{yy}=\rho\upsilon_0\left(\frac{\partial\delta v_y}{\partial x}+\frac{\partial\delta v_x}{\partial y}\right).\quad(6)$$

The parameter υ_0 has the dimensions of the kinematic viscosity and called as magnetic viscosity defined as $\upsilon_0=\Omega_L R_L^2/4$, where R_L is the ion-Larmor radius and Ω_L is the ion gyration frequency.

Let us assume the perturbation of all the quantities vary as $\exp\{i\omega t+i(k_x x+k_z z)\}$

where ω is the frequency of harmonic disturbances and k_x and k_z are the wave number in transverse and longitudinal directions to the magnetic field, with $k^2=k_x^2+k_z^2$.

Now eliminating the perturbation in the linearized equations and using Eqs. (1, 3-6) we may write the following algebraic equations for the components of Eq. (2)

$$\left(\sigma^2+k_x^2V^2+k^2c_A^2\right)\delta v_x+\sigma\left(\upsilon_0(k_x^2+2k_z^2)-2\Omega\right)\delta v_y+k_xk_zV^2\,\delta v_z=0\quad(7)$$

$$\sigma\left(-\upsilon_0(k_x^2+2k_z^2)+2\Omega\right)\delta v_x+\left(\sigma^2+c_A^2k_z^2\right)\delta v_y-2\sigma\upsilon_0k_xk_z\,\delta v_z=0\quad(8)$$

$$k_xk_zV^2\delta v_x+2\sigma\upsilon_0k_xk_z\,\delta v_y+\left(\sigma^2+k_z^2V^2\right)\delta v_z=0.\quad(9)$$

where we have used $c_A^2=B^2/4\pi\rho$, $V^2=c_s^2-4\pi G\rho/k^2+\hbar^2k^2/4m_em_i$ and $i\omega=\sigma$.

The general dispersion relation can be obtained from Eqs. (7) – (9) as

$$\sigma^6+\left[k^2V^2+k_z^2c_A^2+k^2c_A^2+4\upsilon_0^2k_x^2k_z^2+\left(\upsilon_0(k_x^2+2k_z^2)-2\Omega\right)^2\right]\sigma^4+\left[k_z^4c_A^2V^2+4\upsilon_0^2k_x^4k_z^2V^2\right.$$
$$+4\upsilon_0^2k_x^2k_z^2k^2c_A^2+k_z^2c_A^2V^2k^2-k_x^2k_z^2c_A^2V^2+k^2k_z^2c_A^4+k_z^2V^2\left(\upsilon_0(k_x^2+2k_z^2)-2\Omega\right)^2\quad(10)$$
$$\left.-2k_xk_zV^2\upsilon_0k_xk_z\left(\upsilon_0(k_x^2+2k_z^2)-2\Omega\right)^2\right]\sigma^2+k^2k_z^4c_A^4V^2=0.$$

Eq. (10) represents the general dispersion relation for combined influence of rotation, FLR corrections, Alfven velocity and quantum corrections on an infinite homogeneous quantum plasma system. If we ignore the effect of quantum corrections, the above dispersion relation is identical with the dispersion relation obtained by Sharma [3]. Further, by ignoring the rotation in the dispersion relation given by Eq. (10) recovers the dispersion relation obtained by Sharma and Chhajalani [5]. Thus, the dispersion relation in the present analysis is modified due to the combined effect of FLR and rotation in quantum plasma.

3. Discussion of the dispersion relation

3.1 Transverse propagation ($k_x = k, k_z = 0$)

When the perturbations are taken in the perpendicular direction to the magnetic field, the general dispersion relation (10) is reduced and gives

$$\sigma^2 + 4\Omega^2 + k^2V^2 + k^2c_A^2 + \upsilon_0^2k^4 - 4\upsilon_0 k^2\Omega = 0. \tag{11}$$

Eq. (11) represents the dispersion relation for the gravitating mode in transverse propagation that is modified by the combined effects of FLR corrections, rotation, quantum corrections and Alfven velocity. The condition of instability from constant term of Eq. (11) is given by

$$4\Omega^2 + k^2\left(c_s^2 - \frac{4\pi G\rho}{k^2} + \frac{\hbar^2k^2}{4m_em_i}\right) + k^2c_A^2 + \upsilon_0^2k^4 - 4\upsilon_0 k^2\Omega < 0. \tag{12}$$

The above condition of instability depends upon the rotation, FLR and quantum corrections. To discuss the effects of rotation and FLR corrections on the growth rate of Jeans instability we plot the curves between normalized growth rates of Jeans instability versus normalized wave number (normalized in terms of Jeans frequency). Thus by introducing the following non-dimensionless quantities,

$$\sigma^* = \frac{\sigma}{(4\pi G\rho)^{1/2}},\ c_A^* = \frac{c_A}{c_s},\ \upsilon_0^* = \frac{\upsilon_0(4\pi G\rho)^{1/2}}{c_s^2}, k^* = \frac{kc_s}{(4\pi G\rho)^{1/2}},\ H = \frac{\hbar\omega_j}{\sqrt{m_em_i c_s^2}},\ \Omega^* = \frac{\Omega}{(4\pi G\rho)^{1/2}}.$$

We can write the dispersion relation (11) in dimensionless form as

$$\sigma^{*2} + 4\Omega^{*2} + \upsilon_0^{*2}k^{*4} + k^{*2} + \frac{H^2k^{*4}}{4} + k^{*2}c_A^{*2} - 4\upsilon_0^*\Omega^*k^{*2} - 1 = 0 \tag{13}$$

Eq. (13) represents the normalized dispersion relation for the perturbation propagating perpendicular to the magnetic field. Out of the two modes one mode is unstable for which the calculations are presented in figures , where the growth rate σ^* has been plotted against the wave number k^* to show the dependence of the growth rate on the different physical parameters. In Fig. 1 the effect of quantum parameter on the growth rate is shown by depicting the curves between normalized growth rates versus normalized wavenumber. We find that the growth rate decreases with increasing the value of quantum parameter. Hence, the presence of quantum corrections stabilizes the system. In fig. 2 we have depicted the growth rate of Jeans instability against wavenumber for different values of magnetic viscosity (υ_0^*) parameter. It is clear from fig. that growth rate of instability decreases with increasing the magnetic viscosity. Thus, we conclude that the presence of FLR corrections and quantum parameter stabilizes the system.

3.2 Longitudinal propagation ($k_x = 0, k_z = k$)

When the perturbations are assumed along the direction of the magnetic field, the general dispersion relation (10) reduces to

$$\left(\sigma^2 + k^2c_s^2 + \frac{\hbar^2k^4}{4m_em_i} - 4\pi G\rho\right)\left[\sigma^4 + 2k^2c_A^2\sigma^2 + 4\upsilon_0^2k^4\sigma^2 + 4\Omega^2\sigma^2 - 8\upsilon_0\Omega k^2\sigma^2 + V_A^4k^4 + k^4c_A^4\right] = 0 \tag{14}$$

The above dispersion relation (14) shows the combined influence of rotation, Alfven velocity, FLR and quantum corrections. This dispersion relation gives two modes. The first factor of Eq. (14) represents the gravitational mode and second factor gives the Alfven mode. The first factor of Eq. (14) can be written as

$$\left(\sigma^2 + k^2c_s^2 + \frac{\hbar^2k^4}{4m_em_i} - 4\pi G\rho\right) = 0. \tag{15}$$

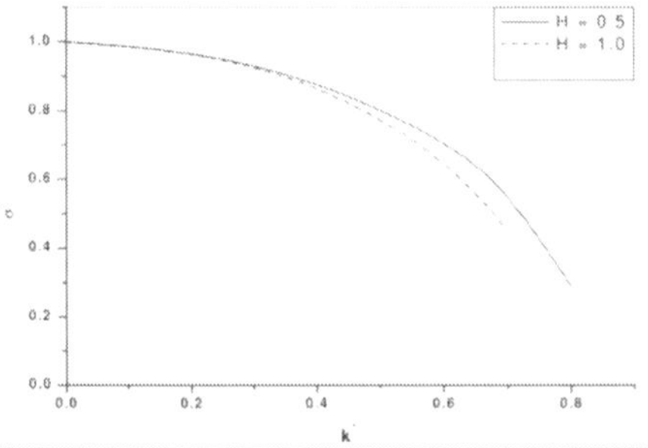

Fig.1. The normalized growth rate of Jeans instability σ^* versus normalized wave number k^* in transverse mode of propagation for different values of the quantum parameter (H). The values of constant parameter are taken as $\upsilon_0^* = 0.5$, $c_A^* = 0.5$, and $\Omega^* = 0.01$

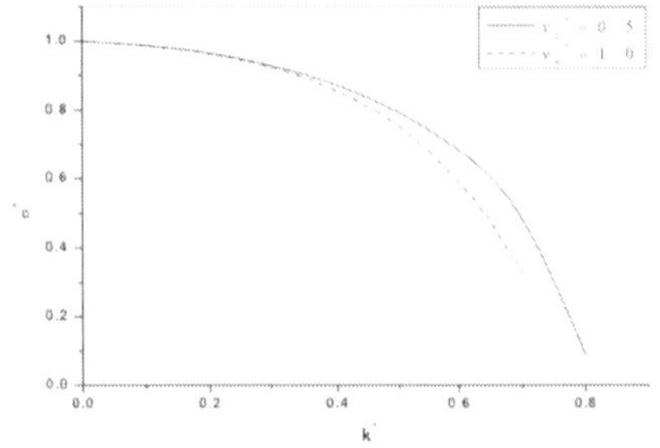

Fig.2. The normalized growth rate of Jeans instability σ^* versus normalized wave number k^* in transverse mode of propagation for different values of the dimensionless magnetic viscosity (υ_0^*). The values of constant parameter are taken as $\Omega^* = 0.01$, $c_A^* = 0.5$, and $H = 1.0$

This is the dispersion relation for gravitating mode modified by quantum corrections. This mode does not depend on rotation, Alfven velocity and FLR corrections. The condition of Jeans instability obtained from the constant term is given by

$$k^2 c_s^2 + \frac{\hbar^2 k^4}{4 m_e m_i} - 4\pi G\rho < 0. \tag{16}$$

The above inequality shows the condition of Jeans instability modified due to quantum corrections. The second factor of Eq. (14) gives

$$\sigma^4 + 2k^2 c_A^2 \sigma^2 + 4\upsilon_0^2 k^4 \sigma^2 + 4\Omega^2 \sigma^2 - 8\upsilon_0 \Omega k^2 \sigma^2 + V_A^4 k^4 + k^4 c_A^4 = 0 \tag{17}$$

This represents the non-gravitational Alfven mode modified due to the rotation and FLR corrections. Thus, it is clear that in transverse propagation the parameters rotation and FLR corrections affect the gravitating mode while in longitudinal propagation these parameters affect the Alfven mode. In the longitudinal mode the gravitational mode is affected only due to the quantum correction term.

4. Conclusions

Thus in the present paper, the combined effects of rotation and FLR corrections on self-gravitational instability of a quantum plasma is investigated. We find that Jeans criterion remains valid and gets modified due to the presence of rotation, quantum and FLR corrections. It is clear that presence of quantum and FLR corrections stabilizes the system in transverse propagation.

Acknowledgement

This work was partially supported by the MPCST Bhopal (M.P.),India.

References

[1] Masood W., Salimullah M., and Shah H. A. 2008, *Phys. Lett. A* **372** 6757.
[2] Wu Z., Ren H., Cao J and Chu P. K. 2010 *Phys. Plasmas* **17** 064503.
[3] Sharma R. C. 1974 *Astrophys. Space Sci.* **29** L1.
[4] Bhatia P. K. and Chhonkar R. P. S. 1985 *Astrophys. Space Sci.* **115** 327.
[5] Sharma P and Chhajlani R. K. 2013 *Phys. Plasmas* **20** 092101.

Hydrodynamic instability of compressible fluid in porous medium

Shraddha Argal[1], Anita Tiwari[1], R P Prajapati [2]and P K Sharma[1]
[1] Barkatullah University Institute of Technology, Barkatullah University, Bhopal- 462026 India.
[2] Department of Physics, Guru Ghasidas Central University, Bilaspur (C.G.), India

E-mail: pks_buit30@yahoo.com, shraddhaargal@gmail.com

Abstract. The hydrodynamic Rayleigh –Taylor instability of two superposed compressible fluids in porous medium has been studied. The dispersion relation is derived for such a medium by using normal mode analysis. The RT instability is discussed for various simplified configuration. The effect of porosity and dynamic viscosity has been analyzed and it is observed that porosity and dynamic viscosity have stabilizing effect on the Rayleigh- Taylor instability of compressible fluids.

1. Introduction

The Rayleigh Taylor instability problems attained considerable interest in the last few years, mainly because of its applications in ICF (inertial confinement fusion), space and astrophysical plasma physics. Rayleigh Taylor instability in the compressible medium has been discussed by many authors. Chandrasekhar [1] has given comprehensive treatise which contains solutions of the classical problem of Rayleigh Taylor instability under varying assumptions of hydromagnetics in the incompressible medium. Vandervroot [2] has discussed Rayleigh- Taylor instability for the compressible medium of the fluid. Verma and Verma [3] have investigated the effect of horizontal magnetic field on Rayleigh Taylor instability of compressible fluids. Shivamoggi [4] used barotropy assumption to study the Rayleigh Taylor instability of compressible fluids. Prajapati and Chhajlani [5] have discussed Rayleigh Taylor instability and Kelvin Helmholtz instability with various parameters in porous medium for incompressible fluids. Kango [6] studied Hydrodynamic and hydromagnetic stability of two superposed incompressible walters' B' viscoelastic fluids in porous medium. In view of the importance of porous and compressible fluids in petroleum engineering, geophysical research and space and astrophysical problems we have discussed the problem of Rayleigh Taylor instability of two superposed compressible fluids in porous medium in the present paper.

2. Linearized equations

We consider that two compressible fluids are separated by a horizontal boundary z = 0 in the porous medium with porosity (ε) and permeability (k_1) in presence of gravitational force **g** (0, 0, -g). The relevant equations of the considered problem are

$$\left[\frac{\rho}{\varepsilon}\frac{\partial}{\partial t}+\frac{\mu}{k_1}\right]\boldsymbol{u} = -\nabla\,\delta p + g\,\delta\rho \tag{1}$$

$$\varepsilon\frac{\partial\delta\rho}{\partial t}+\rho(\nabla.\boldsymbol{u})+(\boldsymbol{u}.\nabla)\rho=0 \tag{2}$$

$$\frac{\partial\delta p}{\partial t}+\frac{(\boldsymbol{u}.\nabla)p}{\varepsilon}=V^2\left[\frac{\partial\delta\rho}{\partial t}+(\boldsymbol{u}.\nabla)\rho\right] \tag{3}$$

Where notations have their usual meaning and $V^2 = p/\rho$ is representing the speed of sound.
To obtain the solution of linearized equations we assume space and time dependent perturbation

$$exp\left(ik_x x + ik_y y + nt\right)$$

Here, n is the growth rate and $k^2 = k_x^2 + k_y^2$, k_x and k_y are the wave numbers along x and y axis.
Eq. (1) to (3) becomes in component form as

$$\left[\frac{\rho}{\varepsilon}n+\frac{\mu}{k_1}\right]u=-ik_x\delta p \tag{4}$$

$$\left[\frac{\rho}{\varepsilon}n+\frac{\mu}{k_1}\right]v=-ik_y\delta p \tag{5}$$

$$\left[\frac{\rho}{\varepsilon}n+\frac{\mu}{k_1}\right]w=-D\delta p-g\delta\rho \tag{6}$$

$$n\delta\rho=-\frac{\rho}{\varepsilon}\left(ik_x u+ik_y v+Dw\right)-\frac{wD\rho}{\varepsilon} \tag{7}$$

$$n\delta p=V^2\left[-\frac{\rho}{\varepsilon}\left(ik_x u+ik_y v+Dw\right)\right]+\frac{gw\rho}{\varepsilon} \tag{8}$$

Substituting the values of u and v from the equation (4) and equation (5) in equations (7) and (8) we get the values of δρ and δp then, using equation (6) we get the following equation

$$\frac{n^2}{\varepsilon}D\left[\rho sQ^{-1}Dw\right]+\frac{gw}{\varepsilon}D\rho-gn^2D\left(\frac{s\rho wQ^{-1}}{V^2\varepsilon}\right)$$
$$-n^2\rho sw+\frac{g^2k^2\rho Q^{-1}w}{V^2\varepsilon^2}+\frac{gn^2\rho sQ^{-1}}{V^2\varepsilon}Dw=0 \tag{9}$$

Here $Q=\frac{n^2s}{V^2}+\frac{k^2}{\varepsilon}$ and $s=\frac{1}{\varepsilon}+\frac{\mu}{k_1n\rho}$

Equation (9) is a general deferential equation for the RT instability of compressible fluids.

3. Dispersion relation

Now we consider that a horizontal boundary separates two uniform fluids at the interface such that the densities of fluids and speed of sound in fluids are constant at the interface.

$$\left(D^2-q^2\right)w=0 \tag{10}$$

$$\text{Where } q^2=\varepsilon Q-\frac{g^2k^2}{\varepsilon sn^2V^2} \tag{11}$$

The general solution of above equation is of the form

$$w = Ae^{q_1 z}+Ae^{-q_2 z}$$

Here A is an arbitrary constant. Applying following boundary conditions on the system of two superposed compressible fluids
(i) The velocity (w) must vanish when z tends to infinity and continuous across the interface.

$w_1 = w_2 = w_0$ at $z = 0$

(ii) The total pressure is continuous at the interface.
Now we integrate equation (9) with respect to z across the interface to satisfy second boundary condition of the system and obtain as follows

$$\frac{n^2}{\varepsilon}\Delta_s\left[\rho sQ^{-1}Dw\right]+\frac{gw}{\varepsilon}\Delta_s\rho - gn^2\Delta_s\left(\frac{s\rho wQ^{-1}}{V^2\varepsilon}\right)=0 \qquad (12)$$

We obtained a characteristic equation given below.

$$\begin{aligned} &n^2\rho_2 s_2\left(\frac{k^2}{\varepsilon}+\frac{s_2 n^2}{V_2^2}\right)^{-1}\left(k^2+\frac{\varepsilon s_2 n^2}{V_2^2}-\frac{g^2k^2}{V_2^2\varepsilon s_2 n^2}\right)^{1/2} \\ &+n^2\rho_1 s_1\left(\frac{k^2}{\varepsilon}+\frac{s_1 n^2}{V_1^2}\right)^{-1}\left(k^2+\frac{\varepsilon s_1 n^2}{V_1^2}-\frac{g^2k^2}{V_1^2\varepsilon s_1 n^2}\right)^{1/2} \\ &=\frac{gk^2\rho_2}{\varepsilon}\left(\frac{k^2}{\varepsilon}+\frac{s_2 n^2}{V_2^2}\right)^{-1}-\frac{gk^2\rho_1}{\varepsilon}\left(\frac{k^2}{\varepsilon}+\frac{s_1 n^2}{V_1^2}\right)^{-1} \end{aligned} \qquad (13)$$

Here $s_1 = \frac{1}{\varepsilon}+\frac{\mu}{k_1 n\rho_1}$ $\quad s_2 = \frac{1}{\varepsilon}+\frac{\mu}{k_1 n\rho_2}$

Subscript 1 and 2 represent lower fluid and upper fluid respectively. The dispersion relation (13) is matched with Vandervroot [2] and Verma and Verma [3] and it is found that the dispersion relation is modified due to porosity and permeability of porous medium.

4. Discussion

The dispersion relation (13) is cumbersome to discuss, therefore, the equation is solved by making the assumption that the superposed fluids are highly compressible. This type of assumption has been considered by El- Sayed [7]. Using binomial expansion on dispersion relation (13) we obtain the dispersion relation as.

$$\begin{aligned} &n^6+n^5\left[\frac{4\varepsilon v'}{k_1}+\frac{\varepsilon v'}{k_1}\left(\beta'+\frac{1}{\beta'}\right)\right]+n^4\left[4V'^2k^2+V'^2k^2\left(\beta'+\frac{1}{\beta'}\right)+\frac{3\varepsilon^2v'^2}{k_1^2}\left(\beta'+\frac{1}{\beta'}+2\right)\right] \\ &+n^3\left[\frac{6\varepsilon v'V'^2k^2}{k_1}\left(\beta'+\frac{1}{\beta'}+2\right)+\frac{2\varepsilon^3v'^3}{k_1^3}\left(\beta'+\frac{1}{\beta'}+2\right)\right] \\ &+n^2\left[\frac{6\varepsilon^2v'^2V'^2k^2}{k_1^2}\left(\beta'+\frac{1}{\beta'}+2\right)+2V'^4k^4\left(\beta'+\frac{1}{\beta'}+2\right)-g^2k^2\right] \\ &+n\left[-\frac{g^2k^2\varepsilon v'}{k_1}\left(\beta'+\frac{1}{\beta'}\right)+\frac{2g\varepsilon v'V'^2k^3}{k_1}\left(\beta'-\frac{1}{\beta'}\right)+\frac{4\varepsilon v'V'^4k^4}{k_1}\left(\beta'+\frac{1}{\beta'}+2\right)\right] \\ &+\left[2gV'^4k^5\left(\beta'-\frac{1}{\beta'}\right)-g^2V'^2k^4\left(\beta'+\frac{1}{\beta'}\right)\right]=0 \end{aligned} \qquad (14)$$

Here $v'=\frac{\mu}{(\rho_1+\rho_2)}, V'^2=\frac{p}{(\rho_1+\rho_2)}, \beta'=\frac{\rho_1}{\rho_2}$

To carry out the effect of porosity and dynamic viscosity on the growth rate of RT instability of compressible fluids in porous medium the dispersion relation is found in dimensionless form by using following substitution in equation (14)

$$n^*=\frac{n}{\sqrt{gk}},\ V'^*=\frac{V'}{gk},\ k^*=k\sqrt{gk}, v'^*=\frac{v'}{\sqrt{gk}}$$

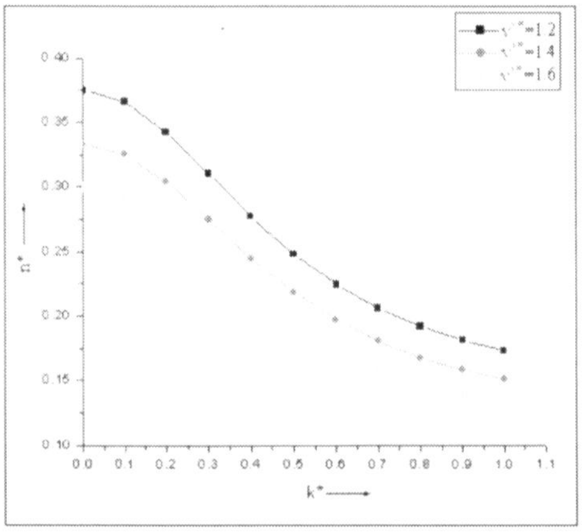

Figure 1. The growth rate (n*) verses wave number (k*) at different values of dynamic viscosity (v'*).

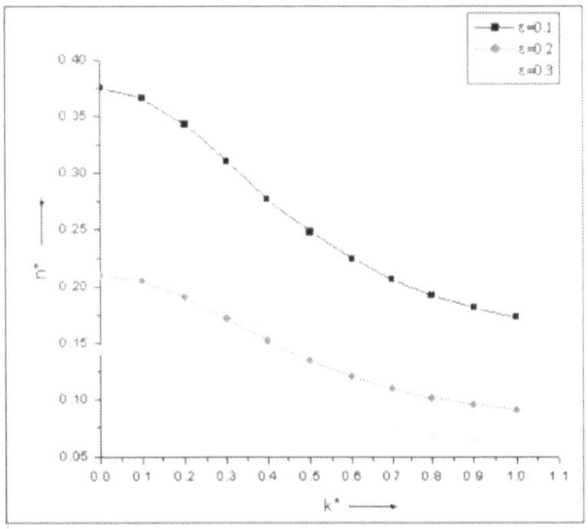

Figure 2. The growth rate (n*) verses wave number (k*) at different values of medium porosity (ε).

$$
\begin{aligned}
& n^{*6} + n^{*5}\left[\frac{4\varepsilon v'^{*}}{k_1} + \frac{\varepsilon v'^{*}}{k_1}\left(\beta' + \frac{1}{\beta'}\right)\right] + n^{*4}\left[4V'^{*2}k^{*2} + V'^{*2}k^{*2}\left(\beta' + \frac{1}{\beta'}\right) + \frac{3\varepsilon^2 v'^{*2}}{k_1^2}\left(\beta' + \frac{1}{\beta'} + 2\right)\right] \\
& + n^{*3}\left[\frac{6\varepsilon v'^{*} V'^{*2} k^{*2}}{k_1}\left(\beta' + \frac{1}{\beta'} + 2\right) + \frac{2\varepsilon^3 v'^{*3}}{k_1^3}\left(\beta' + \frac{1}{\beta'} + 2\right)\right] \\
& + n^{*2}\left[\frac{6\varepsilon^2 v'^{*2} V'^{*2} k^{*2}}{k_1^2}\left(\beta' + \frac{1}{\beta'} + 2\right) + 2V'^{*4}k^{*4}\left(\beta' + \frac{1}{\beta'} + 2\right) - 1\right] \qquad (15) \\
& + n^{*}\left[-\frac{\varepsilon v'^{*}}{k_1}\left(\beta' + \frac{1}{\beta'}\right) + \frac{2\varepsilon v'^{*} V'^{*2} k^{*2}}{k_1}\left(\beta' - \frac{1}{\beta'}\right) + \frac{4\varepsilon v'^{*} V'^{*4} k^{*4}}{k_1}\left(\beta' + \frac{1}{\beta'} + 2\right)\right] \\
& + \left[2V'^{*4}k^{*4}\left(\beta' - \frac{1}{\beta'}\right) - V'^{*2}k^{*2}\left(\beta' + \frac{1}{\beta'}\right)\right] = 0
\end{aligned}
$$

We have solved dimensionless dispersion assuming arbitrary values of parameters and plotted graphs between growth rate (n*) and wave number (k*) Rayleigh Taylor instability in variation of porosity and dynamic viscosity

The behavior of dynamic viscosity is observed through the figure (1) which has been plotted by taking $k_1 = 0.1$, $\varepsilon = 0.1$, $V^* = 1.4$, $\beta'_1 = 0.5$ and $v'^* = 0.1$, 0.2, and 0.3 and it has been found that the growth rate decreases on increasing the value of dynamic viscosity that indicate that the dynamic viscosity has stabilizing influence on the growth rate of Rayleigh Taylor instability.

From Figure 2 we have seen the effect of porosity on the unstable mode of Rayleigh Taylor instability of compressible fluids. The figure is plotted by putting $k_1 = 0.1$, $v'^* = 1.2$, $k_1 = 0.1$, $V^* = 1.4$, $\beta' = 0.5$ and $\varepsilon = 0.1$, 0.2, and 0.3 in dimensionless equation. It has been noticed that porosity has stabilizing effect.

References

[1] Chandrasekhar S 1961 *Hydrodynamic and Hydromagnetic stability* (Oxford: Clarendon Press)
[2] Vandervroot P O 1961 *Astrophys. J.* **134** 699
[3] Verma Y K and Verma Pratibha 1963 *Proc. Nat. Inst. Sci. India* **A 29** 309
[4] Shivamoggi B K 2012 *Z. Angew. Math. and Phys.* **63** 521
[5] Prajapati R P and Chhajlani R K 2010 *J. Porous Media* **13(9)** 765
[6] Kango S K 2011 *Adv. Theor. Appl. Mech.* **4** 113
[7] El-Sayed M F 2003 *Eur. Phys. J. D* **23** 391

Propagation of magnetosonic wave in ion implanted semiconductor: Effects of nano-sized grains

Giriraj Sharma[1], S Ghosh [2] and R C Dad [3]
[1] SRJ Government Girls' College, Neemuch (M P) 458 441, India
[2] School of Studies in Physics, Vikram University, Ujjain (M P) 452001, India
[3] Government Post Graduate College, Mandsaur (M P) 458001, India

E-mail: grsharma@gmail.com

Abstract. We have derived a linear dispersion relation for magnetosonic wave (MSW) in compensated semiconductor plasma, like Ge, embedded with nano-sized grains (NSGs) of ions. These NSGs are bombarded by electrons and holes in the plasma medium and usually acquire net negative charge on account of higher mobility of electrons as compared to that of the holes [1]. The process of charging of NSGs depletes the electron density and creates a charge imbalance, which modifies the propagation characteristics of MSW even if NSGs do not participate in wave perturbation.

1. Introduction

The MSW is a low-frequency electromagnetic wave that propagates perpendicular to magnetic field at the frequency well below the ion-cyclotron frequency. Herlofsen [2] was the first to discover and study MSW in strongly magnetized plasma.

In semiconductor, if ions are implanted at high fluency, they may agglomerate to form nano-sized grains (NSGs). In recent years, there has been growing interest on the study of linear and nonlinear optical properties of NSGs of implanted ions in solids [4, 5-8]. These NSGs can be synthesized within the host material by ion implantation followed by an appropriate annealing process [9]. The shape and size of the NSGs is determined by various factors such as ion fluence, annealing treatment, etc. The metallic NSGs buried in glasses are promising bistable materials because they exhibit ultrafast nonlinear response at plasmon frequency [8], which make them potentially attractive candidates for device applications in the various fields including photonics, nonlinear optics, optoelectronics, etc [5, 6, 9, 10].

Like dust in dusty plasmas, the NSGs in semiconductors might acquire a net negative charge that would change the balance between electron and hole densities in an otherwise compensated semiconductor. This, in turn, gives rise to several novel phenomena. Salimullah et. Al. [11], first time, predicted the possible lattice formation in piezoelectric semiconductors. Later, in the same medium, the role of electron-phonon coupling in the formation of plasma crystal was also reported [12].

A survey of the available literature reveals that the study on properties of NSGs in solids has been hotly pursued in the recent years; however, their effects on the properties of host material have not been explored. We firmly believe that by interacting with the carriers and by creating charge imbalance of the carrier concentration, the NSGs would modify properties of host material. Since a study on wave propagation through host material can provide tremendous insight regarding its properties, comprehensive efforts in this direction are needed.

Motivated by this in the present paper, we have focused our attention on the study of propagation properties of MSW in the ion-implanted-group-IV semiconductor. The choice for the MSW stems from the fact that it is a fundamental electromagnetic mode observed in laboratory and space plasmas and can render an effective tool for material diagnosis [12, 13].

An MSW propagates perpendicular to the magnetic field in gaseous plasma (without any dust contamination) and obey the dispersion relation

$$\frac{\omega^2}{k^2} = c^2 \frac{v_A^2 + v_s^2}{v_A^2 + c^2}, \tag{1}$$

where v_s is the speed of acoustic wave, v_A is the speed of Alfven wave and c is the speed of light.

The paper is organized in the following manner. In section 2, we outline the basic equations describing MSW propagation. In section 3, we derive a dispersion relation for the MSW in the implanted semiconductor plasma. In section 4, we present analytical investigation of the dispersion relation for the MSW followed by discussions.

2. The basic equations

We consider compensated semiconductor plasma i.e. $n_{e0} = n_{h0} = n_0$ embedded with NSGs. The electrons and holes from all directions bombard on the NSGs and get stick onto them. However, owing to high mobility of electrons, greater number of electrons bombards the NSGs as compared to holes in given time interval. If we consider equal probability of sticking of electrons and holes onto the NSGs, the later acquire net negative charge and consequently, the electron density depletes. Thus, the charge neutrality condition reads

$$n_{h0} = n_0 = \bar{n}_{e0} + zn_n, \tag{2}$$

with $\bar{n}_{e0}$ being the depleted electron density, z the number of electrons that reside onto an NSG and n_n their density in the medium. The charge imbalance so created may be measured in terms of parameter

$$\delta = (\bar{n}_{e0} / n_{0h}) = (\bar{n}_{e0} / n_0). \tag{3}$$

Let the semiconductor be immersed in a strong magnetic field $\mathbf{B}_0 = \mathbf{z}\mathrm{B}_0$ and MSW propagates across the external magnetic field, so let $\mathbf{k} = \mathbf{y}\mathrm{k}$. For MSW, the term $\mathbf{E} \times \mathbf{B_0}$ is along $\mathbf{k}$ so that plasma is compressed and released during oscillations in the course of wave propagation. Therefore, we keep pressure gradient term via the density gradient term in our analysis. The propagation of MSW in the medium may be described by Maxwell's equations ignoring the displacement current and the dynamics of electrons and holes in semiconductor plasma with immobile NSGs may be described by the respective continuity and momentum equations for the species $\alpha(=e, h)$ as

$$\frac{\partial n_\alpha}{\partial t} + \nabla.(n_\alpha \mathbf{v}_\alpha) = 0, \tag{4}$$

and $$\frac{\partial \mathbf{v}_\alpha}{\partial t} + (\mathbf{v}_\alpha.\nabla)\mathbf{v}_\alpha = \frac{q}{m_\alpha}(\mathbf{E} + \mathbf{v}_\alpha \times \mathbf{B}_0) - \gamma_\alpha K T_\alpha \nabla n_\alpha, \tag{5}$$

here, m_α, γ_α, T_α and $\mathbf{v}_\alpha$ are the mass, specific heat ratio, temperature and the first-order velocity of the species.

3. Dispersion relation

A dispersion relation for MSW may be deduced by linearizing Eqs. (4-5), which leads us to expressions for x-components of first-order perturbed velocities of electrons and holes:

$$v_{ex}\left(1 - \frac{\omega_c^2/\omega^2}{1-G}\right) = \frac{-ie}{m_e\omega}E_x, \tag{6a}$$

and $$v_{hx}\left(1-\frac{\Omega_c^2/\omega^2}{1-S}\right)=\frac{ie}{m_h\omega}E_x. \tag{6b}$$

here, $\omega_c=\frac{-eB_0}{m_e}$ and $\Omega_c=\frac{eB_0}{m_h}$ are the respective electron and hole cyclotron frequencies, with $G=\frac{k^2}{\omega^2}.\frac{KT_e}{m_e}$ and $S=\frac{k^2}{\omega^2}.\frac{KT_h}{m_h}$ being the dimensionless parameters.

Now, using expressions for v_{ex} and v_{hx} in Maxwell's equations, we get

$$\omega^2-c^2k^2=\omega_p^2\left[\frac{1-G}{1-G-\left(\omega_c^2/\omega^2\right)}\right]+\Omega_p^2\left[\frac{1-S}{1-S-\left(\Omega_c^2/\omega^2\right)}\right], \tag{7}$$

With $\omega_p=\left(\frac{n_0e^2}{m_e\varepsilon_0}\right)^{0.5}$ and $\Omega_p=\left(\frac{n_0e^2}{m_h\varepsilon_0}\right)^{0.5}$, being the electron and hole plasma frequencies.

Under the assumption $\omega<\omega_c,\Omega_c$, and neglecting the factors $(1-G)$ and $(1-S)$ relative to ω_c^2/ω^2 and Ω_c^2/ω^2, respectively. With these assumptions, Eq. (10) reduces to dispersion relation for MSW in compensated semiconductor plasma embedded with immobile NSGs as

$$\frac{\omega^2}{k^2}=c^2\left[\frac{v_A^2(1+m\delta)+v_s^2}{\left(v_A^2+c^2\right)(1+m\delta)}\right], \tag{8}$$

where $m=(m_e/m_h)$, the electron to hole mass ratio, and $v_A=\frac{B_0}{\sqrt{\mu_0n_0m_h(1+m\delta)}}$, the Alfven velocity and v_s, the sound speed. We notice from above equation that the presence of charged NSGs in compensated semiconductor plasma alters the Alfven velocity and modifies the dispersion relation for MSW through charge imbalance parameter δ in the medium. It is evident that, with rise in temperature, the acoustic velocity v_s increases and so does the influences of charged NSGs on the dispersion characteristics of the MSW. Also, by considering positive ions in place of holes and dust in place of NSGs i.e. $m=(m_e/m_i)\rightarrow 0$ in above formulation, we can recover dispersion relation (Eq. (1)) form Eq. (8).

4. Results and discussion

In this section, by employing the dispersion relation Eq. (8), we present analytical investigation on the effects of the NSGs on the propagation characteristics of MSW. The set of parameters used are given in [14]. The results are displayed in the form of graphs in Figures 1 and 2.

Figure 1 illustrates dispersion characteristics of the MSW for different values of charge imbalance parameter δ (=1, 0.8 and 0.6) in group IV semiconductor. The graph reveals that MSW propagates without any dispersion, irrespective of the amount of charge residing on the NSGs. As the charge on the NSGs increases, the electron concentration falls. This reduces inertia of the plasma fluid, which results in slightly increase in phase velocity of MSW. The effect of charged NSGs is negligible for small values of k but increases as k assumes higher values.

Figure 2 displays variation of phase velocity of MSW as a function of hole density in semiconductor plasma for magnetic field B_0=1, 2 and 3 Tesla. An increase in hole density of plasma increases inertia of the plasma fluid and thus, the phase velocity decreases. It is also evident that with increase in magnetic field, the strength of compression and release of plasma increases during course of oscillations and so does the consequent phase velocity of MSW. Therefore, the MSW travels at higher phase velocity in a strongly magnetized plasma with lower hole concentration.

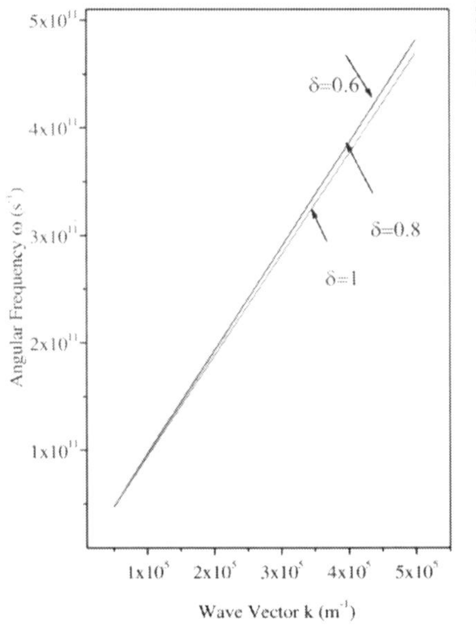

Figure 1. Dispersion characteristics of magnetosonic wave for charge imbalance parameter δ=1, 0.8 and 0.6.

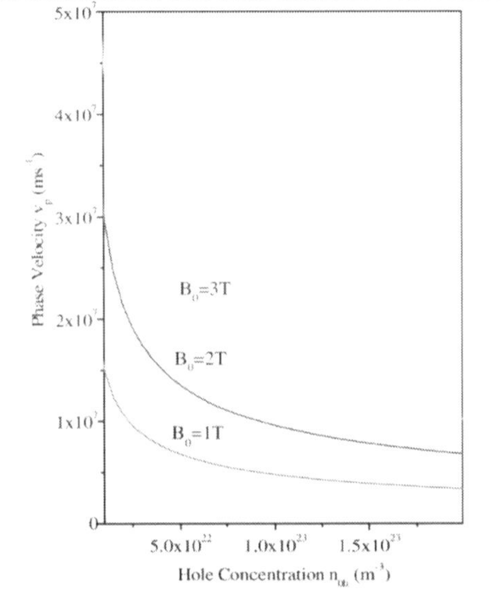

Figure 2. Variation of phase velocity v_p of magnetosonic wave with hole concentration n_{0h} in semiconductor plasma for magnetic field B0=1, 2 and 3T.

References

[1] Shukla P K and Mamun A A 2002 *Introduction to Dusty Plasma Physics* (Bristol: IOP Publishing)
[2] Herlofsen N 1950 *Nature* **165** 1020
[3] Goree J 1994 *Plasma Sources Sci. Technol.* **3** 400
[4] Fortov V E, Ivlev AV, Khrapak S A, Khrapak AG and Morfill G E 2005 *Phys. Reports* **421** 1
[5] White C W, Budal J D, Zhu J G and Withrow S P 1996 *Appl. Phys. Lett.* **68** 2389
[6] Haglund Jr R F, Yang L, Magrunder III R H, Witting J E, Becker K and Zuhr R A 1993 *Opt. Lett.* **18** 373
[7] Okubo N, Umeda N, Takeda Y and Kishimoto N 2003 *Nucl. Instrum. Methods* **B206** 610
[8] Ila D, Wu Z, Smith C C, Poker D B, Hensley D K, Klatt C and Kalbitzer S 1997 *Nucl. Instrum. Methods* B **127/128** 570
[9] Sze S M 2002 *Semiconductor Devices: Physics and Technology* (Wiley)
[10] Cheang-Wong J C, Oliver A, Roiz J, Hernandez J M, Rodrignez-Fernandez L, Morales J G and Crespo-Sosa A 2001 *Nucl. Instrum. Methods* B **175–177** 490
[11] Salimullah M, Ghosh S and Amin M R 2000 *Pramana* **54** 785
[12] Salimullah M, Shukla P K, Ghosh S K, Nitta H and Hayashi Y 2003 *J. Phys. D.: Appl. Phys.* **36** 958
[13] Schunk R and Nagy A 2009 *Ionospheres: Physics, Plasma Physics and Chemistry* (London: Cambridge University Press)
[14] Salimullah M, Khurshed Alam S M and Alam M N 1989 *Phys. Rev.* B **39** 3771

Trapping sub-micron Size Particles in Holographic Optical Tweezers

G. Dwivedi,[1] **A. Gupta,**[1,2] **M. Shukla,**[1,3] **S. Kanaujia,**[1] **S. Yede**[1,4] **and J. T. Andrews**[1,*]

[1] Applied Photonics Laboratory, Department of Applied Physics,
Shri G. S. Institute of Technology & Science, Indore - 452003 MP India.
[2] Dept. Elex. & Comm. Engg., RKDF College of Engg., Bhopal - 462027 MP India.
[3] Bioceramic & Coating Division, CGCRI, Kolkatta - 700032 WB India.
[4] Dept. Elex. & Comm. Engg., S. P. College of Engg. & Tech., Balaghat - 481001 MP India.

E-mail: *Corresponding author: jtandrews@sgsits.ac.in

Abstract. Trapping of sub-micron size particles is of interest to the biological community as well as to nanoelectronic research and industry. We have employed spatially modified Gaussian beam to generate narrow optical traps within diffraction limitation. A spatial light modulator is addressed with the spatial frequencies of the required optical traps. The inverse Fourier transform is obtained at the trap plane of the optical tweezers. We have demonstrated the trapping of sub-micron particles in multiples traps, patterned numerically which is addressed to a spatial light modulator. The trap is vary stable and the particles are trapped for more than 120 seconds.

1. Introduction

Optical tweezers are tools that use optical pressure in trapping micro objects including living cells and micro-organisms, and also in directionally rotating artificial micro-objects fabricated by micro-machining. Given their non-contact and non-invasive nature, optical tweezers are useful particularly in biological processes. The field of optical tweezers has enjoyed a wide range of applications since its invention in the early 1970s. By using light to trap microscopic objects non-invasively, optical tweezers provide a unique tool for ultrafine positioning, measurement, and control. Particle positioning and detection capabilities are on a spatial scale of micrometers down to few angstroms [1, 2].

The emerging applications of laser based optical traps are quite diverse and extensive, ranging from atomic physics to the medical sciences. As a result, optical tweezers have been a focal point for interdisciplinary science. Trapping apparatus ranges from simple, lens-based traps to complex instrumentation integrating multiple optical technologies [3]. A variety of noble techniques have been developed for rapid position detection, trap stiffness determination, and applying controlled, calibrated forces. Instrument advances, such as the use of multiple laser beams, computerized automation of laser beams and sample positioning, and optical tweezers used in combination with other methodologies, such as fluorescence spectroscopy, micro-pipettes, and optical micro-beams, have all helped to make optical tweezers an extremely versatile tool. Owing to their exquisitely controllable force exerting properties, optical tweezers are useful for a variety of nano-mechanical measurements, particularly those with biological applications.

Objects such as bio-polymers e.g., micro-tubules, DNA molecules, lipid membranes, intact or fractionated cells, and single biological macromolecules have all been studied successfully with optical tweezers [4].

In an optical tweezers, a highly focused laser beam creates a large gradient in an optical field, resulting in stable particle trapping at the beam focus. Since the trapping occurs only at the focal point, true 3 - D trapping can occur. Particles in scale from cells (μm) to individual molecules (nm) are routinely studied with this technique. However, due to the large amounts of optical power necessary to create traps of adequate stiffness, large optical intensities are necessary ($\cong 10^6$ W/cm^2). This makes it difficult to create and control multiple traps on the same device hence multiple high powered lasers and/or holography techniques must be employed.

2. Theory

For dielectric particles larger than the wavelength we use simple ray optics to study radiation pressure forces exerted on the particle from scattering of light momentum. The early models for trapping were either all optical two beam traps or single beam levitation traps which required gravity or electrostatic force for their stability.

For particles in the Rayleigh regime, where the size (a) is much less than the wavelength (λ), the particle acts as a dipole. The force acting on dipole divides itself into two components: one is the scattering force that acts in the direction of incident light beam and other is the gradient force component pointing towards the maximum intensity region of light. The balance between the two forces results in trapping of the particle. Optical tweezers was originally designed to work in Rayleigh regime. Conceptually and practically single beam laser traps are one of the simplest lasers trapping model. Its stability in Rayleigh regime is the result of dominance of gradient force that pulls the particle towards the focus and maximum intensity region, over the scattering force that tries to push the particle away from the focus in the direction of incident light beam.

For Rayleigh particle, the scattering force component is given as [5]

$$F_{sca} = \frac{I_0}{c}\frac{128\pi^5 r^6}{3\lambda^4}\frac{m^2-1}{m^2+2}n_b \tag{1}$$

with r is the size of the particle, m is the ratio of refractive indies of medium and particle. The gradient force is given as

$$F_{grad} = -\frac{n_b^3 r^3}{2}\frac{m^2-1}{m^2-2}\nabla|E|^2 \tag{2}$$

The figure shows the trapping forces acting on the particle. The particle is placed at different positions from the focus of incident laser beam. Each time the resultant force acts on the particle such that the particle is displaced towards the focus of the laser beam and when the forces are balanced the particle gets trapped inside the trap volume.

3. Holographic Optical Tweezers

Computer-generated holograms can transform a single laser beam into number of independent optical traps, each with different spatial characteristics, arrayed in arbitrary two dimensional configurations. Holographic optical tweezers use a computer designed diffractive optical element (DOE) to split a single collimated laser beam into several separate beams, each of which is focused into an optical tweezers by a strongly converging lens. Originally demonstrated with micro-fabricated DOEs, holographic optical tweezers have since been implemented with computer addressed liquid crystal transmissive spatial light modulators (SLM). Projecting a sequence of computer designed holograms reconfigures the resulting pattern of traps.

The gradient force same as exploited in single beam optical tweezers also worked in holographic optical tweezers. The dielectric particle, approaching towards the focus of the beam, is pulled into the region of maximum intensity by gradient force. The scattering force competes with this gradient force and tries to displace the trapped particle along the beam axis. Hence, optical tweezers are usually designed around microscope objective lenses whose large numerical apertures and minimal aberrations optimize axial intensity gradients. An optical trap can be placed anywhere within the objective lens' focal volume by appropriately selecting the input beam's propagation direction and degree of collimation [8, 7].

Multiple beams simultaneously entering the lens' input each form optical traps in the focal volume, each at a location determined by its degree of collimation and angle of incidence. This is the principle behind holographic optical tweezers. The figure shows the custom designed optical traps demonstrated using SLM. The first column is the design; middle column shows the Fourier Transform of the objects in first column. The last column shows the trap positions generated using the SLM at trap plane. The particles are trapped, manipulated in three dimensions using the setup.

4. Experiment and Results

The experiment setup consists of three parts, viz., (i) Generating of FFT of desired spatial profile numerically (ii) design of single beam optical tweezers and (iii) guiding the spatially modified beam profile through the beamline without violating Helmholtz equation.

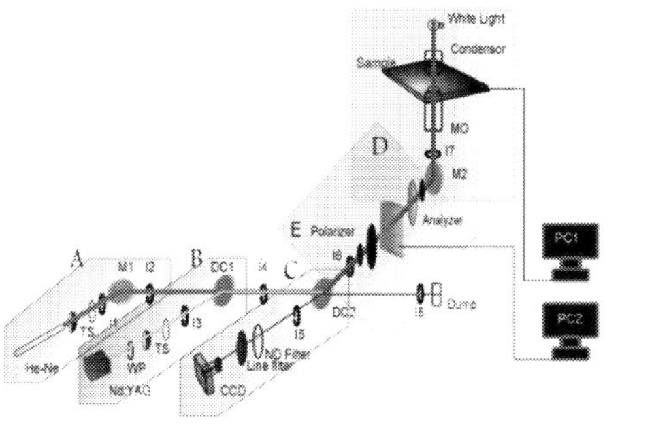

Figure 1. Schematic of experimental setup of Optical Tweezers used to trap particles in a manipulated optical trap.

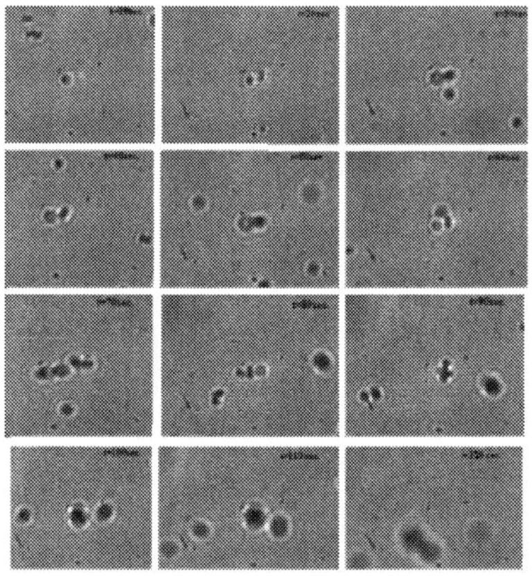

Figure 2. Snaps of trapped particle taken every 10s using setup shown in Fig. 1.

The spatial profile of the desired geometry are designed using appropriate software with 2 bit encoding (B/W) having image size of 256×256 pixels. The image is Fourier transformed using software code written using LabVIEW. The phase information of FFT are distributed spatially. The amplitude part of the FFT are ignored and phase only is considered. The phase is unwrapped suitable to remove any overflow of phase information [6].

The setup of single beam optical Tweezers is shown in Fig. 1. The setup has five ports represented by A-E. The ports A and B guides a He-Ne laser and a Nd:YAG laser to the optical trap. The port C is a imaging port having a 4 mega pixel CCD (Apogee). The port D is the optical tweezers, where two lasers and a white light source illuminate the particle while the backscattered light detected at port C. The port E has the spatial light modification using a setup of inverted telescopes and a SLM. In Fig. 2, snaps taken at an interval of 10 s are displayed.

A particle of sub-micron size (160nm) is shown in the pictures. The trap is found to be stable for more than two minutes times. However, thermodynamic changes and coagulation of particles makes it difficult to keep the trap stable for longer time durations.

In Figs. 3 and 4, we exhibit the manipulation of submicron sized particles in a pre-decided patterned manner. The The First column shows the 2-bit spatial profile of the pattern to be generated. The SLM is address with FFT of information obtained from data in the first column, the FFT are shown in second column. The third column shows the reconstructed profiles at the trap plane.

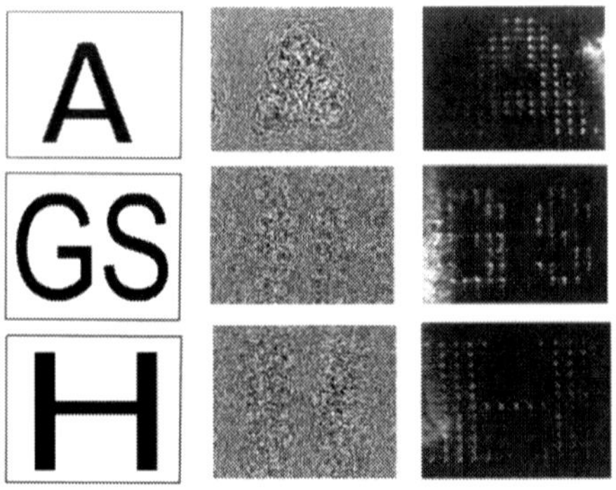

Figure 3. The original image, FFT and reconstructed profile are shown in first, second and third columns, respectively.

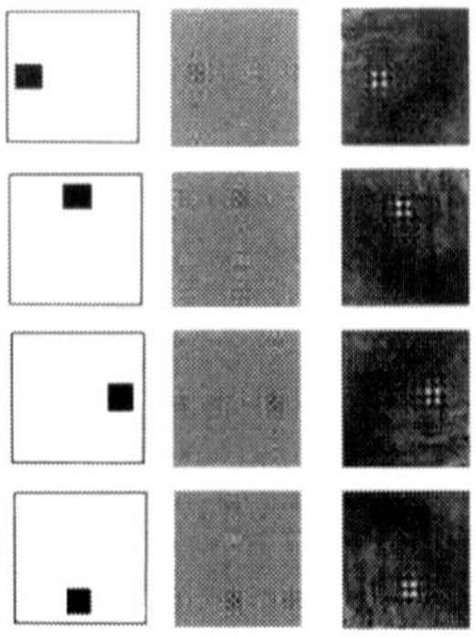

Figure 4. Same as Fig. 3, but the particles are manipulated in a circular path.

The demonstration of optical traps in a predefined pattern promises the utilization of the present setup for various applications. Addition of suitable spectrograph to the system makes the possibility of a micro-photoluminescence studies at single atom level (single electron devices), appropriate manipulation of the beam positions may be used to maneuver DNA and sequencing studies.

5. Conclusions

A single beam optical tweezers is modified to have multiple traps. The spatial profile of the TEM_{00} mode of the laser beam is spatially modified with the FFT of the required pattern. Particles are trapped at the pre-decided positions. Polystyrene spheres of sizes $1\mu m$ and 160nm have been successfully trapped using current Optical Tweezers setup. The trapped particles are manipulated in 3-Dimensional positions using SLM.

[1] A. Ashkin, J. M. Dziedzic, J. E. Bjorkholm and S. Chu, "Observation of a single beam gradient force optical trap for dielectric particles", Opt. Lett., 11, 288-290, (1986).
[2] A. Ashkin and J. M. Dziedzic, "Optical trapping and manipulation of viruses and bacteria", Science, 235, 1517-1520, (1987).
[3] O. M. Marag, P. H. Jones, P. G. Gucciardi, G. Volpe and A. C. Ferrari. "Optical trapping and manipulation of nanostructures", Nature Nanotechnology 8 807-819 (2013).
[4] I. Heller , T.P. Hoekstra , G.A. King , E.J.G. Peterman and G.J.L. Wuite, "Optical Tweezers Analysis of DNAProtein Complexes", Chem. Rev., cr4003006 (2014).
[5] K. C. Neuman and S. M. Block, "Optical trapping", Rev. Sci. Instrum., 75, (2004).
[6] E. R. Dufresne, G. C. Spalding, M. T. Dearing, S. A. Sheets and D. G. Grier, "Computer generated holographic optical tweezers arrays", Rev. Sci. Instrum., 72, 1810-1816, (2001).
[7] Curtis J E, Koss B A and Grier D G, "Dynamic holographic optical tweezers", Opt. Commun., 207 , 69-175, (2002).
[8] L. Novotny and B. Hecht, Principles of Nano Optics (Cambridge Press, Cambridge 2012) pp. 378-418.
[9] B.E.A. Saleh and M.C. Teich, "Fundamentals of photonics", 2nd Edition, Wiley Inter science, (2007).

Study on wavelength of maximum absorbance for phenyl-thiourea derivatives: A topological and non-conventional physicochemical approach

Suprajnya Thakur[a], Ashutosh.Mishra[b], Mamta Thakur[c], Abhilash Thakur[d]
[a] Department of Physics, Shri Vaishnav Institute of Technology and Science, Indore
[b] School of Physics, Devi Ahilya Vishvvidyalaya, Indore
[c] Department of Chemistry, Softvision College, Indore
[d] Department of Applied Sciences, NITTTR, Bhopal

E-mail: suprajnya@gmail.com

Abstract: - In present study efforts have been made to analyze the role of different structural/ topological and non-conventional physicochemical features on the X-ray absorption property wavelength of maximum absorption $\lambda_{m.}$ Efforts are also made to compare the magnitude of various parameters for optimization of the features mainly responsible to characterize the wavelength of maximum absorbance λ_m in X-ray absorption. For the purpose multiple linear regression method is used and on the basis of regression and correlation value suitable model have been developed.

1. Introduction

X-ray absorption spectroscopy is a powerful tool to visualize the molicular arrengment in metal complexes, inorganic compounds, proteins and enzyme systems. XANES parameter λ_m provides the vital information about the overlap between metal and the ligand orbitals. The continuous development of structural descriptors and statistical equations transformed Quantitative Structure-Activity Relationships (QSAR) and Quantitative Structure- Property Relationships (QSPR) into commanding and widely used models for the prediction of physical, chemical, and biological properties [1-6]. The main hypothesis in the QSAR approach is that all properties of a complex are statistically related to its molecular structure. The structure of organic compounds or ligands is represented in a numerical form by various theoretical descriptors exploring the various structural features; these descriptors are used in a statistical model for the computation of the desired molecular property or function. Present study is performed with the use of different parameters exploring the size, shape, branching, connectivity in the molecule or ligand along with the physicochemical properties of ligands representing the steric nature, volume, density, inter and intra molecular interactions.[7] Computational algorithm involves three major steps: (1) selecting the most informative variable; (2) building polyhedra around the variables which are close in space; and (3) merging together polyhedra that contain similar information.[8, 9]

2. Experimental

Studies on Wave Length of Maximum Absorbance for Phenyl-Thiourea derivatives have been performed. The X-ray absorption property or function Wave Length of Maximum Absorbance λ_m is investigated by mean of multiple linear regression (MLR) technique. Considering the relevant

Journal of Physics: Conference Series **534** (2014) 012060 doi:10.1088/1742-6596/534/1/012060

descriptors obtained by stepwise procedure in multiple linear regression technique. Topological descriptors used in present study are calculated. Three separate descriptors were used namely 2D-topological descriptors, physicochemical properties and non-conventional physichochemical properties. 2D-topological descriptors such as Wiener index(W) [10], Randic connectivity index(χ) [11], Balaban Jindex(J) [12], Szeged index(Sz) [13], Shultz molecular topological index(MTI) [14] and Electrotopological index(S) [15] were tested. Similarly in case of physicochemical properties non-conventional physicochemical properties. Approximate surface area and surface area grid are tested along with indicator parameters.

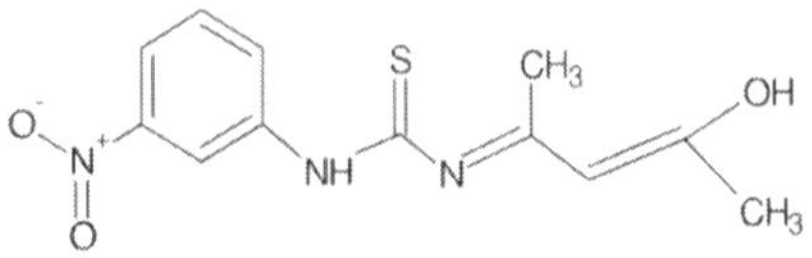

Fig. 1 Molecular structure of Comp.

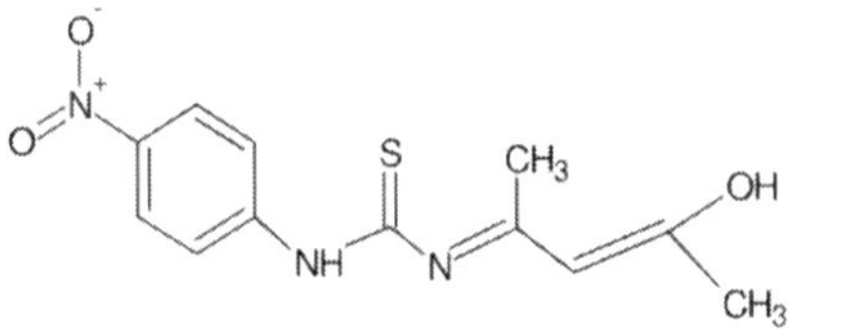

Fig. 2 Molecular structure of Comp. 2.

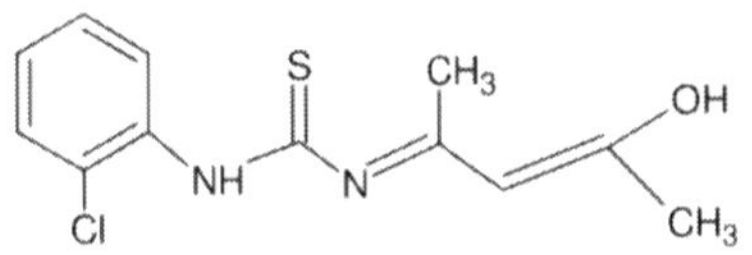

Fig. 3 Molecular structure of Comp 3.

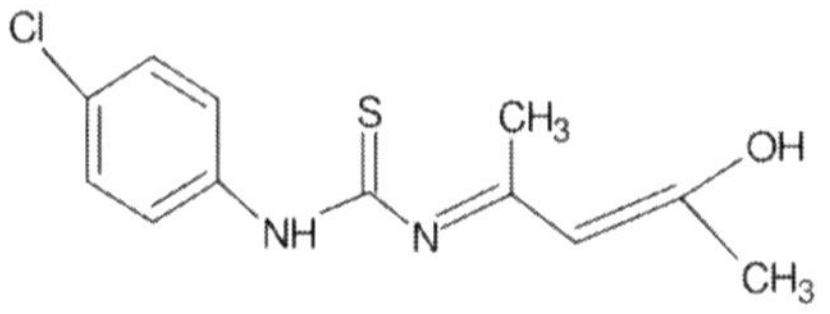

Fig. 4 Molecular structure of Comp 4.

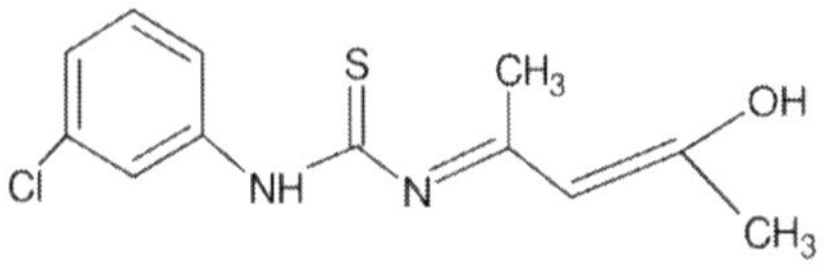

Fig. 5 Molecular structure of Comp. 5.

3. Result and discussion

The set of 5 ligands derived from Phenyl-Thiourea adopted from the literature [16] with their functions or properties i.e., the X-ray absorption property wavelength of maximum absorbance. The studied X-ray absorption property specifically wavelength of maximum absorbance have been expressed as λm, presented in Table 1 and the ligands are presented in figure 1-5 respectively. As may be seen from Table 1, no degeneracy is present in the wave length of maximum absorbance. Therefore, it becomes necessary to examine the degeneracy in the molecular descriptors also. A perusal of Table 2, which contains the topological indices represent the size and volume of the compound, Randic connectivity indices and Balaban Type indices along physicochemical properties with indicator parameter (exploring the substitution on different position in ligands) calculated for the Phenyl-Thiourea ligands used to form metal complexes with Co^{++} having the very low level degeneracy that may be considered as zero level. Balaban has shown that these indices, in spite of their negligible or higher degeneracy, can be used successfully in developing statistically significant mathematical models. From the perusal of statistics for various topological, physicochemical and indicator parameters, it is observed that out of tested 23 parameters 16 parameters having the statistically significant correlation value. Analysis of correlation data and magnitude of the parameters advise that the volumetric properties of ligand or parameters represent the size of the ligand playing the dominating role in theoretical modeling of X-ray absorption property λ_m (wavelength of maximum absorption). The models obtained from the 2D parameters are as follows.

λ_m = -4.3814 (±1.6984) χ_2 + 7753.4102 (1)
n = 5;Se = 1.7476; r = -0.8302; F = 6.655; Q = 0.475

λ_m = 47.9606 (±17.4381) $Jhet_P$ + 7619.2454 (2)
n = 5;Se = 1.6707; r = 0.8462; F = 7.564; Q = 0.506

λ_m = 5.225 (±1.7851) I_O + 7718.2750 (3)
n = 5;Se = 1.5966; r = 0.8606; F = 8.568; Q = 0.539

λ_m = -0.3131 (±0.095) SAG + 7865.8569 (4)
n = 5;Se = 1.4587; r = 0.8853; F = 10.876;Q = 0.607

λ_m = -0.1677 (±0.0454) ASA + 7793.6554 (5)
n = 5;Se = 1.3305; r = -0.9055; F = 13.658;Q = 0.681

All the models obtained from topological or physicochemical parameters exhibits the dominance of volumetric properties over the branching and connectivity. As we pass from the eq. 1 to 5 it is observed that the value of regression coefficient is increasing along with the decrease in the value of standard error of estimation. Increase in the value of R is marginal between to consecutive equations but having the significant difference from 1 to 5 Along with the size or volume, Ortho substitution play the positive role in absorption phenonmenon for wavelength of X-ray. Branching in the ligand with hetero atoms in respect to the polarizability favors the phenonmenon, i.e., branching of any chemical entity responsible to increase the polarizability of ligand favor the absorption phenonmenon. Comparision of magnitude in all the equations shows the dominating role of Surface area grid and approximate surface area over the substitution, connectivity and branching. Highest magnitude of ASA and the maximum 'r' value from eq. 5 demonstrate the role of volume and surface properties in absorption function of X-rays. The analysis of eq. 2, 3 and 5 explore that the presence of any substitution at ortho position, responsible to increase in polarizability and surface area of the ligand is favorable to characterize the X-ray absorption property λ_m. On the basis of mathematical models, calculated values for λ_m are presented in Table 3 and correlation is graphically presented in figure 6.

Table 1. Ligand and their respective functions λ_m used in present study

Comp. No.	Name of Ligand	λ_m
1	1-[(2*E*)-4-hydroxypent-3-en-2-ylidene]-3-(3-nitrophenyl)thiourea	7717.1
2	1-[(2*E*)-4-hydroxypent-3-en-2-ylidene]-3-(4-nitrophenyl)thiourea	7716.
3	1-[(2*E*)-4-hydroxypent-3-en-2-ylidene]-3-(2-chlorophenyl)thiourea	7723.5
4	1-[(2*E*)-4-hydroxypent-3-en-2-ylidene]-3-(3-chlorophenyl)thiourea	7718.
5	1-[(2*E*)-4-hydroxypent-3-en-2-ylidene]-3-(4-chlorophenyl)thiourea	7720.3

Table 2. Topological indices tested in present study.

Comp.No.	χ^2	Jhetp	ASA	I_{Ortho}	SAG
1	8.348	2.061	457.65	0	476.19
2	8.336	2.019	459.83	0	476.19
3	7.333	2.141	426.62	1	459.9
4	7.449	2.117	435.7	0	463.76
5	7.437	2.095	436.62	0	463.76

χ^2 = Second order connectivity index, Jhet p = Balaban type index for polarizability weighted distance matrix, ASA = Approximate Surface Area, SAG= Surface Area Grid, I_{Ortho} (accounted for substitution on ortho position)

Table 3. Calculated values for wave length of maximum absorption (λ_m) using Eq. 5.

Comp. No.	λ_m (Obs.)	λ_m^A
1	7717.1	7716.9
2	7716.9	7716.5
3	7723.5	7722.1
4	7718.8	7720.6
5	7720.3	7720.4

*A = Calculated value of λ_m from Eq. 5.

Figure 6 Graph between experimental and calculated values of λ_m from eq. 5.

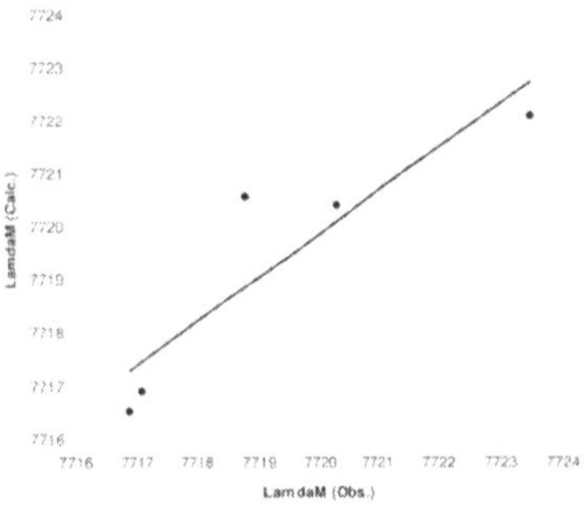

4. Conclusion

The study helps to conclude that the ligands are partially structure specific for the function λ_m. X-ray absorption properties in terms of λ_m affected with the place of substitution and connectivity which may lead to increase in the size or volume of the ligand. Any substitution in the ligand from ortho position increases the polarizability or volume favor the absorption phenomenon λ_m.

References

[1] Wang P., Yang L., Gao D and Long M. 2001 *J. Harbin Inst. Technol.* (N.S.), **8** 394.
[2] Khadikar P. V., Karmarkar S. , V. K. Agrawal, M. Mandloi, and S. Joshi 2000 *Natl. Acad. Sci. Lett.* **23** 50.
[3] Besal´u E., Gallegos A. and Carb´o-Dorca R. 2001 *Match* **44** 41.
[4] Li X.and Lin J. 2003 *J. Math. Chem.*, **33** 81.
[5] Castro E. A., Fern´andez F. M. and Duchowicz P. R.2005 *J. Math. Chem.***37** 433.
[6] Borth D. M., McKay R. J. and Elliott J. 1985 *Technometrics* **27** 25.
[7] Thakur Abhilash, Thakur Mamta, Bharadwaj Arpan and Suprajnya Thakur 2008 *Eur. J. Med. Chem.* **43** 471.
[8] Floersheim, P.,. Nozulak, J. and Weber, H.P. 1993 *Experience with comparative molecular fields analysis*, in Wermuth. C.G. (Ed.) *Trends in QSAR and molecular modeling* (Leiden: ESCOM,), pp. 227.
[9] Boobbyer D.N.A., Goodford, P.J. and McWhinnie, P.M.1989 *J. Med. Chem.* **32** 1083.
[10] Wiener, H.1947 *J. Am. Chem. Soc.* **69** 17.
[11] Randic´, M. 1993 *Croat. Chem. Acta* **66** 289.
[12] Balaban, A. T. 1997 *Selected scientific publications of Alexandru T. Balaban* (New York: Plenum) pp 1-24.
[13] Diudea M. V. 1999 *Croat. Chem. Acta* **72** 835.
[14] Schultz, H.P.; Schultz, E.B. and Schultz, T.P 1990 *J. Chem. Inf. Comput. Sci.* **30** 27.
[15] Kier, L. B. and Hall, L. H.1976 *Molecular Connectivity in Chemistry and Drug Research*; (New York: Academic Press).
[16] Soni Namrata 2011 *Ph.D. thesis* Devi Ahilya University, Indore.

Gibb's energy and intermolecular free length of *'Borassus Flabellifier'* (BF) and *Adansonia digitata* (AnD) aqueous binary mixture

Sushil Phadke[1,*], Bhakt Darshan Shrivastava[2], S. K Ujle[3], Ashutosh Mishra[4] and N. Dagaonkar[5]
[1]Department of Physics, Government Girls College Dhar, MP, India
[2]Department of Physics, Government P.G. College Biora, MP, India
[3]Department of Physics, Government P.G. College Jhabua, MP, India
[4]School of studies in Physics, DAVV, Indore MP, India
[5]Department of Physics, Government P.G. College Dhar, MP, India

Email: sushilphadke5@gmail.com

Abstract. One of the potential driving forces behind a chemical reaction is favourable a new quantity known as the Gibbs free energy (*G*) of the system, which reflects the balance between these forces. Ultrasonic velocity and absorption measurements in liquids and liquid mixtures find extensive application to study the nature of intermolecular forces. Ultrasonic velocity measurements have been successfully employed to detect weak and strong molecular interactions present in binary and ternary liquid mixtures. After measuring the density and ultrasonic velocity of aqueous solution of *'Borassus Flabellifier'* BF and *Adansonia digitata* And, we calculated Gibb's energy and intermolecular free length. The velocity of ultrasonic waves was measured, using a multi-frequency ultrasonic interferometer with a high degree of accuracy operating Model M-84 by M/s Mittal Enterprises, New Delhi, at a fixed frequency of 2MHz. Natural sample *'Borassus Flabellifier'* BF fruit pulp and *Adansonia digitata AnD* powder was collected from Dhar, District of MP, India for this study.

1. Introduction

The measurements of ultrasonic speed in liquid mixture enables accurate determination of some useful acoustical and thermo dynamical parameters are highly sensitive to molecular interactions in their mixtures. Acoustic and thermodynamic parameters have been used to understand different kinds of association, the molecular packing, molecular motion and various types of intermolecular interactions and their strengths influenced by the size in pure components and in the mixtures [1-5]. These parameters play a vital role in assessing the compactness due to molecular arrangement and the extent of molecular interactions in the liquid mixtures through charge transfer, dipole-dipole and dipole induced dipole interactions [6]. Their dependence on concentration, temperature and frequency are of importance in many fields of applied research and also find applications in many important chemical, industrial and biological processes [7]. The aim of this work is to ascertain the nutritive values of BF and AnD fruit, so as to encourage its use and inclusion in human food. The AnD fruit consists of a hard, woody outer shell with a dry, powdery substance inside that covers the hard, black kidney-shaped seeds Phytic acid (myo-inositol hexa phosphate) is a theromolabile rachitogenic agent (Aletor, 1993). It is an important constituent of certain legumes, cereals and forage plants which is capable of chelating divalent cationic minerals like calcium, iron, magnesium and zinc (Liener, 1983). BF fruit as the major species has many traditional uses. They have been used in treating gonorrhea, dysentery, and respiratory disease. BF fruit sap is prized as a tonic, diuretic, stimulant, laxative, anti-phlegmatic and amebicide [8]. Sugar from the sap is a typical agent used to counteract poisoning and prescribed in treating liver disorders, and the pulp of the matured fruits is known to relieve dermatitis. In thermodynamics the Gibbs free energy is a thermodynamic potential that measures the "useful" or process-initiating work obtainable from an isothermal, isobaric thermodynamic system. Just as in mechanics, where potential energy is defined as capacity to do work, similarly different potentials have different meanings. Gibbs energy is the capacity of a system to do non-mechanical work. Attractive or repulsive interactions that occur between all atoms and molecules. Intermolecular forces become significant at molecular separations of about one nano meter or less, but are much weaker than the forces associated with chemical bonding. They are important, however, because they are responsible for many of the physical properties of solids, liquids, and gases. These forces are also largely responsible for the three-dimensional arrangements of biological molecules and polymers.

2. Material and Method

The fresh fruits of *Borassus flabellifer* BF *and Adansonia digitata* And were bought from market Kukshi and Mandu of district Dhar (M.P.). The fresh fruits were washed and air dried. Pods/bark of the fruit were removed, cut into pieces and blended. Aqueous extract of each was obtained, filtered and stored in a clean sample container in the fridge until needed for analysis. The velocity of ultrasonic waves was measured, using a multi-frequency ultrasonic interferometer with a high degree of accuracy operating Model M-84 by M/s Mittal Enterprises, New Delhi, at a fixed frequency of 2MHz [9] and density by density bottle. After measuring density and ultrasonic velocity and density [10], we calculated Gibb's free energy and inter molecular free length [11], of BF and AnD Aqueous Binary Mixture.

3. Result and Discussion

Intermolecular free length and Gibb's free energy increases with the concentration double distil water in both the binary mixture of BF and AnD fruit pulp. The increase in Gibb's free energy suggests shorter time for rearrangement of molecules.

This also indicates existence of molecular association between the components of the liquid mixture. Absorption coefficient, which is a characteristic of the medium, increases In order to understand more about the nature of the interaction between the components of the liquid mixtures, showed in Table 1

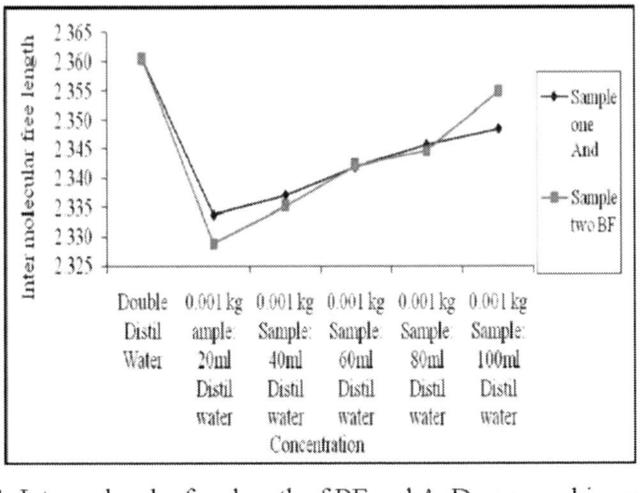

Figure 1. Intermolecular free length of BF and AnD aqueous binary mixture

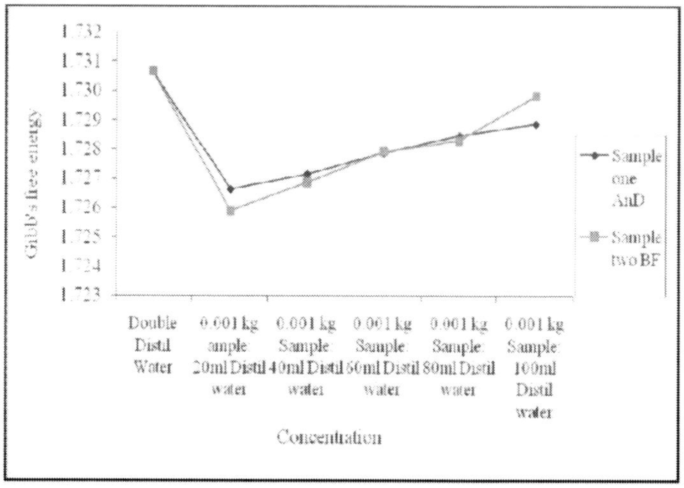

Figure 2. Gibb's energy of BF and AnD aqueous binary mixture

unlike molecules give negative. The value of intermolecular free length follows the same trend and figure 1 and figure 2. Weak interaction between unlike molecules gives in general positive excess values where as dipole-dipole, dipole-induced dipole, charge transfer and hydrogen bonding between. Gibbs' free energy is positive and increases slowly. This is attributed to hydrogen bonding and charge transfer between the molecules of the mixture. As the mixture consists of polar and non-polar liquids, in addition to dipole-dipole interactions there is a dipole-induced dipole interaction.

Table 1. Comparisons of intermolecular free length and Gibb's free energy for BF and AnD. (in SI units)

S.No.	Sample	Concentration	Inter molecular free length $L_f X 10^{-11}$	Gibb's free energy $GX10^{20}$
1	AnD Powder With Distil Water	Double Distil Water	2.3604	1.7307
		0.001 kg ample: 20ml Distil water	2.3337	1.7266
		0.001 kg Sample: 40ml Distil water	2.3371	1.7271
		0.001 kg Sample: 60ml Distil water	2.3419	1.7279
		0.001 kg Sample: 80ml Distil water	2.3458	1.7285
		0.001 kg Sample: 100ml Distil water	2.3485	1.7289
2	BF Fruit Pulp With Distil Water	Double Distil Water	2.3604	1.7307
		0.001 kg ample: 20ml Distil water	2.3288	1.7259
		0.001 kg Sample: 40ml Distil water	2.3352	1.7269
		0.001 kg Sample: 60ml Distil water	2.3421	1.7279
		0.001 kg Sample: 80ml Distil water	2.3446	1.7283
		0.001 kg Sample: 100ml Distil water	2.3548	1.7298

4. Conclusion

From the above study it is seen that there exists molecular association between the components of the binary mixture due to dipole-dipole, dipole-induced dipole, charge transfer and hydrogen bonding interactions, which varies with the change in the frequency of the ultrasonic wave. The presence of Saponin, flavonoid, reducing compound, free-anthraquinone, terpenoids and alkaloids in BF and AnD subjected to phytochemical screening suggest potential source for useful drugs and enhancer of health status to its users.

Acknowledgment

One of the authors B.D.Shrivastava is highly thankful to University Grants Commission, New Delhi for awarding a Project. Our sincere thanks to Dr. S. Alawa Principal Govt. P.G. College, Dhar (MP) and Dr. S. N. Mandloi Principal Govt. Girls College, Dhar (MP) for providing lab facility and maximum cooperation.

References

[1] Sumathi T and Maheswari U 2009 *Ind. J. Pure Appl. Phys.* **47** 782.
[2] Rama Rao G V. 2004 *Ind. J. Pure Appl. Phys.* **42** 820.
[3] Narendra K. 2010 *E-Journal of Chemistry* **7** 927.
[4] Ali A and Nain A K. 2001 *Ind. J. Pure Appl. Phys.* **39** 421.
[5] Ali Anwar 2004 *J. Chin. Chem. Soc.* **51** 477 .
[6] Tadkalkar Alka P 2012 *Archives of Phys. Res.* **3** 287.
[7] Ali A, Abida A K Nain and Hyder S 2003 *J Sol Chem.* **32** 865.
[8] Sofowara. A 1993. *Spectrum books ltd.,* Ibadan, Nigeria Pp 289 -300.
[9] Ferrara L 2005, *Ingredients alimentary* **4** 13.
[10] Phadke S., Shrivastava B.D. & .Mishra A 2009 *Ultra Scientist of Physical Sciences* **21** 499.
[11] Ujle S. K. , Phadke S., Shrivastava B.D., Mishra A., Dagaonkar N., *Manuscript No. ISCA-ISC-2012-15PhyS-17.*

Interaction of electromagnetic fields and biological tissues

Bhakt Darshan Shrivastava [1], Ravindra Barde [2], Ashutosh Mishra[3], S Phadke [4]

[1]Department of Physics, Govt. P. G. College, Biaora (M.P.) India
[2]Department of Physics, S B N Govt. P. G. College Barwani (M.P.) India
[3]School of Physics Devi Ahilya University, Indore (M.P.) India
[4]Department of Physics, Government Girls College, Dhar (M.P.) India

Email: ravibarde88@gmail.com

Abstract. This paper deals with the electromagnetic field interact in biological tissues. It is actually one of the important challenges for the electromagnetic field for the recent years. The experimental techniques are use in Broad-band Dielectric Measurement (BDM) with LCR meters. The authors used Bones and scales of Fish taken from Narmada River (Rajghat Dist. Barwani) as biological tissues. Experimental work carried out done in inter-university consortium (IUC) Indore. The major difficulties that appear are related to the material properties, to the effect of the electromagnetic problem and to the thermal model of the biological tissues.

1. Introduction

A large number of equipment used everyday are electric or electronic ones, and thus generate electromagnetic fields. People are more and more concerned with the consequences of the exposure to the electromagnetic fields. The electromagnetic phenomenon into the cells to the electromagnetic properties values of the tissues.The daily exposure to an electromagnetic environment raises the question of the effects of electromagnetic fields on human health. The accurate assessment of the currents induced in the human body by an electromagnetic field is a major issue, not only for its relevance in medical research, but also for its implications on the definition of industrial standards [1]. Two types of applications may be highlighted. On one hand, electromagnetic fields can be considered as harmful to the health. Using the results of epidemiological studies and by application of the ALARA (As Low As Reasonably Achievable) principle, the from these restrictions using measurement techniques performed on very simple models. Unfortunately governments have imposed some limitations to the authorized radiated fields by the power systems. It has been proposed a set of maximum values of current density or specific absorption rate (SAR), according to the frequency. These values are called the basic restrictions, and reference levels for the fields are derived these reference levels are only external values. They cannot take into account the way the field develops inside the body and they do not take into account the environment of the exposed person. It is now necessary to increase the knowledge of the distribution of the fields inside the body in order to give a more acceptable limit to these radiated fields. On the other hand, electromagnetic fields are used for medical diagnosis (medical scanning MRI) or for medical treatment. As example, hyperthermia is used in oncology treatment to treat localized cancerous tumor [2]. The evaluation of temperature in the *Bone and Scale* is obtained by fish body tissues to a radiofrequency (RF) electromagnetic field. The focalization of the heat inside the *Bone and Scale* is obtained by using several RF sources having specific phases and amplitude. The objective of this paper is to show the present difficulties related to the interaction of induced electromagnetic quantities in biological tissues [3]. The first part deals with the biological material properties. Second part deals with the effect of the electromagnetic problem.

2. Materials and method

The biological tissues with low water content (Dry) have been measured by performance dielectric probe kit. This performance dielectric probe kit is based on broad band measurement (BDM). To quickly and accurately estimate plant biological tissues in situ may provide producers with essential information for making this production. The power source for this method is impedance analyzer and LCR meters with high frequency resolution. The complete system is based on impedance analyzer. This measures the materials response to RF and microwave energy. The samples (*Bones and Scales*) were performed by different biological tissues in low water content (Dry). The two different fish species (catla-catla and common carp) of biological tissues (*Bones and Scales*) were taken from Narmada river Rajghat distt. Barwani (M.P.). First Bones and Scales body tissues remove from fish body. The remove biological tissues dried within 7 days in sunlight heat and preservations. The dried body tissues sample low water content (Dry).The samples prepare proper shape and size at different frequency ranges. The experimental works carried out done in inter university consortium UGC-DAE Indore (M.P.). The Specific Absorption Rate (SAR) distribution has to be determined since it gives a measure of the energy absorption which can be manifested as heat, and since it gives a measure of the internal fields which could affect the biological system without heating (non-thermal interactions). The SAR (W/kg) is defined by SAR = $\sigma E^2 / P_m$ where σ is the conductivity (S/m), E is the root-mean-square magnitude of the electric field (V/m) at the considered point, and P_m is the mass density (kg/m3) of the tissue at that point.

3. Results and discussion

Compared to the material usually used in classical electromagnetic systems, the biological tissues are made of a large number of materials, each of them having specific properties. These properties have been extensively studied in the last fifty years from 10 Hz to almost 10 GHz [4]. In a parametric model able to represent them is presented together with the values of the model parameters for different tissues. Since biological tissues mainly consist of water, they behave neither as a conductor nor a dielectric, but as a dielectric with losses. It is also shown that electromagnetic properties values are highly depending on the frequency. Compared to classical dielectric materials, the dielectric permittivity is high: for example, at 1 KHz, the relative permittivity of the bone is 56. It generally decreases with the frequency. The electric conductivity is low but not zero. As example, the conductivity of the bone varies from 0.3211 to 0.9982 in the 1KHz-1GHz frequency band. The wetter a tissue is, the lossiest it is; the drier it is, the less lossy it is. On the other hand, the permeability of biological tissues [5] is that of free space. 'Figure 1', 'Figure 2', 'Figure 3', 'Figure 4', shows the variation of electromagnetic properties for three representative tissues in the range 1kHz-1GHz: bone (low water content) and scale (low water content). These non-usual values of the electromagnetic properties change the way that Maxwell's equations are dealt with. According to the frequency-domain Ampere's law, the ratio between conduction current and displacement current densities is given by the ratio $\omega\varepsilon/\sigma$. 'Figure 5', shows the variation of this ratio for the three previous tissues in the range 1kHz -10MHz (bone and scale) [6]. 'Figure 6', shows a zoom in the range 1 kHz-10MHz. Clearly, displacement currents cannot be neglected at low frequencies. Dosimetric studies are performed to quantify the interactions of electromagnetic fields with biological tissues. Whether they are numerical or experimental, determining the electromagnetic properties values of biological tissues is a critical step when calculating the SAR. There is not a real consensus on these values: straight comparison of the present data shows significant differences among permittivity values for the same tissue type. Variability in reported values for a single organ may result from different reasons, such as the heterogeneous nature of the biological tissues [7], the freshness of the sample, the tissue preparation procedure, or metabolic changes in samples that occur after death. This poses the question of the effect of these values on calculated SAR values in biological systems. In, it is clear that the dependence of SAR on permittivity changes is important [8].

We have obtained slightly different results in the effect of the Bones and Scale tissues (which are actually not well known) on SAR distribution in local hyperthermia is analyzed. It is shown that neither the permittivity nor the conductivity has a great effect on the mean electric field inside the bones and scale (Table 1). On the other hand, since the SAR is directly proportional to the conductivity, the SAR value in the Bones and Scale is affected by the conductivity value. The temperature will consequently be modified: a SAR value of 20.62 & 10.4 W/kg induces an elevation temperature of 30°C, while a SAR value of 48 &111.75 W/kg increases the temperature in the bones and scale by 100°C.

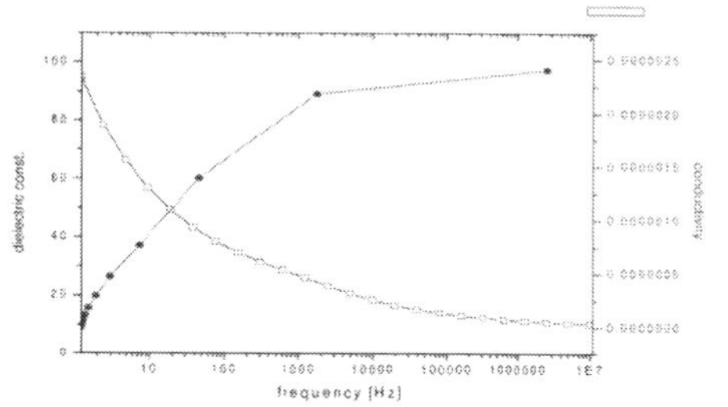

Figure 1. Bone catla-catla

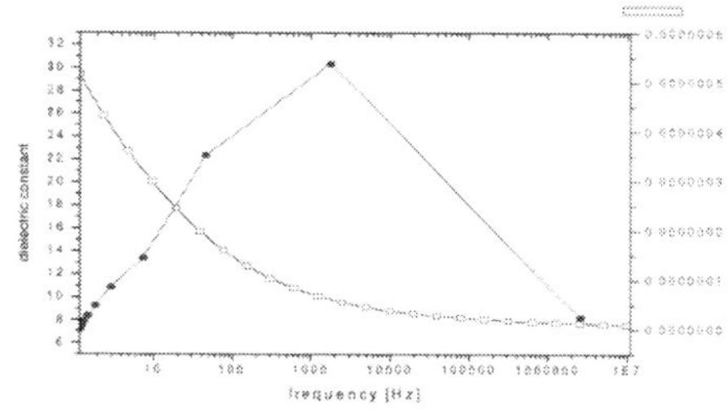

Figure2. Bone common carp

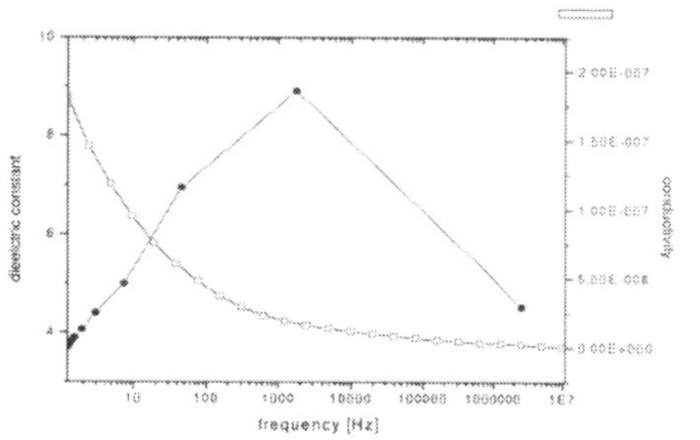

Figure 3. Scale catla-catla

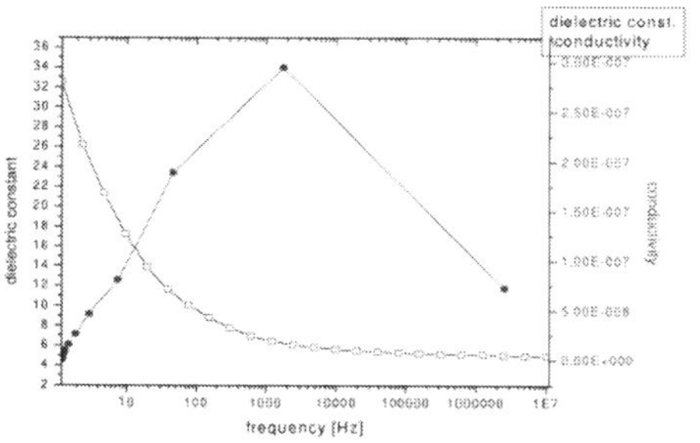

Figure 4. Scale common carp

Figure 1 - 4. Variation with the frequency of the electromagnetic properties values of two representative fish body tissues.

A biological tissue is represented as a heterogeneous and lossy dielectric, whose macroscopic dielectric properties are described by complex permittivity. Interfacial processes, such as Maxwell-Wagner effects, dipolar relaxation effects or counter ion relaxation effects, play an important role in the dielectric properties of tissues. This poses the problem of modeling assemblies of biological cells exposed to electric fields. When exposed to electric fields, a potential difference is induced across the cell membrane. This trans membrane potential (TMP) may be of sufficient magnitude to be biologically significant.

Table 1. The effect of different body tissues on the SAR calculate values

Body tissues	ELF [Hz]	30 ^{0}C	50 ^{0}C	100 ^{0}C	150 ^{0}C
Bone	ε_r	147	254	666	799
Bone	σ	1.65×10^{-10}	5.69×10^{-10}	8.94×10^{-9}	1.50×10^{-7}
Bone	**SAR**	20.62	71.12	111.75	18.75
Scale	ε_r	196	37.5	9.2	6.36
Scale	σ	4.16×10^{-9}	7.95×10^{-9}	1.93×10^{-9}	2.46×10^{-9}
Scale	**SAR**	10.4	19.80	48.75	61.2

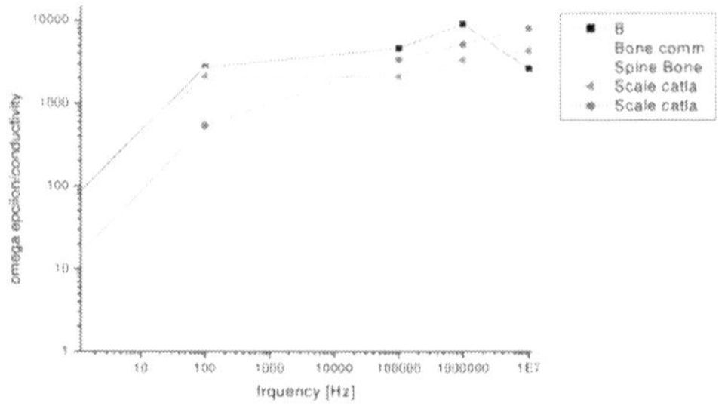

Figure 5. Variation with the frequency of the ε/σ of two fish body tissues: Bones and scales

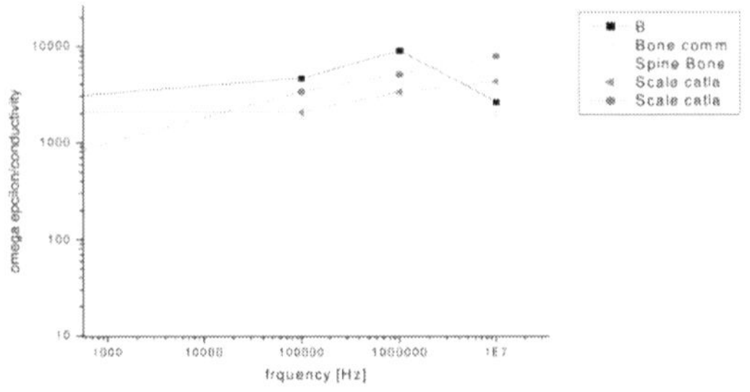

Figure 6. Variation with the frequency of the ratio ω ratio ω ε/σ of two fish body tissues: Bones and scales-zoom on 1Hz - 10MHz band

4. Conclusions

An increasing use of the electromagnetic fields in medical applications can be forecast. Modelling the electromagnetic field distribution in these devices will allow designing optimized systems. On the other hand, a large number of equipment used everyday are electric or electronic ones, and thus generate electromagnetic fields. People are more and more concerned with the consequences of the exposure to the electromagnetic fields. The electromagnetic phenomena into the cells to the electromagnetic properties values of the tissues, to get accurate results on the whole frequency range, validate the electromagnetic-thermal coupling. But in any case, the control of the interaction between electromagnetic fields and biological tissues is one of the challenges for the recent years for which the Compumag community may play an important part.

References

[1] ICNIRP, 1998 *Health Phys.* **74** 494
[2] Nielsen O S, Horsman M and Overgaard J 2001 *Eur. J. Cancer* **37** 1587
[3] SchwanH P 1963 *Biophysik* **1** 198
[4] Stuchly M A and Stuchly SS 1980 *J. Microwave Power* **15** 19
[5] Gabriel C, Gabriel S and Corthout C 1996 *Phys. Med. Biol.* **41** 2231
[6] Gabriel S Lau RW Gabriel C 1996 *Phys. Med. Biol.* **41** 2251
[7] Gabriel S Lau RW Gabriel C 1996 *Phys. Med. Biol.* **41** 2271
[8] Stavroulakis P 2003 *Biological Effects of Electromagnetic Fields* (Springer)

Dielectric Behavior of Biomaterials at Different Frequencies on Room Temperature

B.D. Shrivastava [1,*], Ravindra Barde [2,*], A. Mishra [3], S. Phadke [4]

[1]Department of Physics, Govt. P. G. College Byawara (M.P.) INDIA
[2]Department of Physics, Shaheed Bhima Nayak, Govt. P. G. College, Barwani (M.P.), INDIA
[3]School of Physics D. A. University, Indore - 452001 (M.P.), INDIA
[4]Department of Physics, Government Girls College, Dhar (M.P.), INDI

E-mail: bdshrivastava@gmail.com & ravibarde88@gmail.com

Abstract. Propagation of electromagnetic (EM) waves in radiofrequency (RF) and microwave systems is described mathematically by Maxwell's equations with corresponding boundary conditions. Dielectric properties of lossless and lossy materials influence EM field distribution. For a better understanding of the physical processes associated with various RF and microwave devices, it is necessary to know the dielectric properties of media that interact with EM waves. For telecommunication and radar devices, variations of complex dielectric permittivity (referring to the dielectric property) over a wide frequency range are important. For RF and microwave applicators intended for thermal treatments of different materials at ISM (industrial, scientific, medical) frequencies, one needs to study temperature and moisture content dependencies of the Permittivity of the treated materials. Many techniques have been developed for the measurement of materials. In the present paper authors used Bones and scales of Fish taken from Narmada River (Rajghat Dist. Barwani) as biomaterials. Dielectric properties of Biomaterials with the frequency range from 1Hz to 10 MHz at room temperature with low water content were measured by in-situ performance dielectric kit. Analysis has been done by Alpha high performance impedance analyzer and LCR meters. The experimental work were carried out in Inter University Consortium UGC-DAE, CSR center Indore MP. Measured value indicates the dielectric constant (ε') dielectric loss (ε'') decreases with increasing frequency while conductivity (σ) increases with frequency increased.
Key words: dielectric constant, dielectric loss, conductivity, biomaterials (Bones and Scales)

1 Introduction

The behavior of a dielectric material in the presence of an electromagnetic field is entirely different from that in presence of a direct current field. In order to understand the behavior of dielectric materials under the action of electromagnetic fields, one has to investigate its interaction in such electromagnetic field. In the presence of an alternative electric field, the dielectric materials get polarized along the field direction. The degree of polarization depends on the applied electromagnetic power and the nature of the material itself [1]. The static dielectric constant and dipole moment values are a measure of the polarization of the material at low frequencies. However, in the presence of a high frequency field, there exist a time lag in the attainment of equilibrium in a system with changing field and hence an anomalous dispersion (dielectric constant decrease within increase in frequency) takes place [2, 3]. The different biomaterials (Bones and Scales) sample in low water (Dry) content gives rise to a large variation in the dielectric constant. Thus, knowledge of the variation of the dielectric constant of the biomaterials (*Bones and Scales*) at different frequency for the room temperature. Porosity of the Bones and Scales greatly helps to judge the moisture movement within the biomaterial. Macro pores allow readily movement of air and water. It does not hold water under normal condition. The porosity of soil is easily changed. The dielectric properties of most materials vary with several different factors. In hygroscopic materials such as agricultural products, the amount of water in the materials is generally a dominant factor. The dielectric properties also depend on the frequency of the applied alternating electric field, the temperature of the materials, and on the density and structure of the materials. In granular or particulate materials, the bulk density of the air-particle mixture is another factor that influences their dielectric properties. The dielectric properties of materials are dependent on their chemical composition and particularly molecules in the materials [4]. Researchers have tested different techniques of using the dielectric properties of biomaterials to estimate water content. These studies include transmission line techniques such as waveguide (coaxial and free-space), impedance and cavity methods [5]. Dielectric properties are, by definition , a measure of the polarizability of a material when subjected to an electric field [6]. For lossy materials, the relative complex permittivity, $\varepsilon = \varepsilon' - j\varepsilon''$, represents the dielectric properties. The dielectric constant [7], ε', describes the materials ability to store energy, the dielectric loss factor ε'', describes the materials ability to dissipate the electric field energy, and j is the imaginary root of -1 [8]. The dielectric properties of many materials depend on frequency, moisture content, bulk density, temperature, chemical composition, and the permanent dipole moments association with water and other constituent molecules. With the exception of some extremely low-loss materials, that is, materials that absorb essentially no energy from RF and microwave fields, the dielectric properties vary considerably with the frequency of the applied electric fields. An important phenomenon contributing to the frequency dependence of the dielectric properties is the polarization arising from orientation with the imposed electric field of molecules which have permanent dipole moments. However water in biomaterials exists both as liquid water and as water bound to the inner structure of the biomaterial. The dielectric properties of bound water lie somewhere between those of ice and those of liquid water depending on how tightly the water is bound. Therefore, testing of biomaterials has been conducted comparing potential difference measurements of elements with knows permittivity to the test material or by correlating the transmission potential difference between two or more quantities of same material with the water content.

2 Experiment details

Ribbons Dielectric properties of biomaterials (*Bones and Scales*) with low water content have been measured by performance dielectric probe kit. This performance dielectric probe kit is based on non-destructive method. To quickly and accurately estimate plant biomaterials in situ may provide producers with essential information for making this production. The power source for this method is impedance analyzer and LCR meters with high frequency resolution [9]. The measurement made by simply immersing the probe into the *Bones and Scales* sample. There is no requirement of special fixtures and connectors. The complete system is based on impedance analyzer. This measures the materials response to RF and microwave energy. The samples were performed by different biomaterials (Bones and Scales) in low water content. The two different fish species (*cyprinid and cyprinus carpio*) of biomaterial (*Bones and Scales*) were taken from Narmada river Rajghat distt. Barwani (M.P.). First Bones and Scales body tissues remove from fish body. The remove biomaterials dried within 7 days in sunlight heat and preservations. The dried body tissues sample low water content (Dry).The samples prepare proper shape and size at different frequency ranges. The experimental works carried out done in inter university consortium UGC-DAE Indore (M.P.). The mathematical formulation developed by Debye to describe this process for pure polar materials can be expressed as $\boldsymbol{\varepsilon} = \boldsymbol{\varepsilon_0} + \boldsymbol{\varepsilon_s} - \boldsymbol{\varepsilon_\infty} / \mathbf{1+i\omega\tau}$ where $\boldsymbol{\varepsilon_\infty}$, represents the dielectric constant at frequencies so high that molecular orientation does not have time to contribute to the polarization, $\boldsymbol{\varepsilon_s}$ represents the static dielectric constant, that is, the value at zero frequency (dc value), and $\boldsymbol{\tau}$ is the relaxation time, the period associated with the time for the dipoles to revert to random orientation when the electric field is removed.

Table1. The experimental values of observed biomaterials with different frequencies (1 Hz to 10 MHz) at room temperature.

S.N.	Frequency (Hz)	Biomaterials	Dielectric constant (ε')	Dielectric loss (ε'')	Conductivity σ (S/m)
1	1 Hz	Bone catla	99.9	14.01	8.08×10^{-11}
2	1 Hz	Bone common	30.3	294.0	7.86×10^{-12}
3	1 Hz	Spine catla	147.0	145.0	1.63×10^{-10}
4	1 Hz	Scale catla	34.03	52.04	2.91×10^{-11}
5	1 Hz	Scale common	9.05	7.82	4.35×10^{-12}
6	100 Hz	Bone catla	34.7	293.0	8.05×10^{-10}
7	100 Hz	Bone common	12.7	16.6	2.48×10^{-10}
8	100 Hz	Spine catla	36.9	9.49	9.24×10^{-10}
9	100 Hz	Scale catla	8.82	3.12	2.65×10^{-10}
10	100 Hz	Scale common	4.74	-6.55	5.56×10^{-11}
11	10 KHz	Bone catla	13.2	-3.34	1.78×10^{-7}
12	10 KHz	Bone common	8.04	1.44	2.90×10^{-8}
13	10 KHz	Spine catla	16.7	2.05	7.98×10^{-8}
14	10 KHz	Scale catla	5.21	-1.82	1.58×10^{-8}
15	10 KHz	Scale common	3.85	-82.8	7.19×10^{-9}
16	10 KHz	Bone catla	11.04	-2.09	7.68×10^{-7}
17	10 KHz	Bone common	7.76	-9.73	1.46×10^{-7}
18	10 KHz	Spine catla	15.5	1.10	3.26×10^{-7}
19	10 KHz	Scale catla	5.07	-11.6	8.07×10^{-8}
20	10 KHz	Scale common	3.79	-665	4.62×10^{-8}
21	10 MHz	Bone catla	10.05	-435	2.41×10^{-6}
22	10 MHz	Bone common	7.56	-93.07	2.42×10^{-8}
23	10 MHz	Spine catla	14.04	-43.03	1.81×10^{-6}
24	10 MHz	Scale catla	4.96	-131	7.29×10^{-8}
25	10 MHz	Scale common	3.71	-528	2.94×10^{-8}

3 Results and discussions

The experimental results show the effect of frequency variation on dielectric properties of biomaterials with low water content. The biomaterials give the effective response with different frequencies (1Hz to 10 MHz at room temperature **(Table 1)**. Their frequency dependence study has been qualitatively explained. The value of dielectric constant, dielectric loss and conductivity of low water content and biomaterials are measured and plotted in '**Figure 1**', **Figure 2**, and '**Figure 3**'. By measurements, it is observed that there is a variation in the dielectric constant of different biomaterials at room temperature. The value of dielectric constant (ε') and dielectric loss (ε'') decreases with increasing frequency while the value of conductivity (σ) increases with frequency increases. Such behaviour may be in light of the model of multiple conductive relaxation modes.

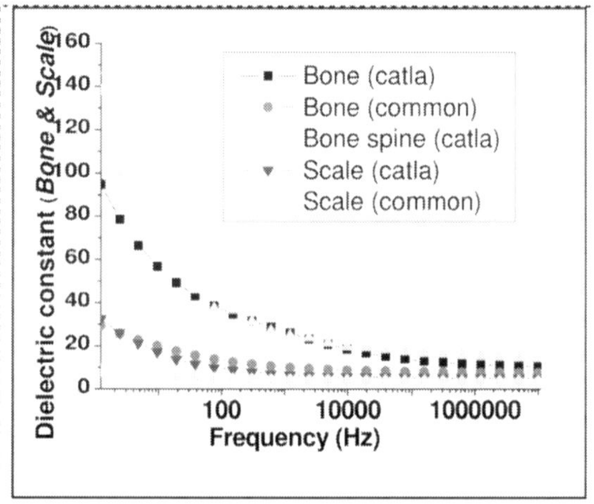

Figure 1: Frequency response dielectric constant (ε') of biomaterials at 30 ^{0}C.

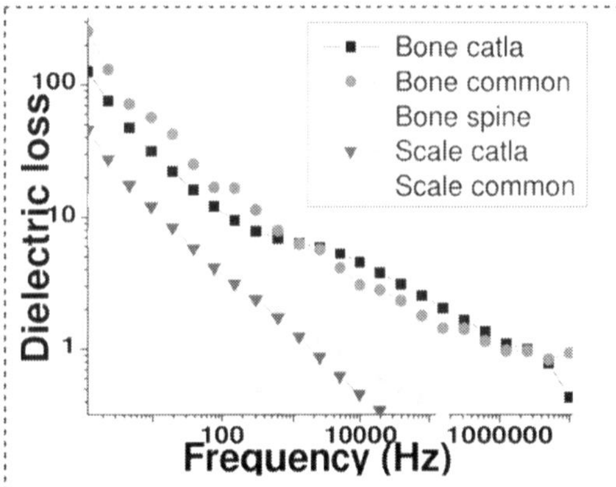

Figure 2: Frequency response dielectric loss (ε'') of biomaterials at 30 ^{0}C.

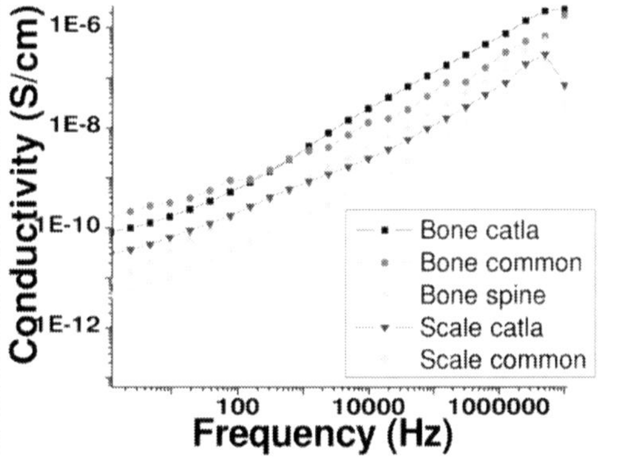

Figure 3. Frequency response Conductivity (S/cm) of biomaterials at 30 ^{0}C.

4 Conclusions

The dielectric properties of biomaterials different frequency ranges good results are obtained for the dielectric constant and dielectric loss. The dielectric properties of the bulk materials in the form of biomaterials with low water content are useful in understanding the effect of biological tissues as biomaterials. The results such studies are important for an understanding of the fundamental nature of the response of the particularly biomaterials to the high frequency electromagnetic fields.

5 Acknowledgments

The authors recognize and appreciate support through the provided laboratory work facility. Department of electromagnetic and thermodynamic laboratories, inter university consortium, UGC-DAE, CSR, center Indore (M.P.), INDIA.

References

[1] Bottcher CJF 1952 *Theory of electric Polarization* (Elsevier Pub. Comp. Ams Hoston London p.415.
[2] Dakin TW Works CN 1947 *J. Appl. Phys. 18(a)*
[3] Frohilch H 1949 Theory of dielectrics Oxford University Press, London
[4] Fedrick TR Thomson LM 1993 Soil and Soil fertility Oxford University Press pp. 16.
[5] Nelson S 1973 *Electrical properties of agricultural products - A critical review*
[6] Ja Von Hippel A R 1954 *Dielectric Properties and Applications* (New York) John Wiley and Sons Trans. ASAE 16 (2):384-400
[7] Trabelsi S and Nelson S 2003 *Measurement Science and Technology 14:589-600*
[8] Nelson, S 1994 *(Journal of Food Engineering 21:365-384)*
[9] Nelson S 1981 (*Review of factors influencing the dielectric properties of cereal grains) Cereal Chem. 58 (6):487-492*

Young's modulus of BF wood material by longitudinal vibration

Sushil Phadke[1,*], Bhakt Darshan Shrivastava[2], Ashutosh Mishra[3] and N. Dagaonkar[4]
[1]Department of Physics Govt. Girls College, Dhar (M.P.), India
[2]Department of Physics Government P.G. College Biora, (M.P.) India
[3]School of Physics, Devi Ahilya University Indore (M.P.) India.
[4]Department of Physics Government P.G. College Dhar (M.P.), India

Email: sushilphadke5@gmail.com

Abstract. All engineered structures are designed and built with consideration of resisting the same fundamental forces of tension, compression, shear, bending and torsion. Structural design is a balance of these internal and external forces. So, it is interesting to calculate the Young's moduli of *Borassus Flabellifier* BF wood are quite important from the application point of view. The ultrasonic waves are closely related with the elastic and inelastic properties of the materials. In the present study, we measured longitudinal wave ultrasonic velocities in BF wood material by longitudinal vibration method. After measuring ultrasonic velocity in BF wood material, we calculated Young's modulus of *Borassus Flabellifier* BF wood material. We used ultrasonic interferometer for measuring longitudinal wave ultrasonic velocity in BF wood material made by Mittal Enterprises, New Delhi, India in our laboratory. ***Borassus Flabellifier*** BF wood material was collected from Dhar district of Madhya Pradesh, India.

1. Introduction

Longitudinal waves are broadly used for the purposes of non-destructive evaluation of materials and for generation and sensing of acoustic vibration [1] of surrounding. Many mathematical models describing longitudinal wave propagation in solids have been derived in order to analyse the effects of different materials and geometries on vibration characteristics without the need for costly experimental studies. The increasing use of piezoelectric quartz [2] for the stabilisation of radio frequencies has promoted many investigations of the vibration of the quartz. In the last few years a mass of information has been accumulated disclosing the complexities of vibratory modes and types which may exist simultaneously in one and the same crystal plate or rod. Along with the longitudinal, flexural and tensional oscillations may exist, as well as overtones of any or of all. All engineered structures are designed and built with consideration of resisting the same fundamental forces of tension, compression, shear, bending and torsion. Structural design is a balance of these internal and external forces. At the same time, materials selection, component specification, joining techniques, individual member design and assembly must be also considered. Knowledge of the mechanical properties of materials and how they perform is a must for designers, engineers and architects. It is this knowledge of a material's performance relative to standardized tests and benchmarks that gives designers greater certainty in the structural integrity of their designs. Determination of the elastic proprieties of materials by means of measurements of ultrasonic velocity has a long history for different materials [3-4]. Application of Ultrasonic techniques to wood appears to have been pioneered by Hearmon [5]. There is direct and continuing application of these analyses to the design and construction of musical instruments [6-7]. The determination of the elastic properties of different wood species is also of importance in the more mundane aspects of their use in the building industry. The ultrasonic bulk wave velocity measurements, require the hypothesis of orthotropic symmetry, are used to determine these properties. The ultrasonic modes considered are longitudinal waves [8], and shear waves with particle motion along the direction of the applied stress. Compressive and tensile stresses are applied in the longitudinal direction of small clear wood specimen, and ultrasonic waves are propagated through the radial direction of the wood specimen. Stress-induced Velocity changes of the ultrasonic waves are measured, and acoustoelastic constants are also determined. Additionally, stress distributions in bending of wood beam specimen are estimated by measuring ultrasonic velocity.

2. Material and Method

The longitudinal axis of each specimen coincided with the longitudinal direction of the wood. The test specimens of BF Female and Male tree leaf stem were kept under air-dried condition. Dried samples were cut in desired shape. We used ultrasonic interferometer for measuring longitudinal wave ultrasonic velocity in BF wood material made by Mittal Enterprises, New Delhi, India in our laboratory. *Borassus Flabellifier* BF Female and Male tree leaf stem wood Material was collected from Dhar district of Madhya Pradesh, India. For longitudinal vibration's the velocity of longitudinal Vibration given by $V_L = (C_{11}/\rho)^{\frac{1}{2}}$ and $Y=V_L^2\rho$, Where Young's modulus $Y= C_{11}$ and ρ density [9].

3. Result and Discussion

Table 1 shows typical experimental results indicating the relationships between the stress, strain, and the changes in the ultrasonic velocity. The stress-strain relationships are represented by generally recognized curves for longitudinal compression of softwood species.

The ultrasonic velocity was large in male BF wood material then female BF wood material. Longitudinal wave velocity decreased with increasing density of wood material which affects the young's modulus of the wood material. The elastic constant of male BF wood material was greater then female BF wood material. compressive stress immediately after the natural state, as in figure 1& figure 2.

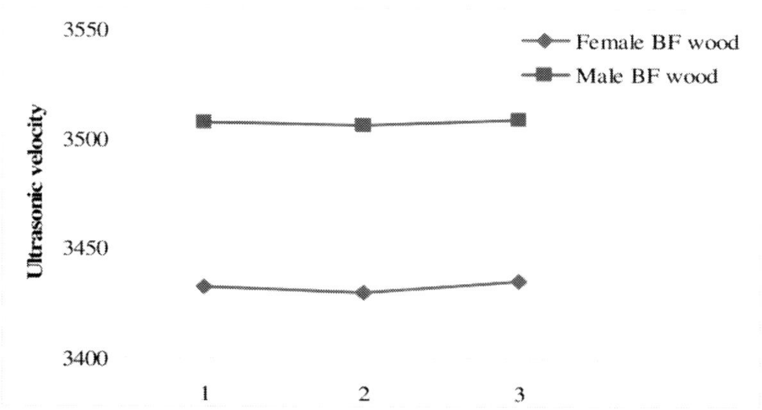

Figure 1 Comparisons of ultrasonic velocity in Female and male wood material.

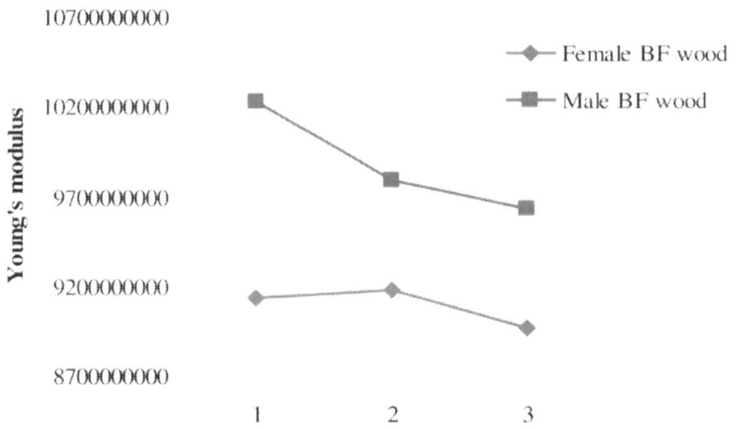

Figure 2 Comparisons of Young's modulus of Female and male wood material.

Table 1-The measured and calculated values of mass, density, ultrasonic velocity and elastic constant of BF (in MKS units)

Mass of the quartz rod mq= 0.000924 Kg, Natural frequency of quartz rod fq =79670 Hz Thickness of specimen=0.004 m and Breath of specimen= 0.004 m								
S. No	Mass of specimen m_s	length of specimen L	Volume of specimen V	Density ρ	Frequency of composite system fc	Frequency of specimen fs	Ultrasonic velocity V_L	Elastic constant E
For Female BF								
1	0.000278	0.0224	3.58E-07	775.67	78146	76621	3432.621	9.1E+09
2	0.000483	0.0387	6.19E-07	780.039	61995	76621	3430.291	9.2E+09
3	0.000652	0.0536	8.58E-07	760.261	55859	76621	3435.438	9E+09
For Male BF								
1	0.000298	0.0224	3.58E-07	831.473	78986	78301	3507.885	1E+10
2	0.000493	0.0387	6.19E-07	796.189	62485	45299	3506.143	9.8E+09
3	0.000671	0.0536	8.58E-07	782.416	56203	32735	3509.192	9.6E+09

4. Conclusion

There exist obvious linear relationships .The changes in the velocities of these materials are small. The changes in the propagation velocities of ultrasonic waves have been accounted for by the changes in the densities and elastic modulii of the materials. As a result of the application of stress to an elastic material, the density and elastic modulus of the material change. This change is considered to lead to a change in the propagation velocity. Such phenomena that density or elastic modulii change due to applied stresses or deformations. In addition to this, the phenomenon obtained for wood is considered to relate to its cellular structure complexity. This suggests the existence of a relationship between the acoustoelastic phenomena and the anatomical structure of wood. The changes in the velocities of propagation of ultrasonic waves were given as functions of the applied stress. The relationships between the velocity and stress at the initial level, and those between the velocity and strain at the range of large deformation were straight lines.

Acknowledgement
One of the authors Sushil Phadke is highly thankful to University Grants Commission, New Delhi for awarding a Project. Our sincere thanks to Dr. S. Alawa Principal Govt. P.G. College, Dhar (MP) and Dr. S. N. Mandloi Principal Govt. Girls College, Dhar (MP) for providing lab facility and maximum cooperation.

References

[1] Fedotov I. A., Polyanin A. D. and Shatalov M.Yu. 2007 *Doklady Physics* **52** 607.
[2] Boyle R W and Sproule D O *Nature* **123** 13
[3] Hearmon R. F. S. 1961 *An introduction to applied anisotropic elasticity* (Oxford University Press, Oxford).
[4] Musgrave M. J. P. 1970 *Crystal acoustics.* (San Francisco: Holden Day).
[5] Hearmon R. F. S. 1948 *The elasticity of wood and plywood.* (London: His Majesty's Stationery Office).
[6] Schelleng J. C. 1969 *Catgut Acoust. Soc. Newsletter* **11** 18.
[7] Schumacher R. T. 1988 *J. Acoust. Soc. Amer.* **84** 1223.
[8] Yasutoshi Sasaki and Kosei Ando 1999 *Acoustoelastic Phenomena of Wood NDT.net* **4** 11.
[9] Shrivastava B. D., Mishra A and Phadke Sushil 2012 *Defect and Diffusion Forum* **326-328** 105

Jeans instability of self-gravitating rotating radiative plasma with finite Larmor radius corrections

Sachin Kaothekar[1] and R K Chhajlani[2]
[1] Department of Engineering Physics, Mahakal Institute of Technology and Science, Ujjain-456664, M.P., India
[2]School of Studies in Physics, Vikram University, Ujjain-456010, M.P., India

E-mail: sac_kaothekar@rediffmail.com

Abstract. The problem of Jeans instability is investigated for a self-gravitating, rotating, radiative plasma with finite Larmor radius corrections, which has connection in astrophysical condensation. A general dispersion relation is derived using normal mode analysis method with the help of relevant linearized perturbation equations of the problem. We find that the classical Jeans result regarding the size of initial break-up is considerably modified due to rotation, finite Larmor radius corrections and radiative heat-loss function. Numerical calculations have been performed to discuss the effect of various physical parameters on the growth rate of the Jeans gravitational instability.

1. Introduction

In the process of formation of astrophysical objects, the most outstanding phenomenon is interstellar gas fragmentation. These fragments of interstellar gaseous clouds are gravitationally contracted and as a final result the astrophysical objects are formed. In this context, Khan and Bhatia [1] have carried out the effects of rotation, Hall current and finite electrical conductivity no Jeans gravitational instability of plasma. Vyas and Chhajlani [2] have investigated the problem of gravitational instability with rotation, magnetic field, viscosity, finite electrical resistivity and thermal conductivity. Recently Chakrabarti et al. [3] have discussed the problem of nonlinear Jeans instability in an uniformly rotating gas. In addition to this, it is a well-known fact that thermal and radiative processes play a significant role in the instability investigation in gravitating plasma medium. Aggarwal and Talwar [4] have pointed out the role of radiative heat-loss function in the study of magnetothermal instability in a rotating gravitating fluid. Prajapati et al. [5] have investigated the self-gravitational instability of rotating viscous Hall plasma with arbitrary radiative heat-loss function and electron inertia. Recently Kaothekar et al. [6] have carried out the effect of radiative heat-loss function on self-gravitational instability of viscous thermally conducting partially-ionized plasma. Along with this, in several plasma situations such as in solar corona, interplanetary and interstellar plasmas the FLR plays an important role in stability investigations. Sharma [7] has investigated the stabilizing role of FLR with rotation and magnetic field on gravitational instability of plasma. Kaothekar and Chhajlani [8] have investigated the effect of radiative heat-loss function and finite Larmor radius corrections on Jeans instability of viscous thermally conducting self-gravitating astrophysical plasma. Recently Sharma and Chhajlani [9] have carried out the effect of FLR corrections on Jeans instability of quantum plasma. Thus in the present work, we have investigated the effect of rotation, radiative heat-loss function, FLR corrections, thermal conductivity and finite electrical resistivity on Jeans instability of self-gravitating plasma. The above work is helpful in understanding the structure formation in interstellar medium.

2. Linearized Perturbation equations

Let us consider an infinite homogeneous, self-gravitating, thermally conducting, radiating plasma of finite electrical resistivity in the presence of magnetic field $\boldsymbol{H}$ (0, 0, H). The linearized perturbation equations of motion of the problem with these effects are written as-

$$\rho\frac{\partial \boldsymbol{u}}{\partial t} = -\nabla\delta p - \nabla\cdot\boldsymbol{P} + \rho\nabla\delta U + \frac{1}{4\pi}(\nabla\times\boldsymbol{h})\times\boldsymbol{H} + 2\rho(\boldsymbol{u}\times\boldsymbol{\Omega}), \quad (1)$$

$$\frac{1}{\gamma-1}\frac{\partial\delta p}{\partial t} - \frac{\gamma}{\gamma-1}\frac{p}{\rho}\frac{\partial\delta\rho}{\partial t} + \rho\left[\delta\rho\left(\frac{\partial L}{\partial\rho}\right)_T + \delta T\left(\frac{\partial L}{\partial T}\right)_\rho\right] - \lambda\nabla^2\delta T = 0, \ (2), \quad \frac{\partial\boldsymbol{h}}{\partial t} = \nabla\times(\boldsymbol{u}\times\boldsymbol{H}) + \eta\nabla^2\boldsymbol{h}, \quad (3)$$

$$\frac{\delta p}{p} = \frac{\delta T}{T} + \frac{\delta\rho}{\rho}, \ (4), \quad \frac{\partial\rho}{\partial t} = -\rho\nabla.\boldsymbol{u}, \ (5), \quad \nabla^2\delta U = -4\pi G\delta\rho, \ (6), \quad \nabla.\boldsymbol{h} = 0, \quad (7)$$

where $\rho, p, T, \boldsymbol{u}(u_x, u_y, u_z)$, $\boldsymbol{\Omega}(0, 0, \Omega_z)$, η, λ, U, G, R, and γ denote the fluid density, pressure, temperature, velocity, rotation, electrical resistivity, thermal conductivity, gravitational potential, gravitational constant, gas constant and ratio of two specific heats respectively. Also here δ is for perturbed quantities and $\boldsymbol{h}$ is perturbed magnetic field. The components of pressure tensor $\boldsymbol{P}$, considering the finite ion gyration radius for the magnetic field along z-axis as given by Roberts and Taylor [10] are

$$P_{xx} = -\rho\upsilon_0\left(\frac{\partial u_y}{\partial x} + \frac{\partial u_x}{\partial y}\right), \quad P_{yy} = \rho\upsilon_0\left(\frac{\partial u_y}{\partial x} + \frac{\partial u_x}{\partial y}\right), \quad P_{zz} = 0,$$
$$P_{xy} = P_{yx} = \rho\upsilon_0\left(\frac{\partial u_x}{\partial x} - \frac{\partial u_y}{\partial y}\right), \quad P_{xz} = P_{zx} = -2\rho\upsilon_0\left(\frac{\partial u_y}{\partial z} + \frac{\partial u_z}{\partial y}\right), \quad P_{yz} = P_{zy} = 2\rho\upsilon_0\left(\frac{\partial u_z}{\partial x} + \frac{\partial u_x}{\partial z}\right). \quad (8)$$

The parameter υ_0 has the dimensions of the kinematics viscosity and defined as $\upsilon_0 = \Omega_L R_L^2/4$, where R_L is the ion-Larmor radius and Ω_L is the ion gyration frequency. We assume the variation of perturbed quantities as $e^{(ik_x x + i\sigma t)}$, (9), where k_x is the wave number of the perturbations along x-axis, and σ is the frequency of harmonic disturbance.

3. Dispersion relation and discussion

The substitution of equation (9) into equations (1)-(7) using equation (8), gives the dispersion relation for transverse mode having axis of rotation along magnetic field (i.e. $k_x = k$ and $\Omega_z = \Omega$) is

$$\omega^3\left\{\omega\left[\omega + \frac{V^2k^2}{\omega+\eta k^2}\right] + \left(\upsilon_0 k^2 - 2\Omega\right)^2 + \frac{k^2(\gamma-1)\left(\frac{\lambda k^2 T}{\rho} + TL_T - \rho L_\rho\right) - 4\pi G\rho(\gamma-1)\left(\frac{\lambda k^2 T}{p} + \frac{T\rho L_T}{p}\right) + \omega\left(c^2k^2 - 4\pi G\rho\right)}{\omega + (\gamma-1)\left(\frac{\lambda k^2 T}{p} + \frac{T\rho L_T}{p}\right)}\right\} = 0. \quad (10)$$

Equation (10) has two independent factors. The first factor of equation (10) gives $\omega^3 = 0$, which is a marginal stable mode. The second factor gives the fourth degree dispersion relation in ω, which is modified by the presence of radiative heat-loss function, self-gravitation, rotation, FLR correction, thermal conductivity, finite electrical resistivity and magnetic field. The condition of instability obtained from the constant term is given as

$$\left\{k^2\left(\frac{\lambda k^2 T}{\rho}+TL_T-\rho L_\rho\right)-\left(\frac{\lambda k^2 T}{p}+\frac{T\rho L_T}{p}\right)\left[4\pi G\rho+4\Omega\upsilon_0 k^2-4\Omega^2-\upsilon_0^2 k^4\right]\right\}<0, \tag{11}$$

From equation (11) we conclude that the fundamental Jeans criterion of gravitational instability is modified due to rotation, FLR correction, thermal conductivity and radiative heat-loss function.

For an infinitely conducting medium in absence of FLR corrections $(\eta=\upsilon_0=0)$, the condition of instability is given as

$$\left\{k^2(\gamma-1)\left(\frac{\lambda k^2 T}{\rho}+TL_T-\rho L_\rho\right)-(\gamma-1)\left(\frac{\lambda k^2 T}{p}+\frac{T\rho L_T}{p}\right)\left[4\pi G\rho-V^2k^2-4\Omega^2\right]\right\}<0, \tag{12}$$

From equation (12) we conclude that rotation and magnetic field stabilizes the radiative instability. The above inequality (12) can be solved to yield the following critical wave number.

$$k_{J1}^2=\frac{1}{2}\left|\left[\left(\frac{4\pi G\rho}{c'^2}+\frac{\rho^2 L_\rho}{\lambda T}-\frac{4\Omega^2}{c'^2}\right)\left(1+\frac{V^2}{c'^2}\right)^{-1}-\frac{\rho L_T}{\lambda}\right]\pm\left\{\left[\left(\frac{4\pi G\rho}{c'^2}+\frac{\rho^2 L_\rho}{\lambda T}-\frac{4\Omega^2}{c'^2}\right)\right.\right.$$
$$\left.\left.\times\left(1+\frac{V^2}{c'^2}\right)^{-1}-\frac{\rho L_T}{\lambda}\right]^2+\frac{16\rho L_T}{\lambda c'^2}\left(\pi G\rho-\Omega^2\left(1+\frac{V^2}{c'^2}\right)^{-1}\right)\right\}^{1/2}\right|. \tag{13}$$

In absence of rotation the above modified critical Jeans wave number reduces to that of Aggrawal and Talwar [4]. Numerical calculations were performed for equation (10) by making it dimensionless in terms of self-gravitation to determine the roots of ω^* as a function of wave number k^* taking $\gamma=5/3$. Out of the four modes only one mode is unstable for which the calculations are presented in figures 1-4, where the growth rate ω^* (positive real value of ω) has been plotted against the wave number k^* to show the dependence of the growth rate on the different physical parameters.

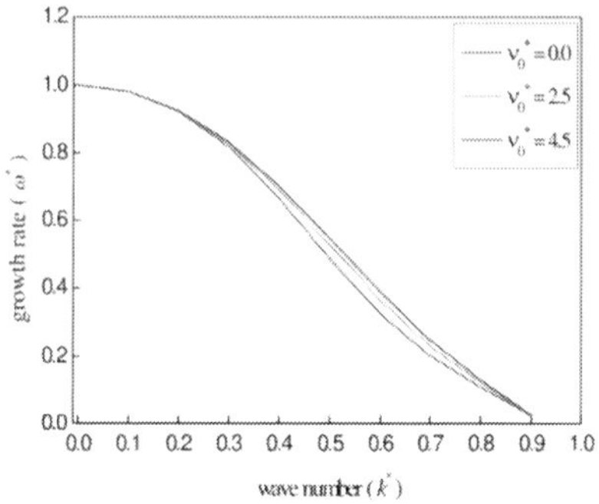

Figure 1. The normalized growth rate v/s normalized wave number with variation in FLR υ_0^* keeping the other parameters fixed.

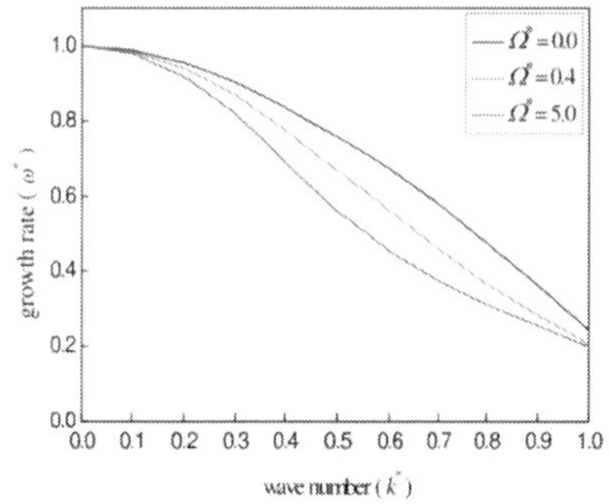

Figure 2. The normalized growth rate v/s normalized wave number with variation in rotation Ω^* keeping the other parameters fixed.

It is clear from figure 1 that growth rate decreases with increasing the value of FLR correction. Thus the effect of FLR correction is stabilizing. From figure 2 we conclude that the growth rate decreases on increasing the value of rotation. Thus the effect of rotation is stabilizing. From figure 3 we conclude that growth rate increases with increasing density dependent heat-loss function. Thus the effect of density dependent heat-loss function is destabilizing. One can observe from figure 4 that the growth rate decreases with increase in temperature dependent heat-loss function. Thus the effect of temperature dependent heat-loss function is stabilizing.

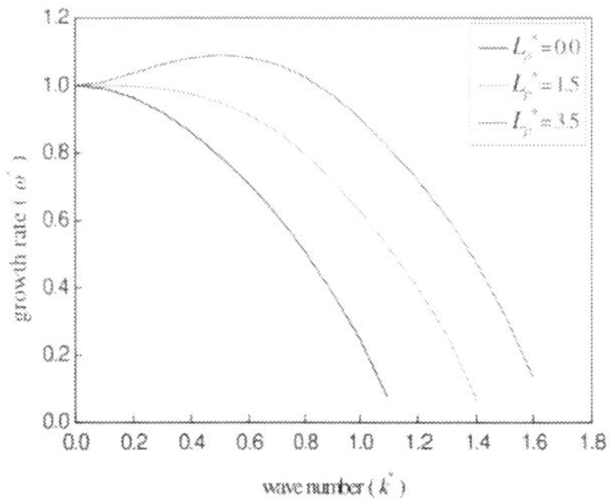

Figure 3. The normalized growth rate v/s normalized wave number with variation in density dependent heat-loss function L_ρ^* keeping the other parameters fixed.

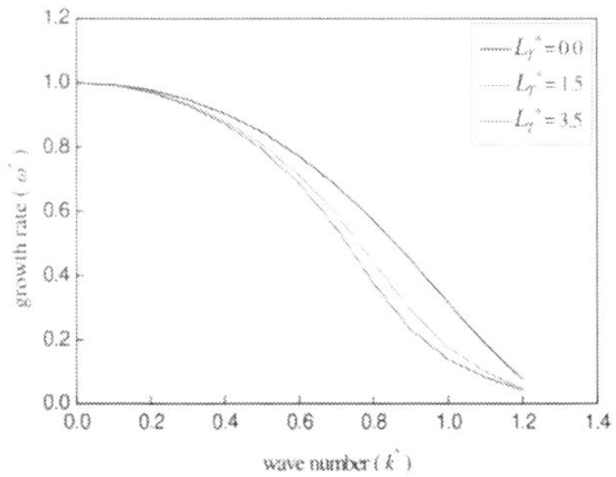

Figure 4. The normalized growth rate v/s normalized wave number with variation in temperature dependent heat-loss function L_T^* keeping the other parameters fixed.

4. Conclusion

Thus in the present paper, the effects of rotation, FLR corrections and radiative heat-loss function on the Jeans gravitational instability is investigated in the presence of thermal conductivity and finite electrical resistivity. The instability conditions are obtained on the basis of Jeans criterion. We find that Jeans criterion of instability gets modified due to the presence of rotation, FLR corrections, thermal conductivity and radiative heat-loss function.

Acknowledgments

The authors are indebted to Dr. S.K. Ghosh, Prof. and Head, S.S. in Physics Vikram University, Ujjain; for his constant encouragements. One author (SK) is grateful to Er. Praveen Vashishtha, Chairman of Mahakal Institute of Technology and Science, Dr. V.S. Ubboveja, Advisor MITS, and Dr. A.C. Shukla Director, MITS Ujjain; for continuous support and guidance.

References

[1] Khan A and Bhatia P K 1993 *Phys. Scr.* **47** 230
[2] Vyas M K and Chhajlani R K 1988 *Astrophys. Space Sci.* **140** 89
[3] Chakrabarti N, Pal B and Krishan V 2008 *Terbulence, Dynamos, Accretion Disc, Pulsar and Collective Plasma Processes Astrophys. Space Sci. Proc.* pp 281-289 (Springer)
[4] Aggarwal M and Talwar S P 1969 *Mon. Not. R. Astron. Soc.* **146** 235
[5] Prajapati R P, Pensia R K, Kaothekar S and Chhajlani R K 2010 *Astrophys. Space Sci.* **327** 139
[6] Kaothekar S, Soni G D and Chhajlani R K 2012 *AIP Advances* **2** 042191
[7] Sharma R C 1974 *Astrophys. Space Sci.* **29** L1
[8] Kaothekar S and Chhajlani R K 2012 *ISRN Astron. Astrophys.* **2012** 420938
[9] Sharma P and Chhajlani R K 2013 *Phys. Plasmas* **20** 092101
[10] Roberts K V and Taylor J B 1962 *Phys. Rev. Lett.* **8** 197

MOEMS optical delay line for optical coherence tomography

Om P. Choudhary[1], S. Chouksey[1], P. K. Sen[1], P. Sen[1], J. Solanki[1, 3], and J. T. Andrews[1],*

[1]National MEMS Design Center, Department of Applied Physics, Shri G S Institute of Technology & Science, Indore – 452003, India.
[2]Laser Bhawan, School of Physics, Devi Ahilya University, Indore - 452007, India.
[3]Department of Biomedical Engg., Texas A&M University, College Station, Texas, USA.

E-mail: jtandrews@sgsits.ac.in

Abstract. Micro-Opto-Electro-Mechanical optical coherence tomography, a lab-on-chip for biomedical applications is designed, studied, fabricated and characterized. To fabricate the device standard PolyMUMPS processes is adopted. We report the utilization of electro-optic modulator for a fast scanning optical delay line for time domain optical coherence tomography. Design optimization are performed using Tanner EDA while simulations are performed using COMSOL. The paper summarizes various results and fabrication methodology adopted. The success of the device promises a future hand-held or endoscopic optical coherence tomography for biomedical applications.

1. Introduction

Electro-optic (EO) materials are shown to exhibit their potential as optical waveguide for precise and continuous laser beam guiding, manipulation, signal processing and controlled light guiding capabilities [1, 2, 3]. Optical modulation is performed using bulk photo-refractive materials and also by adopting electro-optic, magneto-optic, thermo-optic effects or combination of these effects. However, the electro-optic effect is preferred over other optical effects due to its simplicity, linearity, easy availability of large number of material as well as due the size factor. Also, the fast response time of less than few picoseconds and an ultra-wide wavelength bandwidth of over several hundred nanometres make this more attractive [1]. The EO modulators are being used in variety of applications including wave guiding, frequency manipulation, intensity or amplitude modulation, phase generation, polarization splitting, electro-optic switches, filters and other optoelectronic modulation device.

Optical coherence tomography (OCT) is an emerging imaging technique that examines, non-destructively internal structure of the biological tissues. It is based on hetero-dyne mixing of near-infrared light back-scattered from the sample under study, which could carry information. OCT relies on the principle of coherence gating, using the limited coherence length of a broadband source, whose bandwidth basically determines the achievable resolution. As the low coherence light back reflects from sample layers and the reference light, they are recombined and get interference at the detector. The larger band width leads to better transverse resolution and high resolution along image depth [2]. Most of fiber based or free space OCT used mechanically scanned optical component to delay the reference signal over a few millimetres [3, 4]. Except time-domain OCT (TD-OCT), most of the OCT technique does not require optical time delay. To realize a TD-OCT system on chip, we will have to

always consider for optical delay generation. However, it is desirable to scan the reference path with no moving part. Various researchers have performed optical delay generation by mean of various techniques [5, 6] and also proposed complete integrated TD-OCT on a chip as glucose monitoring device [7].
The possibility to modify the refractive index of a material with an applied electric field allow in the optical light modulation for various applications. Lithium Niobate guided wave optical modulator are widely used for long-haul communication systems and optical signal processing systems. Silicon [8] and EO polymer [9], [10] is also a useful candidate for such applications. Currently it appears that significant challenges remain for EO polymer material to be commercial used in long-haul optical communication. The linear EO phenomena, is capability of an anisotropic material, without an inversion symmetry, to linearly change the refractive index with an applied electric field. An applied electric field, parallel to the optical axis of an electro-optic material structure or poled electro-optic doped polymer chromospheres changes its index by a quantity of $\Delta n = -\frac{1}{2} n_0^3 r_{33} E$. Where n_0 is the refractive index of electro optic responded material without applied electric field (E), r_{33} is an element of the linear electro-optic tensor or electro-optic coefficient. An example of various electro-optic materials such as LiNbO3, Doped Polysilicon and DR1 (Dye) doped SU-8 polymer with their refractive index (n_0) 3.2, 1.57, 1.45 and electro-optic coefficient of 30.8×10^{-12}m/V, 25.0×10^{-12}m/V and 30.8×10^{-12}m/V, respectively. It is to be noted that the magnitude of EO coefficient of these materials are very small, however, the beam can be sufficiently modulated with controlled voltage in order to realize light beam modulation. The design and development process is approached systematically using fourfold method viz., (i) To optimize the losses in the waveguides such that wave propagates through the waveguide and reflect back to the detector port (ii) To generate optical delay in the waveguide such that the expected Doppler shift occurs in the medium (iii) finding and designing suitable ports for launching light into the device and extract it to detector and (iv) fabricating the device with optimized values.

2. Design

A Schematic of fiber optics being used by us is shown in figure 1. An equivalent micro-opto-mechanical system (MOEMS) integrated circuit for TD-OCT is shown in figure 2. In the fiber optic setup the reference and sample arms compensated by either by mechanical motion of translational stages or using grating mirror combination. The same in the integrated circuit is carried out electro-optically. The EO induced change refractive index induces path deference in the reference arm, while the back scattered light from the sample arm will be obtained through a fiber optic cable.

Optical modulation is possible by externally applied DC voltage with parallel plate electrode configuration. These characteristics are very attractive in device application for optical beam manipulation in many optoelectronic systems. The designs of the waveguide with suitable electrode are discussed elsewhere [7].

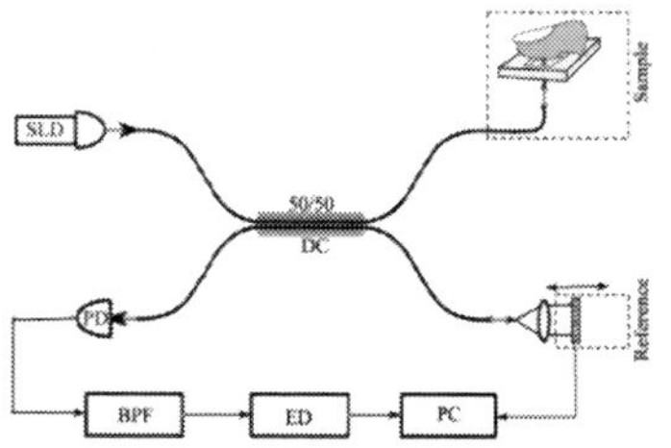

Figure 1. Schematic of fiber-optic OCT setup with additional optical and electronic components.

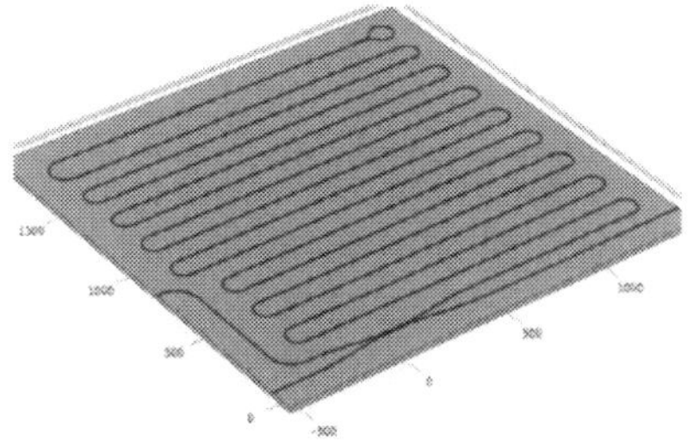

Figure 2. The design of MOEMS_-OCT setup with all optical components integrated into a single chip.

3. Fabrication

The mask designed using the Tanner EDA L-Edit v14.1 for fabrication. The fabrication of designed chip was performed by polyMUMPs (multi user MEMS processes) technology [11]. The process steps (shown in figure 3) followed for the fabrication as:

(1) The substrate is prepared with HF using standard RCA1 and RCA2 cleaning methods.

(2) The MUMPs process is a three-layer poly-silicon surface micro-machining process. The process being with 100mm n-type (100) silicon wafers. Next, a 600nm low-stress LPCVD (low pressure chemical vapour deposition) silicon nitride layer is deposited on the wafers as an electrical isolation layer.

(3) This is followed directly by the deposition of a 500nm LPCVD poly-silicon film Poly0 (see figure 4). Poly0 is then patterned by photo-lithography with mask Poly (see figure 4).

(4) After patterning the photo-resist, the Poly0 layer is then etched in an RIE (Reactive Ion Etch) system.

(5) A 2.0μm phosphosilicate glass (PSG) sacrificial layer is then deposited by LPCVD and annealed at 1050 °C for 1 in argon.

(6) The wafer are then patterned with the third mask layer, ANCHOR1 and RIE. This step provides anchor hole that will be filled by the Poly1 layer.

(7) After etching ANCHOR1, the first structural layer of poly-silicon (Poly1) is deposited at a thickness of 2.0μm and patterned lithographically with the Poly1 mask (see figure 4).

(8) After Poly1is etched, a second PSG layer (second oxide) is deposited and annealed.

(9) The Poly1_Poly2_via provided for etch holes in the second oxide down to the Poly1 layer. This provides a mechanical and electrical connection between the Poly1 and Poly2 layers.

(10) The ANCHOR2 level is provided to etch both the first and second oxide layers in one step, thereby eliminating any misalignment between separately etched holes. More importantly, the ANCHOR2 etch eliminates the need to make a cut in first oxide unrelated to anchoring a Poly1 structure, which need lesser exposes the substrate to subsequent processing that can damage either Poly0 or nitride.

(11) The ANCHOR2 layer is lithographically patterned and etched by RIE in the same way as Poly1_Poly2_via.

(12) The second structural layer Poly2, is then deposited (1.5μm thick) followed by the deposition of 200nm PSG.

(13) Poly2 layer is lithographically patterned with the seventh mask (Poly2) and the PSG and poly-silicon layer are etched by RIE using the same processing conditions as for Poly1.

(14) The final deposited layer in the MUMPs process is a 0.5μm metal layer that provides for probing, bonding and electroding.

(15) The wafer is patterned lithographically with eight masks (Metal) and the metal is deposited and patterned using lift-off method.

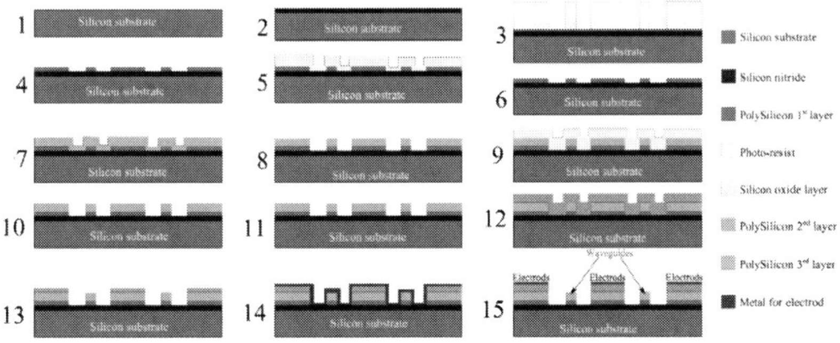

Figure 3. Fabrication process flow followed for the fabrication of the waveguide as shown in figure 2. Details are discussed in text.

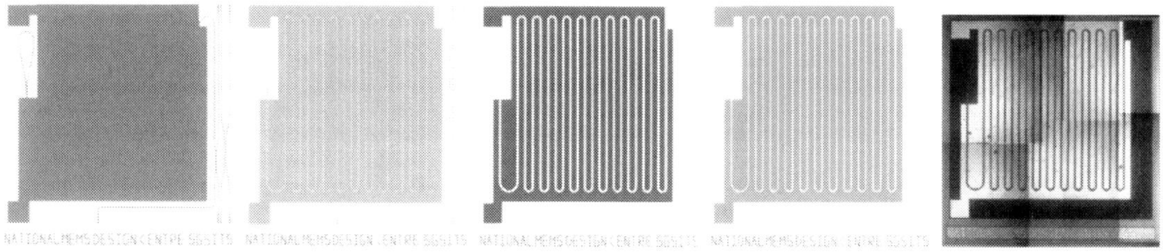

Figure 4. The fabrication method adopted is PolyMUMPs, accordingly, various masks (Poly0, Poly1, Poly2 and Metal masks) are shown in row. The final fabricated product is shown in the last figure.

4. Results and conclusions

We have presented successful fabrication of MOEMS optical coherence tomography. To the aurhors acknowledge, this is the first such device fabrication. The novel miniaturizations have demonstrated the realization of smaller optical coherence tomography setup. This setup could be employed for mobile and portable biomedical applications. Also, the device has the potential to use in clinical and industrial environment.

To conclude, we have designed, simulated and fabricated a MOEMS optical coherence tomography on a chip for biomedical applications. If the various characterization and packaging are successful, it would find huge amount of application in various field of Engineering, Science and Technology.

Acknowledgments

The authors acknowledge the financial support received from ICMR-New Delhi, DBT-New Delhi, NPMASS-Bangalore, and MPCST-Bhopal, India. They also thank CENSQ, Bangalore for carrying out the fabrication work.

References

[1] Wong K K, Ed., 2002, *Properties of Lithium Niobate* (IEE, London, 2002), pp 113-115
[2] Huang D, Swanson E A, Lin C P, Schuman J S, Chang WG, Hee M R, 1991 *Science* **254** 1178
[3] Solanki J, Sen P, Andrews J T and Thareja K, 2012 *J. Mod. Phys.* **3** 64.
[4] Solanki J, Sen P, Andrews J T and Thareja K, 2012 *J. Opt.* **41** 127
[5] Margallo-Balbas E, Geljon M, Pandraud G and French P J, 2010 *Opt. Lett.* **35** 4027.
[6] Jansz P V, Hinckley S, and Wild G, 2011 *6th IEEE international symposium* pp 162.
[7] Andrews J T, Solanki, Choudhary Om P, Chouksey S, Malviya N, Chaturvedi P and Sen P, 2012 *J. Phys. Conf. Ser.* **365** 012004.
[8] Daldosso N and Pavesi L, 2009 *Laser & Photonic Reviews.*, **3** 508.
[9] Chen D, Fetterman H R, Chen A, Steirer W H, Dalton L R and Wang W, 1997 *Appl. Phys. Lett.*, **70** 3335.
[10] Balakrishnan M, Facini M, Diemmer M B J, Verboom W, and Reinhoudt D N, 2006, *IEE Electronic Lett.* **42** 1.
[11] www.memscap.com

A comparative study of PPM and solenoid focusing in multi-beam electron gun

Ashok Kumar Nehra[1,2], **R K Sharma**[1], **and Y Choyal**[2]
[1]Microwave Tubes Division, Council of Scientific and Industrial Research/ Central Electronics Engineering Research Institute (CEERI), Pilani-333031, Rajasthan, India
[2]School of Physics, Devi Ahilya University, Indore-452001, M.P, India

Email: ashokkumarnehra@gmail.com

Abstract. This paper represents the comparison of periodic permanent magnet (PPM) and solenoid focusing for dual anode multi-beam electron gun using OPERA3D code. The electron gun has been operated at 6 kV having 75 mA beam current with 0.45 mm beam waist radius. The design has an additional feature of cathode protection from ion bombardment with the application of extra ion barrier anode.

1. Introduction

As the multi-beam technology has numerous advantages [1]-[3] and therefore it is used in microwave tubes, such as klystrons and traveling wave tubes but due to increasing demand of more compact and lighter weight devices PPM focusing is used than conventional solenoidal focusing. The elimination of the solenoid power supply can also lead to higher overall device efficiency. PPM focusing has been widely used in helix and coupled cavity traveling wave tubes but is much less common in klystrons because the historic applications for klystrons has not required. The weight reduction compared to a solenoid or a uniform field permanent magnet is 1 to 2 orders of magnitude.

A deficiency in the PPM case that is not present in uniform-field focusing is that the rotational energy of the beam increases as the beam becomes more confined. As a result, the energy in the axial motion of the beam and the efficiency decrease. For reasonable values of confinement, a large fraction of the beam energy is utilized in rotation for microperveances above ~0.5. As a result, efficiency is degraded. Thus, the confined flow PPM focusing is limited to low-perveance designs [4]. But in present case of four beam electron gun microperveances per beam-let is 0.16 which can be easily handle with confined flow PPM focusing.

2. Design Approach

A four-beam electron gun on 14mm PCD at an angular distance of 90° from each other has been simulated using OPERA 3-D [5] to study the effect of PPM and solenoidal focusing. The total beam current (300 mA) of multi-beam electron gun is equally divided (75 mA) among each beamlet with laminar beam flow at 6 kV beam voltage. The total beam power is 1.8 kW. The beam waist radius and beam throw of electron gun are 0.45 mm and 15 mm respectively. It has been estimated that a positive potential of the order of 100-150 V with respect to ground is good enough to prevent the flow of positive ions from beam-wave interaction region to the cathode-anode region, thus protecting cathode surface from ion bombardment and deteriorate the emission current. The design parameters of multi-beam electron gun are summarized in the Table I.

Table 1. Design parameters of multi-beam electron gun

Parameter	Value
Cathode voltage	-6 kV
No. of Beam Lets	4
Total Beam Current	300 (75×4) mA
Total Beam Perveance	0.64 μP
Individual Beam Perveance	0.16 μP
Cathode Diameter	3.2 mm
Cathode Emission Loading	0.92 A/cm^2
Beam Waist Radius	0.45 mm
Beam Throw	15 mm

3. Focusing structure

The design of PPM focusing structures, however, is more complex, and beam transmission is not as good as for uniform field focusing; therefore, this focusing method is limited to lower average power. But specially designed pole piece (of suitable thickness, aperture shape, and size) are the major design parameters that are responsible for preventing the electron beams from corkscrewing in order to get the desired beam transmission [6]. When the beam-lets enter the pole-piece aperture, a symmetric magnetic force is experienced by them, their center remains at its own axis, and hence, the beam-lets remain laminar. In both cases i.e. PPM and solenoid same magnetic field (2400 Gauss) is used to guide the beam-lets but less than 2 % scalloping in solenoid focusing while 5.2 % scalloping is observed in PPM focusing and cut views of OPERA 3D simulated solenoid and PPM focused four-beam electron gun are shown in Fig.1 and Fig.2 respectively. Comparison of magnitudes of magnetic field in PPM and solenoid focusing are shown in Fig.3

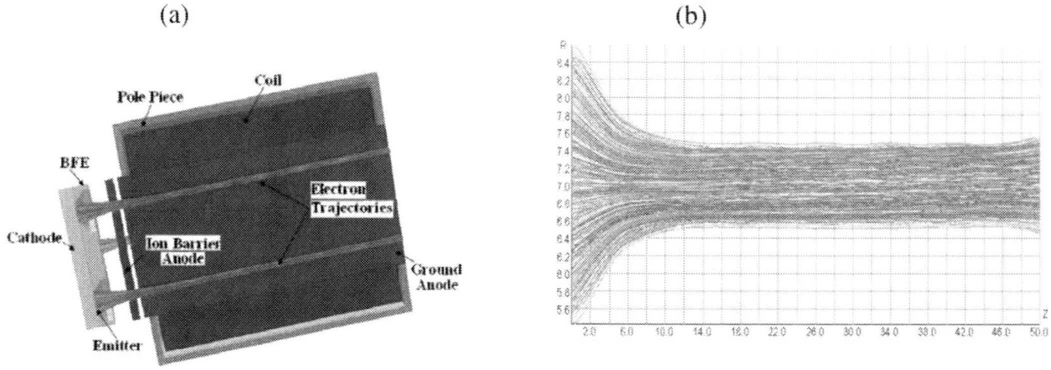

Figure 1. (a) Cut view of OPERA 3D simulated solenoid focused four beam electron gun (b) magnified view of single beam

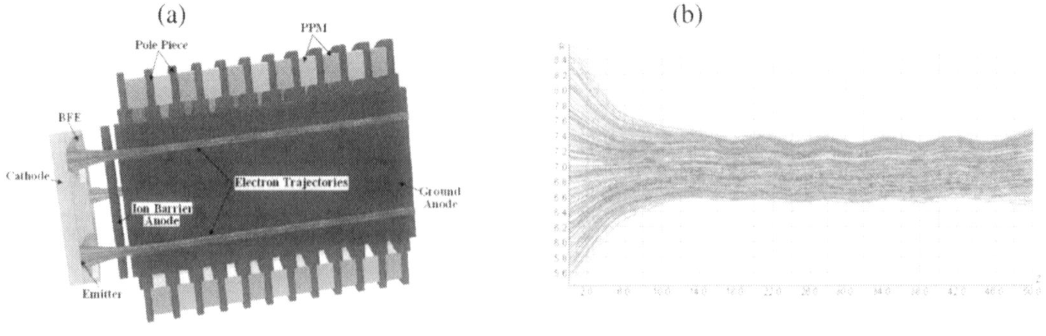

Figure 2. (a) OPERA 3D simulated PPM focused four-beam electron gun (b) magnified view of single beam

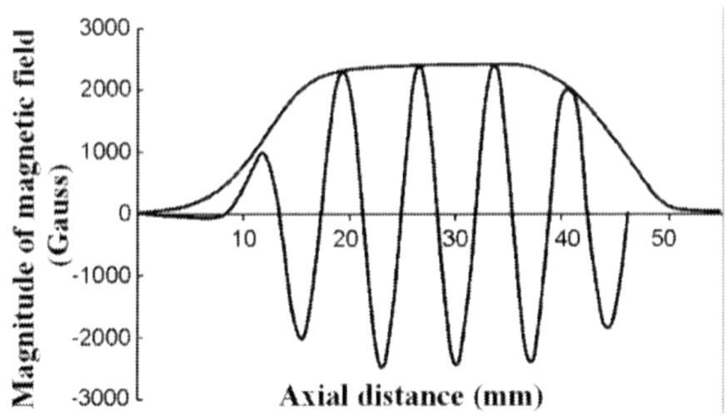

Figure 3. Comparison of magnitudes of magnetic field in PPM and solenoid focusing

Acknowledgements
The authors are thankful to Dr Chandershaker, Director, CEERI, Pilani, for granting permission to publish this paper. Thanks are also due to CSIR for awarding the Senior Research Fellowship to Mr. Ashok Kumar Nehra.

References

[1] Korolyov A. N., Gelvich E. A., Zhary Y. V., Zakurdayev A. D. and Poognin V. I. 2004 *IEEE Trans. Plasma Sci.* **32**, 1109.

[2] Nehra A., Joshi L. M., Gupta R. K., Sharma S., Choyal Y.and Sharma R. K., 2011 *Journal of Infrared Millimeter and Terahertz Waves* **32** 793.

[3] Nehra A.,. Sharma R. K, Kumar D. and Choyal Y, 2012 *IEEE Trans. Plasma Science* **40**, 1672

[4] Gilmour A. S. Jr., 2011 *Klystrons, Traveling Wave Tubes, Magnetrons, Crossed-Field Amplifiers, and Gyrotrons*, (Boston, Artech House).

[5] Opera-3D, Vector fields software, Oxford, U.K.

[6] Nguyen K. T., Pershing D. E., Abe D. K., Levush B., Wood F. N., Calame J. P., Pasour J. A., Petillo J. J., Cusick M., Cattelino M. J. and Wright E. L. 2004 *IEEE Trans. Plasma Sci., vol.* **32** 1212.

Optical nutation in GaAs/AlGaAs core-shell quantum dot

R Solanki[1, 3], P Sen[1] and P K Sen[2]

[1] School of Physics, Devi Ahilya University, Khandwa Road, Indore (M.P.) 452001, India
[2] Department of Applied Physics, Shri G S Institute of Technology and Science, Park Road 23, Indore (M.P.) 452003, India

[3] Corresponding author E-mail address: er.ravisolanki.80@gmail.com

Abstract. Optical nutation in GaAs/$Al_{0.3}Ga_{0.7}As$ core-shell quantum dot (CSQD) shined by femtosecond laser has been analyzed theoretically. The nutation signal intensity, arising due to photo-induced electronic transitions between ground state and first excitonic state, has been calculated using time dependent perturbation theory and rotational wave approximation. The dependency of the properties of optical nutation signal viz. its maximum amplitude and the oscillation frequency of the nutation signal has been studied at various CSQD's core radius and shell thickness.

1. Introduction

Semiconductor nanostructures are proving themselves as potential candidate materials for a wide range of application in nano-photonics, nano-electronics, ultrafast optical devices and logic gates in optical computing and quantum information processing devices[1]. Core-shell quantum dots (CSQDs) are the structures in which the spherical core of lower bandgap material is enclosed within a shell of relatively higher bandgap material. Of late, CSQDs have been shown to be useful in the design of low temperature nano-sensors [2]. Moreover, the properties of the structure become size dependent and have discrete energy levels if its size is of the order of or less than bulk exciton Bohr radius i.e. under strong confinement regime.

The interaction between coherent light pulse and matter on a time scale much shorter than the dipole dephasing time of the material results in coherent transient effects.

Optical nutation is one of such effects in which there is a periodic change (nutation) of the population between two selected states of the system which is manifested in oscillating damped ringing behaviour of the transmitted light. It is an important tool for the study of relaxation mechanism in semiconductor materials and to calculate lifetime of excited state [3]. Analysis of optical nutation in various semiconductor nanostructures like in quantum wells [4], quantum wires [5] and core-shell quantum dots have already been reported in the recent past. One of the groups has analyzed optical nutation between the energy levels inside and outside the core of the CSQDs [6].

In the present communication, we have study the nutation signal obtained due to the transition between ground state and first excitonic state and examined the dependency of maximum amplitude and frequency of the nutation signal on core radius and shell thickness of the CSQDs under strong confinement regime.

Journal of Physics: Conference Series **534** (2014) 012068 doi:10.1088/1742-6596/534/1/012068

2. Mathematical Formulation

The interaction between light and CSQD can be analyzed using semi-classical theory, representing the QD quantum mechanically and light classically. Considering the transition from crystal ground state $|a\rangle$ to the excitonic state $|b\rangle$, we describe the CSQD by two-level wavefunction as

$$\Psi(t) = a(t)e^{i\omega_a t}\,|a\rangle + b(t)e^{i\omega_b t}\,|b\rangle\,. \tag{1}$$

Here, $a(t)$ and $b(t)$ are the probability amplitudes of the crystal ground state and the excited state, respectively. We assume the interaction Hamiltonian to be of dipole type. The equations of motion of the probability amplitudes of states can be obtained using time dependent perturbation technique [7]. The probability amplitudes have been calculated by the standard formalism as

$$a(t) = (iX\sin\theta + \beta\cos\theta)e^{-i\Delta t/2}e^{-\Gamma t/2}/\beta \tag{2}$$

and

$$b(t) = -2i\mu\,E_0/2\hbar\,\sin\theta\,e^{i\Delta t/2}e^{-\Gamma t/2}/\beta\,. \tag{3}$$

Here $\omega_k\,(=\omega_b-\omega_a)$ is the transition frequency. $\Gamma = T_2^{-1}$; T_2 being the collisional dephasing time of the electrons, $\theta = \beta t/2$, and $\Delta\,(=\omega_k-\omega)$ is defined as the detuning. The transition energy is given by

$$\hbar\omega_k = E_{g(bulk)} + E_{nl} - E_{ex}, \tag{4}$$

with E_{nl} as the confinement energy of the carriers confined in core region and E_{ex} as the exciton binding energy in the core material.

$X = \Delta - i\Gamma$ represent complex detuning frequency and $\beta = \sqrt{X^2 + 4(\mu\,E_0/2\hbar\,)^2}$ is the generalized complex Rabi flopping frequency in the presence of detuning and dephasing.

The corresponding induced polarization in CSQD will be

$$P(t) = N(|eP_{ab}(0)/m_0\omega_k|)\,[a(t)b^*(t)e^{i\omega_k t} + b(t)a^*(t)e^{-i\omega_k t}]/2, \tag{5}$$

for N number of QDs per unit volume. Considering single CSQD we take N=1. m_0 is the electron rest mass and $P_{ab}(0)$ is the transition momentum matrix element at the band edge.
The corresponding value of transmitted radiation intensity is obtained using the standard relation

$$I_T = \frac{1}{2}n_0\,\varepsilon_0\,c|E_T|^2, \tag{6}$$

with transmitted field $E_T(t)\,(=-i\omega L\,P(t)\,/c\,\varepsilon_0)$ determined from Maxwell's equations under slowly varying envelope approximation for interaction length L [8]. Now we calculate confinement energy of electron and hole.

The physical structure and energy band diagram of the CSQD under investigation is shown in figure 1. In the figure, r_1 and r_2 are the core and shell radius, respectively and $d\,(=r_2-r_1)$ is the thickness of the shell. The eigenenergy is calculated by solving the 3-dimensional Schrodinger equation for electrons confined in the conduction band and for the holes confined in valence band. We have restricted our analysis for 1s energy state for which $n = 1$ and $l = m = 0$. Using the boundary conditions of the wavefunction, we get the transcendental equation

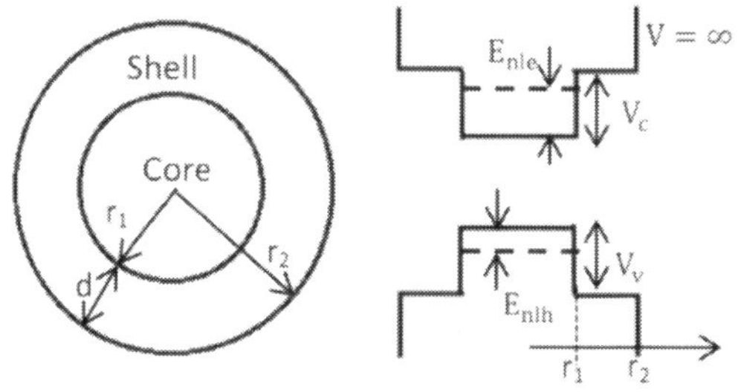

Figure 1. Structure and energy band diagram of the core-shell quantum dot. Dashed line in conduction and valence band represent the confinement eigenenergy level of electron and hole, respectively.

$$(m^*_{e2}/m^*_{e1})(1 - kr_1 \cot kr_1) = 1 + Kr_1 \coth(Kd) \tag{7}$$

using which the allowed value of E_{nle} for $E_{nle} < V_c$ can be obtained. $k = \sqrt{2m^*_{e1}E_{nle}/\hbar^2}$ and $K = \sqrt{2m^*_{e2}(V_c - E_{nle})/\hbar^2}$ are the propagation constants in core and shell region, respectively.

Similarly the eigenenergy E_{nlh} of holes can be calculated by replacing electron effective mass by the hole effective mass and conduction band offset by valence band offset. The value of E_{nl} used in equation (4) is then given by $E_{nl} = E_{nle} + E_{nlh}$.

3. Results and discussions

We have investigated analytically the optical nutation phenomena, induced in GaAs/$Al_{0.3}Ga_{0.7}As$ CSQDs for various core radius r_1 and shell thickness d, on application of the ultra-short coherent light pulse. The CSQD is considered to be excited by a near resonant 150-fs Ti: Sapphire pulsed laser and electric field amplitude of 5 x 10^7 V/m under weak excitation regime at low temperature of about 5 K. For numerical analysis, we take $\Delta = 6 \times 10^{12}\ s^{-1}$. The dephasing time is taken as T_2 = 200 fs. The core and shell material parameters are tabulated in table 1. The refractive index n_0, exciton Bohr radius a_{ex} and exciton binding energy E_{ex} of GaAs core are 3.6, 12 nm and 5.7 meV, respectively.

Table 1. Material parameters of the core and shell region of CSQD used in analysis at temperature 5K.

	E_g (eV)	m^*_e (m_0)	m^*_h (m_0)
GaAs	1.521	0.067	0.45
$Al_{0.3}Ga_{0.7}As$	1.923	0.092	0.585

We take the value of $|P_{ab}(0)|^2/m_0$= 11.44 eV for core material. The conduction and valence band offsets are taken as 0.249 eV and 0.165 eV, respectively [9].

In this paper, we have examined the dependency of maximum intensity and frequency of the nutation signal on core radius and shell thickness. The nutation signal obtained from CSQD of various core radii at constant shell thickness is shown in figure 2 and the signal obtained from CSQD with core radius 3 nm and different shell thickness is given in figure 3. The maximum amplitude and frequency of oscillation increases with the core radius whereas the change in the above same parameters is almost insignificant on changing the shell thickness, at the core radius of 3 nm.

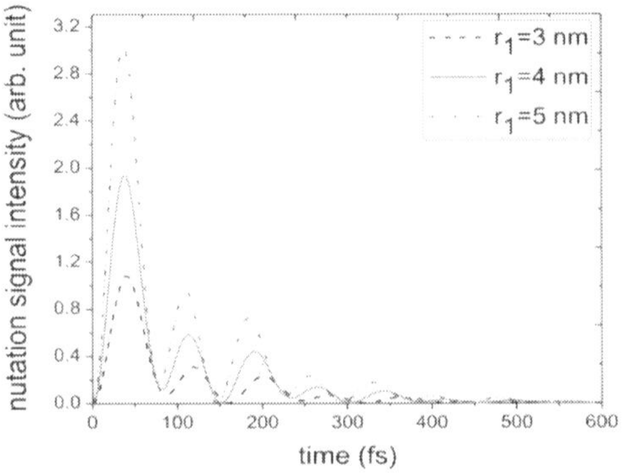

Figure 2. The optical nutation signal of various core radius with shell thickness d = 2 nm. Intensity of nutation signal increases with the core radius.

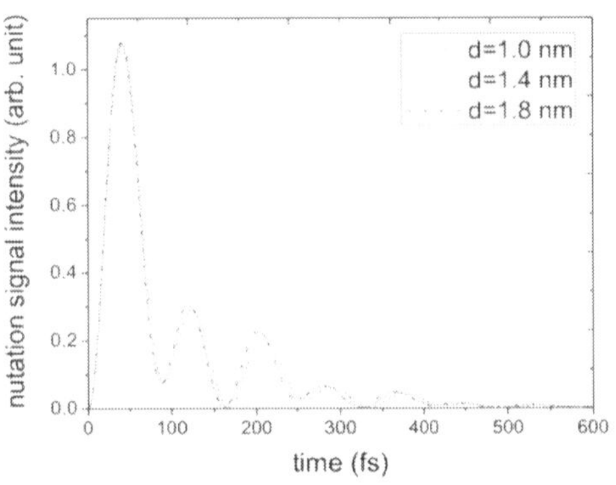

Figure 3. The optical nutation signal of CSQD having core radius r_1= 3 nm and various shell thickness. The nutation signal is almost insensitive to shell thickness.

For a given system of pump and CSQD the electric field amplitude and detuning remains constant, and eventually the factors on which maximum amplitude and frequency of the nutation signal depend is the ensemble average of transition dipole moment $\langle \hat{\mu}(t) \rangle$ and the interaction length L. As we increase the core radius the electrons will move farther from the hole and thus results in increasing dipole moment.

The calculation of wavefunctions and confinement energies for different core radii and shell thicknesses reveal that the probability of electron present in the shell increases with the shell thickness keeping the core radius constant. But the penetration of holes' wavefunction doesn't change because of change in core radius and shell thickness due to their higher effective mass and less confinement energy and therefore they remain confined inside the core. This will also increase the separation between electron and hole and hence the dipole moment also increases with the increase in shell thickness.

Another factor responsible for change with core radius is due to the increase in the interaction length. Since the incident radiation is coherent only with the core region and the shell acts as a transparent medium for it because of its higher bandgap than the core region; the interaction length is equal to the core diameter.

The interaction length thus remains constant when the core radius is kept same and the shell thickness is varied as shown in figure 3. This is the parameter which makes the intensity of nutation signal dependents more on the core radius than the shell thickness.

4. Conclusions

We have studied the optical nutation process in GaAs/$Al_{0.3}Ga_{0.7}As$ CSQDs using effective mass approximation and time dependent perturbation theory. From the results, one may conclude that both the maximum amplitude and Rabi frequency of the nutation signal depends on the core radius whereas its dependency on the shell thickness is insignificant for the bigger core radius CSQDs. The dependency of nutation signal on physical parameters arises due transition dipole moment and the interaction length dependency on physical parameters.

Acknowledgments

Authors kindly acknowledge the financial support from Science and Engineering Research Board (SERB) research project of Department of Science and Technology (DST), New Delhi, India.

References

[1] Baimuratov A S, Rukhlenko I D, Turkov V K, Baranov A V and Fedorov A V 2013 Quantum-dot supercrystals for future nanophotonics *Sci. Rep.* **3** 01727-36

[2] Chattopadhyay S, Sen P, Andrews J T and Sen P K 2012 Semiconductor core-shell quantum dot: a low temperature nano-sensor material *J. Appl. Phys.* **111** 034310-18

[3] Guang S H and Song H L 1999 *Physics of nonlinear optics* (Singapore: World Scientific)

[4] Kapoor S, Kumar J and Sen P K 2004 Optical coherent transient effects in magnetoactive GaAs/AlGaAs quantum well structure *J. Appl. Phys.* **95** 4833-39

[5] Gupta S K and Sen P K 2007 A study of the ultrafast optical response of magnetized GaAs/AlGaAs quantum wire structures *Superlattices Microstruct.* **41** 248-255

[6] Gong S and Yao D 2006 Optical nutation induced by transition between levels inside and outside the well in a core-shell CdSe/ZnS quantum dot *J. Phys.: Condens. Matter* **18** 10989-98

[7] Sargent M, Scully M O and lamb W E 1982 *Laser Physics* (Massachusetts: Addison-Wesley)

[8] Shimoda K 1984 *Introduction to Laser Physics* (Berlin:Springer)

[9] Adachi S 1993 Properties of Aluminium Gallium Arsenide (London:INSPEC)

AUTHOR INDEX

AUTHOR INDEX

Institute of Physics
Dirac House, Temple Back
Bristol BS1 6BE UK

proceedings.com